高等学校“十三五”规划教材

Chromatographic Analysis

色谱分析

（双语版）

张裕平　李　英　崔乘幸　主编

化学工业出版社

·北京·

《色谱分析》（双语版）共分八章，系统介绍了色谱分析的基本理论、薄层色谱、气相色谱、高效液相色谱、离子色谱、尺寸排阻色谱、毛细管电泳和电色谱以及色谱技术中样品前处理技术等内容。本书集理论性与实践性为一体，在材料选择上注重知识的基础性和系统性，同时兼顾色谱技术的最新进展。

《色谱分析》（双语版）可供化学类专业高年级本科生和研究生使用，以提高专业英语水平，为后续的科研工作做好准备；本书也可供科研工作者参考。

图书在版编目（CIP）数据

色谱分析：汉、英/张裕平，李英，崔乘幸主编. —北京：化学工业出版社，2018.5（2024.8重印）
高等学校“十三五”规划教材
ISBN 978-7-122-31741-4

Ⅰ.①色…　Ⅱ.①张…②李…③崔…　Ⅲ.①色谱法-化学分析-双语教学-高等学校-教材-汉、英　Ⅳ.①O657.7

中国版本图书馆CIP数据核字（2018）第049788号

责任编辑：宋林青　　　文字编辑：刘志茹
责任校对：王素芹　　　装帧设计：关　飞

出版发行：化学工业出版社（北京市东城区青年湖南街13号　邮政编码100011）
印　　装：北京科印技术咨询服务有限公司数码印刷分部
787mm×1092mm　1/16　印张 15¾　字数 382 千字　　2024年8月北京第1版第2次印刷

购书咨询：010-64518888　　售后服务：010-64518899
网　　址：http://www.cip.com.cn
凡购买本书，如有缺损质量问题，本社销售中心负责调换。

定　　价：40.00元

前言

随着我国高等教育教学改革进程的持续推进，教育部要求高校培养具有直接接受国际最新科技信息，能用英语交流的本科生或研究生的国际型人才，提高所培养人才的英语水平，适应国家和社会对英语的需求，建立符合中国实际的双语课程教学模式。

色谱分析是分析化学中应用最广泛发展最迅速的研究领域，新技术新方法层出不穷。近年来随着科学技术的突飞猛进，色谱分析从理论到技术也得到了较快发展，极大地拓宽了其研究与应用领域。目前，色谱法已经成为分析化学学科的一个重要分支，在化学、化工、轻工、石油、环保和医药等几乎所有科学领域内得到了广泛应用，为信息科学、生命科学、材料科学、环境科学等新兴学科的发展做出了重要贡献。

本教材由多年从事色谱教学及研究工作，且具有留学经历的老师编写，吸收了国外分析化学及色谱理论相关教材重理论分析和实际应用相结合的特点，信息量大，图标例题丰富，在保证学生理论学习的同时，注重培养学生的实践能力，有利于学生的自主学习。这本教材在正式出版之前曾印成讲义进行了试用，得到了师生的好评。为了保证质量，也邀请了国外的相关学科专家进行了审阅和校勘。

为了便于读者理解内容，同时又使读者减少对中文的依赖性，编者在教材内容的编排上，同一页上采用了两栏编排方式，右栏中列出对读者可能为生僻词的英文音标及释义，以有助于学生理解，参考译文列于对应章节之后。另外，为了帮助学生更好地理解与掌握每个章节的内容，本教材每章之后均给出了相关的练习题，并在书后附有参考答案。

本书由河南科技学院张裕平、李英、崔乘幸主编，张毅军、杨靖华、胡林峰、陈军、陈娜为副主编，白秀芝、李盼盼等参与了部分工作。编写过程得到了化学工业出版社的鼎力支持，在此表示诚挚的感谢。本书可作为高等院校化学、化工、食品、环境、医药类等专业本科生的双语教材，同时也可作为相关专业研究生的教材或参考书。

由于编者能力有限，书中有些内容难免不够妥善，希望读者、同行提出批评和改进意见，以便日后修订完善。

编者

2018.10

Table of Contents

Chapter 2 Thin Layer Chromatography 39

第2章 薄层色谱法 48

第4章 高效液相色谱法 135

Chapter 5 Ion Chromatography 148

Chapter 7 Capillary Electrophoresis and Electrochromatography 189

第7章 毛细管电泳和电色谱 209

Chapter 1
The Invention of Chromatography

Who invented chromatography, one of the most widely used laboratory techniques? This question leads to controversies. In the 1850s, Schonbein used filter paper to partially separate substances in solution. He found that not all solutions reach the same height when set to rise in filter paper. Goppelsroder (in Switzerland) found relations between the height to which a solution climbs in paper and its chemical composition. In 1861, he wrote "I am convinced that this method will prove to be very practical for the rapid determination of the nature of a mixture of dyes."

Even if both of them did valuable work towards the progress of paper chromatography, it is traditional to assign the invention of modern chromatography to Michael S. Tswett, shortly after 1900. Through his successive publications, one can indeed reconstitute his thought processes, which makes of him a pioneer, even if not the inventor, of this significant separative method. His field of research was involved with the biochemistry of plants. At that time, one could extract chlorophyll and other pigments from house plants, usually from the leaves, easily with ethanol. By evaporating this solvent, there remained a blackish extract which could be redissolved in many other solvents and in particular in petroleum ether (now one would say polar or non-polar solvents). However, it was not well understood why this last solvent was unable to directly extract chlorophyll from the leaves. Tswett put forth the assumption that in plants chlorophyll was retained by some molecular forces binding on the leaf substrate, thus preventing extraction by petroleum ether. He foresaw the principle of adsorption here. After drawing this conclusion, and to test this assumption he had the idea to dissolve the pigment extract in petroleum ether and to add filter paper (cellulose), as a substitute for leaf tissue. He realized that paper collected the colour and that by adding ethanol to the mixture one could reextract these same pigments.

As a continuation of his work, he decided to carry out

substance
['sʌbstəns]
n.［物］物质

chlorophyll
['klɒrəfɪl]
n. 叶绿体

blackish
[b'lækɪʃ]
adj. 稍黑的

extract
['ekstrækt]
n. 提取物

redissolve
[ˌriːdɪ'zɒlv]
vt. 溶解

pigment
['pɪgmənt]
n. 颜料；色料

petroleum ether
石油醚

continuation
[kənˌtɪnjuː'eɪʃən]
n.继续

systematic tests with all kinds of powders (organic or inorganic), which he could spread out. To save time he had carried out an assembly which enabled him to do several assays simultaneously. He placed the packed powders to be tested in the narrow tubes and he added to each one of them a solution of the pigments in petroleum ether. That enabled him to observe that in certain tubes the powders produced superimposed rings of different colours, which testified that the force of retention varied with the nature of the pigments present. By rinsing the columns with a selection of suitable solvents he could collect some of these components separately. Modern chromatography had been born. A little later, in 1906, then he wrote the publication, in which he wrote the paragraph generally quoted: Like light rays in the spectrum the different components of a pigment mixture, obeying a law, are resolved on the calcium carbonate column and then can be measured qualitatively and quantitatively. I call such a preparation a chromatogram and the corresponding method the 'chromatographic method'.

powder [ˈpaʊdə(r)] *n*.粉末

assay [əˈseɪ] *v*.分析

ring [rɪŋ] *n*.环

quote [kwəʊt] *vi*.引用

resolve [rɪˈzɒlv] *vi*.处理

calcium carbonate 碳酸钙

qualitatively [ˈkwɒlɪtətɪvlɪ] *adv*.定性地

quantitatively [ˈkwɒntɪtətɪvlɪ] *adv*.定量地

corresponding [ˌkɒrɪˈspɒndɪŋ] *adj*.相当的，对应的

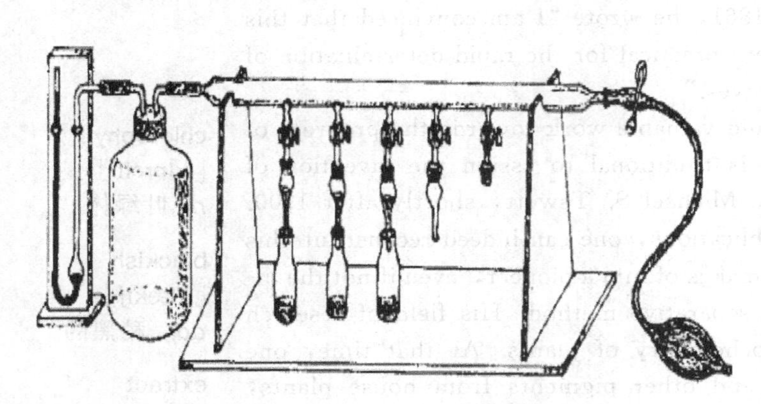

Chromatography, the process by which the components of a mixture can be separated, has become one of the primary analytical methods for the identification and quantification of compounds in the gaseous or liquid state. The basic principle is based on the concentration equilibrium of the components of interest, between two immiscible phases. One is called the stationary phase, because it is immobilized within a column or fixed upon a support, while the second, called the mobile phase, is forced through the first. The phases are chosen such that components of the sample have differing solubility in each phase. The differential migration of compounds leads to their separation. Of all the instrumental analytical techniques, this hydrodynamic procedure is the one with the broadest application. Chromatography occupies a dominant position that all laboratories involved in molecular analysis can confirm.

quantification [ˌkwɒntɪfɪˈkeɪʃn] *n*.量(化)，以数量表示

stationary [ˈsteiʃənəri] *adj*.固定的，静止的；*n*.固定物

instrumental [ˈɪnstruˈmentl] *adj*.仪器的

hydrodynamic [ˌhaɪdrəʊdaɪˈnæmɪk] *adj*.水力的

1.1 General concepts of analytical chromatography

Chromatography is a physico-chemical method of separation of components within mixtures, liquid or gaseous, in the same vein as distillation, crystallization, or the fractionated extraction. The applications of this procedure are therefore numerous since many of heterogeneous mixtures, or those in solid form, can be dissolved by a suitable solvent (which becomes, of course, a supplementary component of the mixture).

A basic chromatographic process may be described as follows (Figure 1.1):

1) A vertical hollow glass tube (the column) is filled with a suitable finely powdered solid, the stationary phase.

2) At the top of this column is placed a small volume of the sample mixture to be separated into individual components.

3) The sample is then taken up by continuous addition of the mobile phase, which goes through the column by gravity, carrying the various constituents of the mixture along with it.

distillation [ˌdɪstɪ'leɪʃn] *n*.蒸馏

crystallization [ˌkrɪstəlaɪ'zeɪʃn] *n*.结晶化

fractionate ['frækʃənˌeɪt] *vt*.分离

extraction [ɪk'strækʃən] *n*.萃取(法)

gravity [ˌgrævɪtɪ] *n*.重力

constituent [kən'stɪtjʊənt] *n*.成分

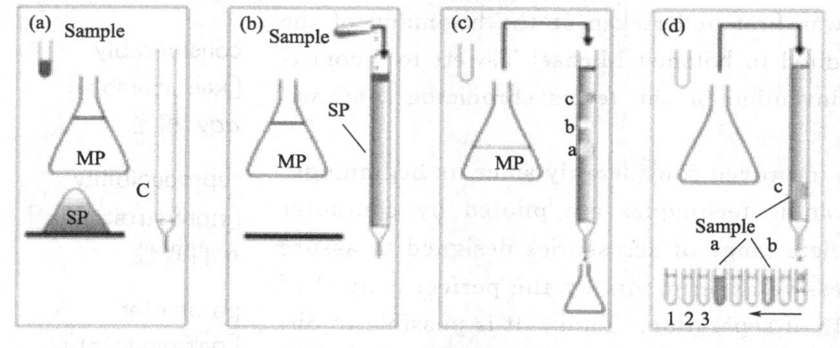

Figure 1.1 A basic experiment in chromatography. (a) The necessary ingredients (C, column; SP, stationary phase and MP, mobile phase; (b) Introduction of the sample; (c) Start of elution; (d) recovery of the products following separation

This process is called elution. If the components migrate at different velocities, they will become separated from each other and can be recovered, mixed with the mobile phase.

This basic procedure, carried out in a column, has been used since its discovery on a large scale for the separation or purification of numerous compounds (preparative column chromatography), but it has also progressed into a stand-alone analytical technique, particularly once the idea of measuring the migration times of the different compounds as a mean to identify them had been conceived, without the need for their collection. To do that, an optical device was placed at the column exit, which indicated the variation of the composition

purification [ˌpjʊərəfɪ'keɪʃən] *n*.洗净,提纯

of the eluting phase with time. This form of chromatography, whose goal is not simply to recover the components but to control their migration, first appeared around 1940 though its development since has been relatively slow.

The identification of a compound by chromatography is achieved by comparison: To identify a compound which may be A or B, a solution of this unknown is run on a column. Next, its retention time is compared with those for the two reference compounds A and B previously recorded using the same apparatus and the same experimental conditions. The choice between A and B for the unknown is done by comparison of the retention times.

In this experiment, a true separation had not been effected (A and B were pure products), but only a comparison of their times of migration was performed. In such an experiment there are, however, three unfavorable points to note: the procedure is fairly slow; absolute identification is unattainable; and the physical contact between the sample and the stationary phase could modify its properties, therefore its retention times and finally the conclusion.

This method of separation, using two immiscible phases in contact with each other, was first undertaken at the beginning of the 20th century and is credited to botanist Michael Tswett to whom is equally attributed the invention of the terms chromatography and chromatogram.

The technique has improved considerably since its beginnings. Nowadays chromatographic techniques are piloted by computer software, which complete range of accessories designed to assure reproducibility of successive experiments by the perfect control of the different parameters of separation. Thus, it is possible to obtain, during successive analyses of the same sample conducted within a few hours, recordings that are reproducible to within a second (Figure 1.2).

The essential recording that is obtained for each separation is called a chromatogram. It corresponds to a two-dimensional diagram traced on a chart paper or a screen that reveals the variations of composition of the eluting mobile phase as it exits the column. To obtain this document, a sensor, of which there exists a great variety, needs to be placed at the outlet of the column. The detector signal appears as the ordinate of the chromatogram while time or alternatively elution volume appears on the abscissa.

The identification of a molecular compound only by its retention time is somewhat arbitrary. A better method consists of associating two different complementary methods, for example, a chromatograph and a second instrument on-line, such as a mass spectrometer or an

migration [maɪ'greɪʃən] *n*.迁移

experimental [ɪks ˌperɪ'mentəl] *adj*.实验的

identification [aɪˌdentɪfɪ'keɪʃn] *n*.鉴定，识别

considerably [kən'sɪdərəblɪ] *adv*.相当

reproducibility [rɪprədju:sə'bɪlɪtɪ] *n*.再现性

parameter [pə'ræmɪtə(r)] *n*.参数

chart [tʃɑ:t] *n*.图表

detector [dɪ'tektə] *n*.检测器

abscissa [æb'sɪsə] *n*.横坐标

arbitrary ['ɑ:bɪtrərɪ] *adj*.主观的

infrared spectrometer. These hyphenated techniques enable the independent collating of two different types of information that are independent (time of migration and "the spectrum"). Therefore, it is possible to determine without ambiguity the composition and concentration of complex mixtures in which the concentration of compounds can be of the order of nanograms.

infrared
[ˌɪnfrə'red]
n.红外线,红外区

spectrometer
[spek'trɔmɪtə]
n.分光计,分光仪

hyphenate
['haɪfəneɪt]
v.联用

ambiguity
[ˌæmbɪ'gju:ətɪ]
adj.未知的

concentration
[ˌkɔnsən'treɪʃən]
n.浓度

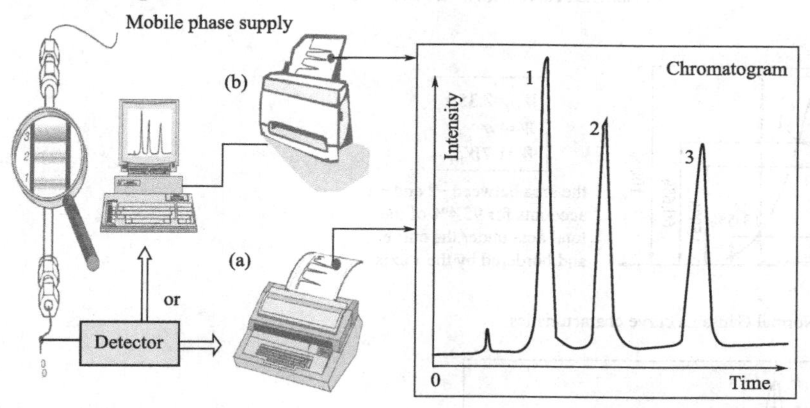

Figure 1.2 The principle of analysis by chromatography. The chromatogram, the essential graph of every chromatographic analysis, describes the passage of components. It is obtained from variations, as a function of time, of an electrical signal emitted by the detector. It is often reconstructed from values that are digitized and stored to a microcomputer for reproduction in a suitable format for the printer, (a) For a long time, the chromatogram was obtained by a simple chart recorder or an integrator; (b) Right, a chromatogram illustrating the separation of a mixture of at least three principal components. Note that the order of appearance of the compounds corresponds to the relative position of each constituent on the column

1.2 The chromatogram

The chromatogram is the representation of the variation, with time (rarely volume), of the amount of the analyte in the mobile phase exiting the chromatographic column. It is a curve that has a baseline which corresponds to the trace obtained in the absence of a compound being eluted. The separation is complete when the chromatogram shows as many chromatographic peaks as there are components in the mixture to be analyzed (Figure 1.3).

A constituent is characterized by its retention time t_R, which represents the time elapsed from the sample introduction to the detection of the peak maximum on the chromatogram. In an ideal case, t_R is independent of the quantity injected.

representation
[ˌreprɪzen'teɪʃən]
n.表现;陈述

variation
[ˌveərɪ'eɪʃn]
n.变化

constituent
[kən'stitjuənt]
n.组分;成分

independent
[ˌɪndɪ'pendənt]
adj.独立自主的

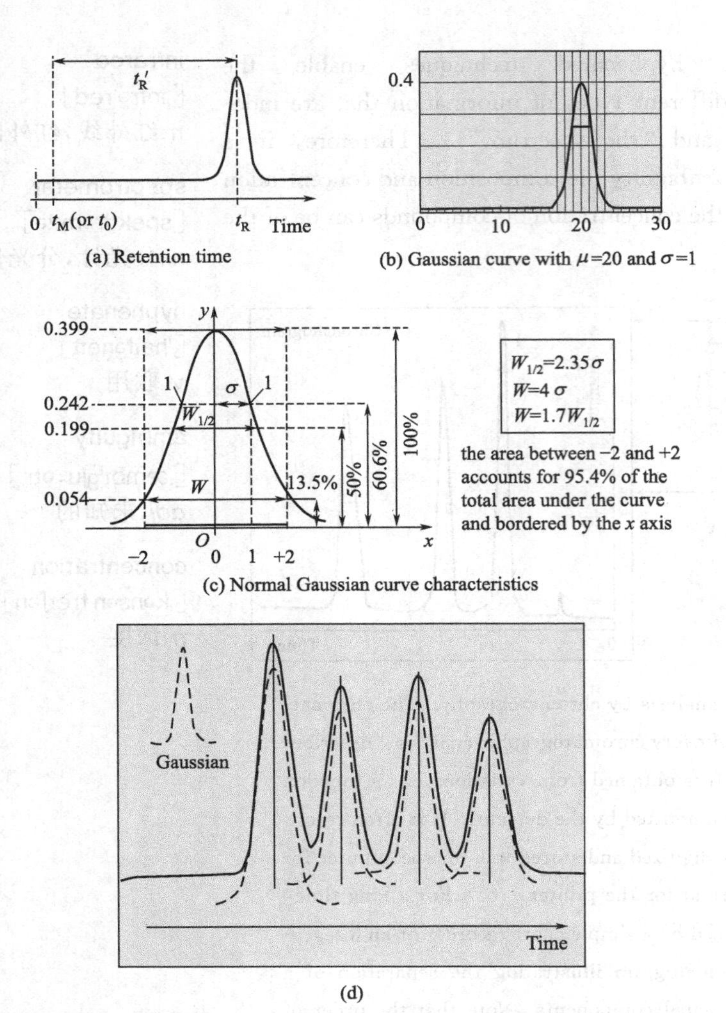

Figure 1.3 Chromatographic peaks. (a) The concept of retention time. The hold-up time t_M is the retention time of an unretained compound in the column (the time it took to make the trip through the column); (b) Anatomy of an ideal peak; (c) Significance of the three basic parameters and a summary of the features of a Gaussian curve; (d) An example of a real chromatogram showing that while travelling along the column, each analyte is assumed to present a Gaussian distribution of concentration

concept
[ˈkɔnsept]
n.观念，概念

significance
[sɪgˈnɪfɪkəns]
n.意义，意思，重要性

distribution
[ˌdɪstrɪˈbjuːʃən]
n.分配，分布

A constituent which is not retained will elute out of the column at time t_M, called the hold-up time or dead time (formerly designated t_0). It is the time required for the mobile phase to pass through the column.

The difference between the retention time and the hold-up time is designated by the adjusted retention time of the compound, t'_R.

If the signal sent by the sensor varies linearly with the concentration of a compound, then the same variation will occur for the area under the corresponding peak on the chromatogram. This is a basic condition to perform quantitative analysis from a chromatogram.

compound
[ˈkɔmpaʊnd]
n.化合物

1.3 Gaussian-shaped elution peaks

On a chromatogram, the perfect elution peak would have same form as the graphical representation of the law of Normal distribution of random errors. In keeping with the classic notation, μ would correspond to the retention time of the eluting peak while σ to the standard deviation of the peak (σ^2 represents the variance). y represents the signal as a function of time x, from the detector located at the outlet of the column (Figure 1.3).

This is why ideal elution peaks are usually described by the probability density function (1.2).

$$Y=\frac{1}{\sqrt[\sigma]{2\pi}}\cdot\exp\left[-\frac{(X-\mu)^2}{2\sigma^2}\right] \tag{1.1}$$

$$Y=\frac{1}{\sqrt{2\pi}}\cdot\exp\left[-\frac{x^2}{2}\right] \tag{1.2}$$

$$a=\frac{b}{f} \tag{1.3}$$

$$TF=\frac{b+f}{2f} \tag{1.4}$$

This function is characterized by a symmetrical curve (maximum for $x=0$, $y=0.3999$) possessing two inflection points at $x=\pm 1$ (Figure 1.3), for which the ordinate value is 0.242 (being 60.6 per cent of the maximum value). The width of the curve at the inflection points is equal to 2σ ($\sigma=1$).

In chromatography, $W_{1/2}$ represents the width of the peak at half-height ($W_{1/2}=2.35\sigma$) and σ^2 the variance of the peak. The width of the peak"at the base"is labeled W and is measured at 13.5 percent of the height. At this position, for the Gaussian curve, $W=4\sigma$ by definition.

Real chromatographic peaks often deviate significantly from the Gaussian ideal aspect. There are several reasons for this. In particular, there are irregularities of concentration in the injection zone, at the head of the column. Moreover, the speed of the mobile phase is zero at the wall of the column and maximum in the centre of the column.

The observed asymmetry of a peak is measured by two parameters, the skewing factor a measured at 10 percent of its height and the tailing factor TF measured at 5 percent (for the definition of these terms, see Figure 1.4).

random error
随机误差

notation
[nəʊ'teɪʃn]
n.符号

variance
['veərɪəns]
n.方差

probability
[ˌprɒbə'bɪlətɪ]
n.概率

inflection
[ɪn'flekʃn]
n.拐点

significantly
[sɪg'nɪfəkəntlɪ]
adv.值得注目地

irregularity
[ɪregjʊ'lærɪtɪ]
n.不规则

asymmetry
[ˌeɪ'sɪmətrɪ]
adj.不对称的

skewing factor
倾斜因子

tailing factor
拖尾因子

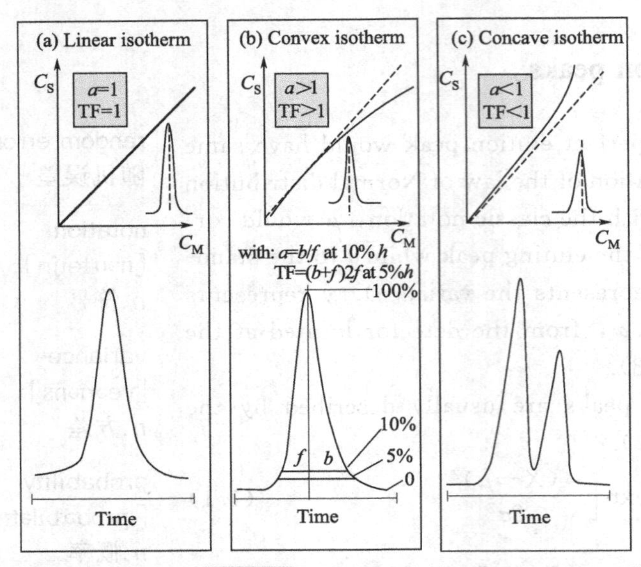

Figure 1.4 Distribution isotherms. (a) The ideal situation corresponding to the invariance of the concentration isotherm; (b) Situation in which the stationary phase is saturated-as a result of which the ascent of the peak is faster than the descent (skewing factor greater than 1); (c) The inverse situation: the constituent is retained too long by the stationary phase, the retention time is therefore extended and the ascent of the peak is slower than the descent apparently normal. For each type of column, the manufacturers indicate the capacity limit expressed in ng/compound, prior to a potential deformation of the corresponding peak. The situations (a), (b) and (c) are illustrated by authentic chromatograms taken out from liquid chromatography technique

isotherm
[ˈaɪsəθɜːm]
n.等温线

1.4 The plate theory

For half a century, different theories have been and continue to be proposed to model chromatography and to explain the migration and separation of analytes in the column. The best known are those employing a statistical approach (stochastic theory), the theoretical plate model or a molecular dynamics approach.

stochastic
[stəˈkæstɪk]
adj.随机的

To explain the mechanism of migration and separation of compounds on the column, the oldest model, known as Craig's *theoretical plate model* is a static approach now judged to be obsolete, but which once offered a simple description of the separation of constituents.

description
[dɪsˈkrɪpʃən]
n.描述,形容

Although chromatography is a dynamic phenomenon, Craig's model considered that each solute moves progressively along a sequence of distinct static steps. In liquid-solid chromatography this elementary process is represented by a cycle of adsorption/desorption.

adsorption
[ædˈsɔːpʃən]
n.吸附(作用)

The continuity of these steps reproduces the migration of the compounds on the column, in a similar fashion to that achieved by a cartoon which gives the illusion of movement through a sequence of fixed images. Each step corresponds to a new state of equilibrium for the *entire* column.

cartoon [kɑːˈtuːn] *n*.动画

These successive equilibria provide the basis of *plate theory* according to which a column of length L is sliced horizontally into N fictitious, small plate-like discs of same height H and numbered from 1 to n. For each of them, the concentration of the solute in the mobile phase is in equilibrium with the concentration of this solute in the stationary phase. At each new equilibrium, the solute has progressed through the column by a distance of one disc (or plate), hence the name *theoretical plate theory*. The height equivalent to a theoretical plate ($HETP$ or H) will be given by equation (1.5):

equilibria [iːkwɪˈlɪbrɪə] *n*.平衡

slice [slaɪs] *vt*.划分

fictitious [fɪkˈtɪʃəs] *adj*.假设的

$$H=\frac{L}{N} \tag{1.5}$$

This employs the polynomial approach to calculate, for a given plate, the mass distributed between the two phases present. At instant I, plate J contains a total mass of analyte m_T which is composed of the quantity m_M of the analyte that has just arrived from plate $J-1$ carried by the mobile phase formerly in equilibrium at instant $I-1$, to which is added the quantity m_S already present in the stationary phase of plate J at time $I-1$ (Figure 1.5).

polynomial [ˌpɒlɪˈnəumjəl] *adj*.多项式的

$$m_T(I,J)=m_M(I-1,J-1)+m_S(I-1,J)$$

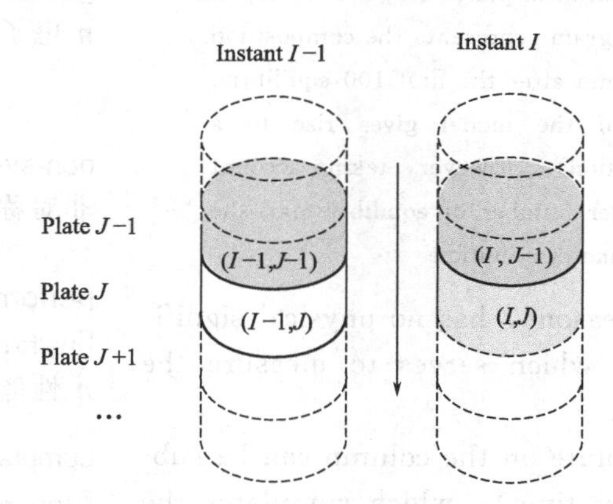

Figure 1.5 Schematic of a column cross-section

If it is assumed for each theoretical plate that: $m_S=K\cdot m_M$ and $m_T=m_M+m_S$, then by a recursive formula, m_T (as well as m_M and m_S), can be calculated. Given that for each plate the analyte is in a concentration equilibrium between the two phases, the total mass of analyte in solution in the volume of the mobile phase V_M of the column remains constant, so long as the analyte has not reached the column outlet. So, the chromatogram corresponds to the mass in

recursive formula 递推公式

constant [ˈkɒnstənt] *adj*.不变的

transit carried by the mobile phase at the $(N+1)th$ plate (Figure 1.6) during successive equilibria. This theory has a major fault in that it does not take into account the dispersion in the column due to the diffusion of the compounds.

diffusion [dɪ'fju:ʒən] n.扩散

The plate theory comes from an early approach by Martin and Synge (Nobel laureates in Chemistry, 1952), to describe chromatography by analogy with distillation and counter current extraction as models.

analogy [ə'nælədʒɪ] n.类似

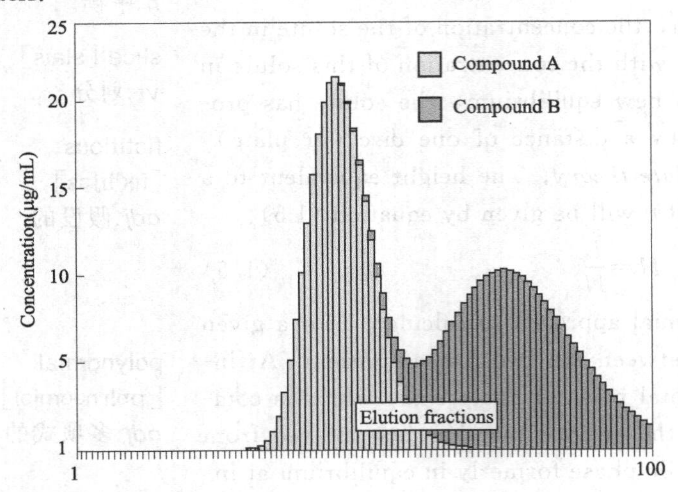

Figure 1.6 Theoretical plate model. Computer simulation, aided by a spreadsheet, of the elution of two compounds A and B, chromatographed on a column of 30 theoretical plates ($K_A=0.6$; $K_B=1.6$, $m_A=300\mu g$; $m_B=300\mu g$). The diagram represents the composition of the mixture at the outlet of the column after the first 100 equilibria. The graph shows that application of the model gives rise to a non-symmetrical peak (Poisson summation). However, taking account of compound diffusion and with a larger number of equilibriums, the peaks look more and more like a Gaussian distribution

spreadsheet ['spred ʃi:t] n.电子制表软件

non-symmetrical 非对称

This term, used for historical reasons, has no physical significance, in contrast to its homonym which serves to measure the performances of a distillation column.

performance [pə'fɔ:məns] n.性能

The retention time t_R, of the solute on the column can be subdivided into two terms: t_M (hold-up time), which cumulates the times during which it is dissolved in the mobile phase and travels at the same speed as this phase, and t_S the cumulative times spent in the stationary phase, during which it is immobile. Between two successive transfers from one phase to the other, it is accepted that the concentrations have the time to reequilibrate.

cumulate ['kju:mjə ˌleɪt] v.累积

successive [sək'sesiv] adj.连续的

In a chromatographic phase system, there are at least three sets of equilibria: solute/mobile phase, solute/stationary phase and mobile phase/stationary phase. In a more recent theory of chromatography, no consideration is given to the idea of molecules immobilized by the stationary phase but rather that were simply slowed down when passing in close proximity.

immobilize
[ɪ'məubəlaɪz]
vt.使不动，使固定

1.5 Nernst partition coefficient (K)

The fundamental physico-chemical parameter of chromatography is the equilibrium constant K, termed the partition coefficient, quantifying the ratio of the concentrations of each compound within the two phases.

$$K=\frac{C_S}{C_M}=\frac{\text{Molar concentration of the solute in the stationary phase}}{\text{Molar concentration of the solute in the mobile phase}} \tag{1.6}$$

Values of K are very variable since they can be large (e.g.1000), when the mobile phase is a gas or small (e.g.2) when the two phases are in the condensed state. Each compound occupies only a limited space on the column, with a variable concentration in each place, therefore the true values of C_M and C_S vary in the column, but their ratio is constant.

constant
[ˈkɒnstənt]
n.常数

Thermodynamic relationships can be applied to the distribution equilibria defined above. K (C_M/C_S), the equilibrium constant relative to the concentrations C of the compound in the mobile phase (M) and stationary phase (S) can be calculated from chromatography experiments. Thus, knowing the temperature of the experiment, the variation of the standard free energy for this transformation can be deduced:

transformation
[ˌtrænsfə'meɪʃən]
n.变化

$$C_M \rightleftharpoons C_S \qquad \Delta G^{\ominus}=-RT\ln K$$

In gas chromatography, where K can be easily determined at two different temperatures, it is possible to obtain the variations in standard enthalpy $\Delta H^{\ominus}$ and entropy $\Delta S^{\ominus}$ (if it accepted that the entropy and the enthalpy have no changed):

enthalpy
[en'θælpɪ]
n.[热]焓

$$\Delta G^{\ominus}=\Delta H^{\ominus}-T\Delta S^{\ominus}$$

The values of these three parameters are all negative, indicating a spontaneous transformation. It is to be expected that the entropy is decreased when the compound moves from the mobile phase to the stationary phase where it is fixed. In the same way, the Van't Hoff equation can be used in a fairly rigorous way to predict the effect of temperature on the retention time of a compound. From this it is

spontaneous
[spɒn'teɪnɪəs]
adj.自发的

rigorous
[ˈrɪgərəs]
adj.严格的

$$\frac{\mathrm{d}\ln K}{\mathrm{d}T}=\frac{\Delta H}{RT^2}$$

clear that for detailed studies in chromatography, classic thermodynamics are applicable.

1.6 Column efficiency

1.6.1 Theoretical efficiency (number of theoretical plates)

As the analyte migrates through column, it occupies a continually expanding zone (Figure 1.6). This linear dispersion σ_1 measured by the variance σ_1^2 increase with the distance of migration. When this distance becomes L, the total column length, the variance will be:

continually
[kən'tɪnjʊəlɪ]
adv.持续地

dispersion
[dɪ'spɜːʃn]
n.分散，分布

$$\sigma_L^2 = HL \tag{1.7}$$

Reminding the plate theory model, this approach also leads to the value of the height equivalent to one theoretical plate H and to the number N, of theoretical plates ($N = L/H$).

Therefore (Figure 1.7), any chromatogram that shows an elution peak with the temporal variance permits the determination of the theoretical efficiency N for the compound under investigation (1.8), and by deduction, of the value of H knowing that $H = L/N$;

temporal
['tempərəl]
adj.短暂的

efficiency
[ɪ'fɪʃənsɪ]
n.效率

investigation
[ɪnˌvestɪ'geɪʃn]
n.(正式的)调查

$$N = \frac{L^2}{\sigma_L^2} \quad \text{or} \quad N = \frac{t_R^2}{\sigma^2} \tag{1.8}$$

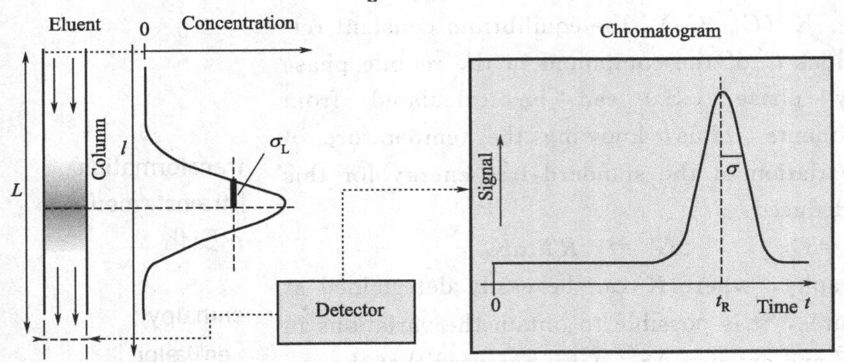

Figure 1.7 Dispersion of a solute in a column and its translation on a chromatogram. Left, graph corresponding to the isochronic image of the concentration of an eluted compound at a particular instant. Right, chromatogram revealing the variation of the concentration at the outlet of the column, as a function of time. t_R and σ are in the same ratio as L and σ_L. In the early days, the efficiency N was calculated from the chromatogram by using a graduated ruler

calculated
['kælkjuˌleɪtɪd]
adj.计算出的

If these two parameters are accessible from the elution peak of the compound, just because t_R and σ are in the same ratio as that of L to σ_L.

On the chromatogram, σ represents the half-width of the peak at 60.6 percent of its height and t_R the retention time of the compound. t_R and σ should be measured in the same units (time, distances or eluted volumes if the flow is constant). If σ is expressed in units of volume (using the flow), then 4σ corresponds to the "volume of the peak", that contains around 95 percent of the injected compound. By consequence of the properties of the Gaussian curve ($W = 4\sigma$ and $W_{1/2} = 2.35\sigma$), Equation 1.9 results. However, because of the distortion of most peaks at their base, expression 1.9 is rarely used and finally Equation 1.10 is preferred.

consequence
[ˈkɔnsɪkwəns]
n.结果,成果

N is relative parameter, since it depends upon both the solute chosen and the operational conditions adopted. Generally, a constituent is selected which appears towards the end of the chromatogram in order to get a reference value, for lack of advance knowledge of whether the column will successfully achieve a given separation.

selected
[sɪˈlektɪd]
adj.挑选出来的

$$N = 16\frac{t_R^2}{W^2} \quad (1.9)$$

$$N = 5.54\frac{t_R^2}{W_{1/2}^2} \quad (1.10)$$

1.6.2 Effective plates number (real efficiency)

In order to compare the performances of columns of different design for a given compound-or to compare, in gas chromatography, the performances between a capillary column and a packed column-more realistic values are obtained by replacing the total retention time, which appears in expressions 1.8-1.10, by the adjusted retention time which does not take into account the hold-up time spent by any compound in the mobile phase ($t'_R = t_R - t_M$). The three preceding expressions become:

hold-up
n.停留,保留

$$N_{eff} = \frac{t_R'^2}{\sigma^2} \quad (1.11)$$

$$N_{eff} = 16\frac{t_R'^2}{W^2} \quad (1.12)$$

$$N_{eff} = 5.54\frac{t_R'^2}{W_{1/2}^2} \quad (1.13)$$

Currently it is considered that these three expressions are not very useful.

1.6.3 Height equivalent to a theoretical plate (HETP)

The equivalent height of a theoretical plate H, as already defined (expression 1.5), is calculated for reference compounds to permit a comparison of columns of different lengths. H does not behave as a constant, its value depends upon the compound chosen and upon the experimental conditions.

For a long time in gas chromatography an adjustment value called the effective height of a theoretical plate H_{eff} was calculated using the true efficiency.

This corresponds to the Equation 1.14;

$$H_{eff}=\frac{L}{N_{eff}} \quad (1.14)$$

In chromatography, in which the mobile phase is a liquid and the column is filled with spherical particles, the adjusted height of the plate h, is often encountered. This parameter takes into account the average diameter d_m of the particles. This eliminates the effect of the particle size. Columns presenting the same ratio (length of the column)/(diameter of the particles) will yield similar performances.

$$h=\frac{H}{d_m}=\frac{L}{Nd_m} \quad (1.15)$$

eliminate [ɪ'lɪmɪneɪt] vt.消除，除去

yield [ji:ld] vt.产生

1.7 Retention parameters

Hold-up times or volumes are used in chromatography for various purposes, particularly to access to retention factor k. and thermodynamic parameters. Only basic expressions are given below.

access ['ækses] n.接近

thermodynamic [ˌθɜ:məʊdaɪ'næmɪks] adj.热力学的

expression [ɪk'spreʃən] n.表示

1.7.1 Retention times

The definition of hold-up time, t_M, retention time, t_R and adjusted retention time, t'_R, have been given previously (Section 1.2).

1.7.2 Retention volume (or elution volume)

The retention volume V_R of an analyte represents the volume of mobile phase necessary to enable its migration throughout the column from the moment of entrance to the moment in which it leaves. To estimate this volume, different methods (direct or indirect) may be used, that depend of the physical state of the mobile phase. On a standard chromatogram with time in abscissa, V_R is calculated from expression 1.16, if the flow rate F is constant,

$$V_R=t_RF \quad (1.16)$$

analyte [ænə'lɪt] n.分析物，被分析物

leave [li:v] vt.离开

chromatogram [krəʊ'mætəˌgræm] 色谱图，色层分离谱

The volume of a peak, V_{peak} corresponds to that volume of the mobile phase in which the compound is diluted when leaving the column. It is defined by:

$$V_{peak}=WF \quad (1.17)$$

dilute [daɪ'lu:t] vt.稀释

1.7.3 Hold-up volume (or dead volume)

The volume of the mobile phase in the column (known as the dead volume), V_M, corresponds to the accessible interstitial volume. It is often calculated from a chromatogram, provided a solute not retained by the stationary phase is present. The dead volume is deduced from t_M and the flow rate F:

interstitial [ˌɪntə'stɪʃl] adj.空隙的，裂缝的

$$V_M = t_M F \quad (1.18)$$

Sometimes, in the simplest cases, the volume of the stationary phase designated by V_S can be calculated by subtracting the dead volume V_M from the total internal volume of the empty column.

designate
[ˈdezɪgneɪt]
vt.固定

1.7.4 Retention (or capacity) factor

When a compound of total mass m_T is introduced onto the column, it separates into two quantities: m_M, the mass in the mobile phase and m_S, the mass in the stationary phase. During the solute's migration down the column, these two quantities remain constant.

Their ratio, called the retention factor k, is constant and independent of m_T:

$$k = \frac{m_S}{m_M} = \frac{C_S}{C_M} \times \frac{V_S}{V_M} = K \frac{V_S}{V_M} \quad (1.19)$$

The retention factor, also known as the capacity factor k, is a very important parameter in chromatography for defining column performances. Though it does not vary with the flow rate or the column length, k is it not a constant as it depends upon the experimental conditions. For this reason, it is sometimes designated by k' rather than k alone.

capacity
[kəˈpæsətɪ]
n.容量

This parameter takes into account the ability, great or small, of the column to retain each compound. Ideally, k should be superior to one but less than five, otherwise the time of analysis is unduly elongated.

An experimental approach of k can be as follows:

elongate
[ˈiːlɒŋgeɪt]
vt.使延长

Suppose the migration of a compound in the column. Recalling Craig's model, each molecule is considered as passing alternately from the mobile phase (in which it progresses down the column), to the stationary phase (in which it is immobilized). The average speed of the progression down the column is slowed if the time periods spent in the stationary phase are long. Extrapolate now to a case which supposes n molecules of this same compound (a sample of mass m_T). If we accept that at each instant, the ratio of the n_S molecules fixed upon the stationary phase (mass n_S) and of the n_M molecules present in the mobile phase (mass n_M), is the same as that of the times (t_S and t_M) spent in each phase for a single molecule, the three rations will therefore have the same value:

immobilize
[ɪˈməubəlaɪz]
v.使固定

extrapolate
[ɪkˈstræpəleɪt]
v.推断

$$\frac{n_S}{n_M} = \frac{m_S}{m_M} = \frac{t_S}{t_M} = k$$

Take the case of a molecule which spends 75 per cent of its time in the stationary phase. Its average speed will be four times slower than if it rested permanently in the mobile phase. As a consequence,

stationary
[ˈsteɪʃənrɪ]
adj.固定的,静止的

if 4μg of such a compound has been introduced onto the column, there will be an average of 1μg permanently in the mobile phase and 3μg in the stationary phase.

Knowing that the retention time of a compound t_R is such that $t_R = t_M + t_S$, the value of k is therefore accessible from the chromatogram ($t_S = t'_R$); see Figure 1.7:

$$k = \frac{t'_R}{t_M} = \frac{t_R - t_M}{t_M} \quad (1.20)$$

This important relation can also be written:

$$t_R = t_M(1+k) \quad (1.21)$$

Bearing in mind the relations (1.16) and (1.18), the retention volume V_R of a solute can be written:

$$V_R = V_M(1+k) \quad (1.22)$$

or

$$V_R = V_M + KV_S \quad (1.23)$$

This final expression linking the experimental parameters to the thermodynamic coefficient of distribution K, is valid for the ideal chromatography.

coefficient
[kəʊɪ'fɪʃnt]
n.[数]系数,率;协同因素

1.8 Separation (or selectivity) factor between two solutes

The separation factor α, (1.24) enables the comparison of two adjacent peaks 1 and 2 present in the same chromatogram (Figure 1.8). Using Equations 1.20 and 1.19, it can be concluded that the separation factor can be expressed by Equation 1.25.

distribution
[dɪstrə'bjuʃən]
n.分布,分配

comparison
[kəm'pærɪsən]
n.比较

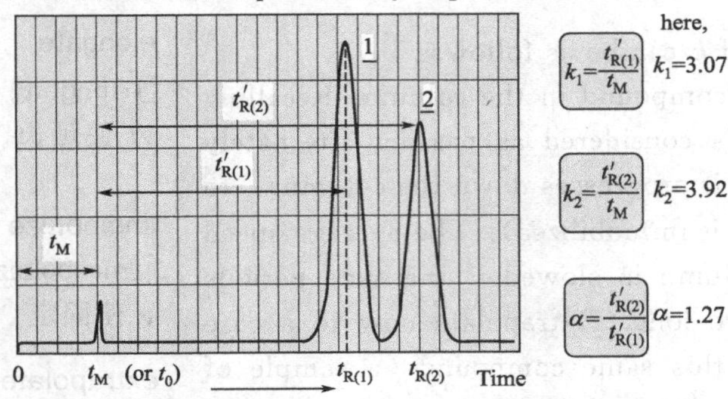

Figure 1.8 Retention factors and separation factor between two compounds. Each compound has its own retention factor. On this figure, the separation factor is around 1.3. The separation factor is also equal to the ratio of the two components

By definition α is greater than unity (species 1 elutes faster than species 2):

$$\alpha = \frac{t'_{R(2)}}{t'_{R(1)}} \quad (1.24)$$

or

$$\alpha = \frac{k_2}{k_1} = \frac{K_2}{K_1} \quad (1.25)$$

For non-adjacent peaks the relative retention factor r, is applied, which is calculated in a similar manner to α.

1.9 Resolution factor between two peaks

To quantify the separation between two compounds, another measure is provided by the resolution factor R. Contrary to the selectivity factor which does not take into account peak widths, the following expression is used to calculate R between two compounds 1 and 2 (Figure 1.9):

calculate
[ˈkælkjuleɪt]
vt.计算

$$R = 2\frac{t'_{R(2)} - t'_{R(1)}}{W_1 + W_2} \qquad (1.26)$$

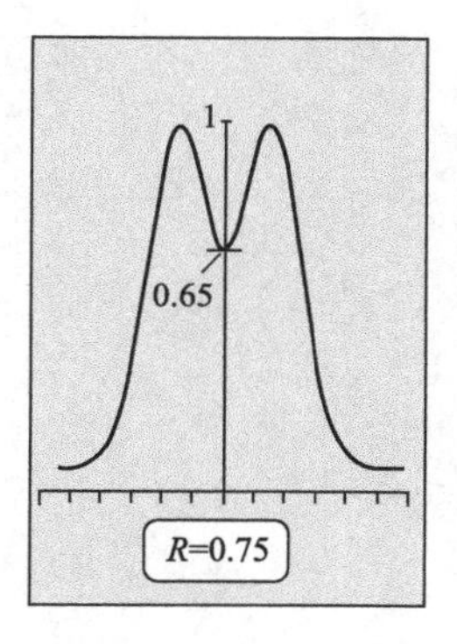

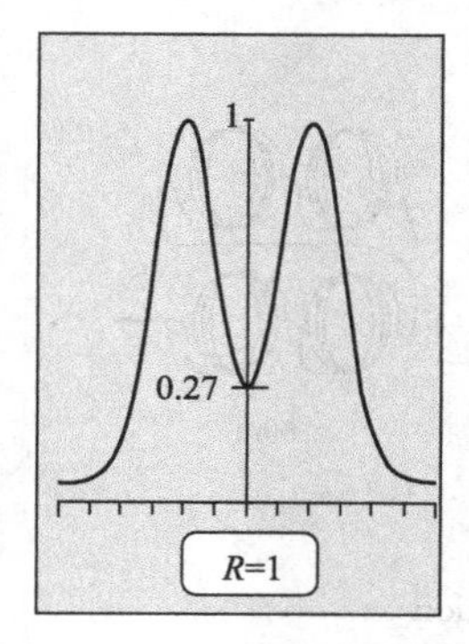

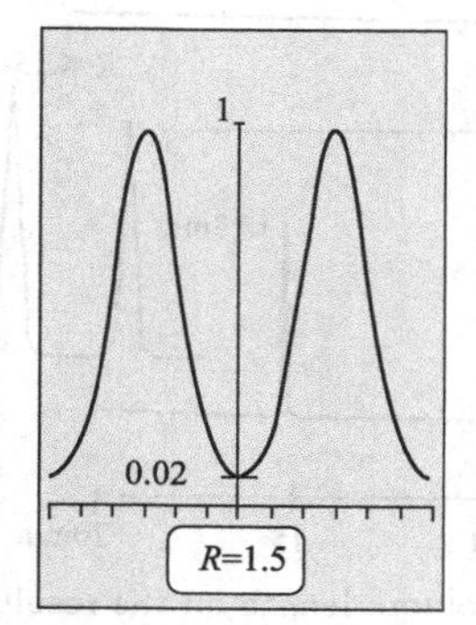

Figure 1.9 Resolution factor. A simulation of chromatographic peaks using two identical Gaussian curves, slowly separating. The visual aspects corresponding to the values of R are indicated on the diagrams. From a value of $R = 1.5$ the peaks can be considered to be baseline resolved, the valley between them being around 2 percent

Other expressions derived from the preceding ones and established with a view to replacing one parameter by another or to accommodate simplifications may also be employed to express the resolution. Therefore expression 1.27 is used in this way.

It is also useful to relate the resolution to the efficiency, the retention factor and the separation factors of the two solutes (expression 1.28, obtained from 1.26 when $W_1 = W_2$). The chromatograms on Figure 1.10 present an experimental verification.

simplification
[ˌsɪmplɪfɪˈkeɪʃn]
n.简化

$$R = 1.77\frac{t_{R(2)} - t_{R(1)}}{\delta_1 + \delta_2} \qquad (1.27)$$

verification
[ˌverɪfɪˈkeɪʃən]
n.证明，验证，核实

$$R = \frac{1}{4}\sqrt{N} \times \frac{\alpha - 1}{\alpha} \times \frac{k_2}{1 + k_2} \qquad (1.28)$$

$$R = \frac{\sqrt{N}}{2} \times \frac{\alpha - 1}{\alpha} \times \frac{k_1 - k_2}{k_1 + k_2 + 2} \qquad (1.29)$$

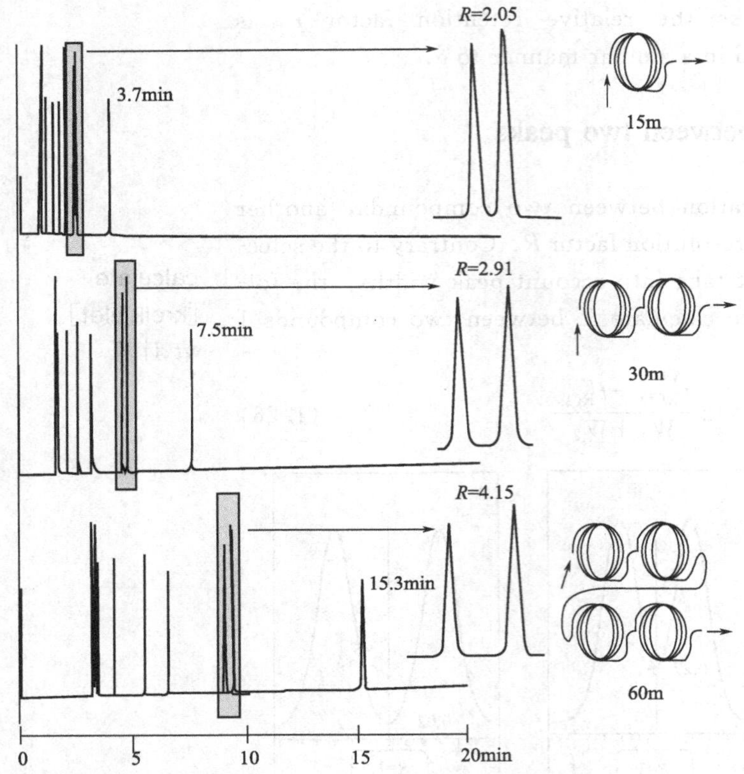

Figure 1.10 Effect of column length on the resolution

Chromatograms obtained with a GC instrument illustrating that by doubling the length of the capillary column, the resolution is multiplied by 1.41 or $\sqrt{2}$ (adapted from a document of SGE Int. Ltd)

illustrate [ɪˈləstreɪt] *vt*.阐明，说明，例证

1.10 The rate theory of chromatography

In all of the previous discussion and particularly in the plate theory, the velocity of the mobile phase in the column and solute diffusion are, perhaps surprisingly, never taken into account. Of all things, the speed should have an influence upon the progression of the analytes down the column, hence their dispersion and by consequence, upon the quality of the analysis undertaken.

Rate theory is a more realistic description of the processes at work inside a column which takes account of the time taken for the solute to equilibrate between the two phases. It is dynamics of the separation process which is concerned. The first kinetic equation for packed columns in gas phase chromatography was proposed by Van Deemter.

dispersion [dɪˈspɜːʒn] *n*.散布

equilibrate [ɪˈkwɪlaɪbreɪt] *vi*.平衡，相称；*vt*.使平衡，与……平衡

dynamics [daɪˈnæmɪks] *n*.动力学

1.10.1 Van Deemter's equation

This equation is based on a Gaussian distribution, similar to that of plate theory. Its simplified form, proposed by Van Deemter in 1956, is well known (expression 1.30). The expression links the plate high H to the average linear velocity of the mobile phase $\overline{u}$ in column (Figure 1.11).

$$H = A + \frac{B}{\overline{u}} + C\overline{u} \tag{1.30}$$

The three experimental basic coefficients A, B and C are related to diverse physico-chemical parameters of the column and to the experimental conditions. If H is expressed in cm, A will also be in cm, B in cm^2 and C in s (where velocity is measured in cm/s).

physico-chemical 物理化学的；物化

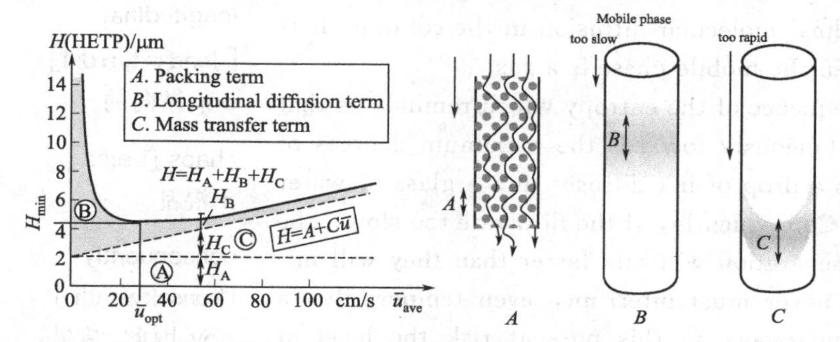

Figure 1.11 Van Deemter's curve in gas chromatography with the domains of parameters A, B and C indicated. There exists an equation similar to that of Van Deemter that considers temperature: $H = A + B/T + CT$

indicated [ˈɪndəˌketɪd] *adj*.表明的，指示的；*v*.表明，指出，显示

This equation reveals that there exists an optimal flow rate for each column, corresponding to the minimum of H, which predicts the curve described by Equation 1.30.

The loss in efficiency as the flow rate increases is obvious, and represents what occurs when an attempt is made to rush the chromatographic separation by increasing the pressure upon the mobile phase.

occur [əˈkɜː] *v*.出现

However, intuition can hardly predict the loss in efficiency that occurs when the flow rate is too slow. To explain this phenomenon, the origins of the terms A, B and C must be recalled. Each of these parameters represents a domain of influence which can be perceived on the graph (Figure 1.11).

intuition [ˌɪntjuˈɪʃn] *n*.直观

perceived [pəˈsiːvd] *v*.感知

The curve that represents the Van Deemter equation is a hyperbola which goes through a minimum (H_{min}) when:

$$\overline{u}_{opt} = \sqrt{\frac{B}{C}} \tag{1.31}$$

hyperbola [haɪˈpɜːbələ] *n*.双曲线

minimum [ˈmɪnɪməm] *n*.最小值，最低限度

(1) Packing related term $A = 2\lambda d_p$

Term A is related to the flow profile of mobile phase passing through the stationary phase. The size of the particles (diameter d_p), their dimensional distribution and the uniformity of the packing (factor characteristic of packing λ) can all be the origin of flow paths of different length which cause broadening of the solute band and improper exchanges between the two phases. This results in turbulent or *Eddy* diffusion, considered to have little importance in liquid chromatography and absent for WCOT capillary columns in GC (Golay's equation without term A, cf. paragraph 1.10.2). For a given column, nothing can be done to reduce the A term.

(2) Gas (mobile phase) term $B=2\gamma D_G$

Term B, which can be expressed from D_G, the diffusion coefficient of the analyte in the gas phase and γ, the above packing factor, is related to the longitudinal molecular diffusion in the column. It is especially important when the mobile phase is a gas.

This term is a consequence of the entropy which reminds us that a system will tend spontaneously towards the maximum degrees of freedom, chaos, just as a drop of ink diffuses into a glass of water into which it has fallen. Consequently, if the flow rate too slow, the compounds undergoing separation will mix faster than they will migrate. This is why one never must interrupt, even temporarily, a chromatography once underway, as this puts at risk the level of efficiency of the experiment.

(3) Liquid (stationary phase) term $C=C_G+C_L$

Term C, which is related to the resistance to mass transfer of the solute between the two phases, becomes dominant when the flow rate is too high for an equilibrium to be attained. Local turbulence within the mobile phase and concentration gradients slow the equilibrium process ($C_S \rightleftharpoons C_M$). The diffusion of solute between the two phases is not instantaneous, so that it will be carried along out of equilibrium. The higher the velocity of mobile phases, the worse the broadening becomes. No simple formula exists which takes into account the different factors integrated in term C. The parameter C_G is dependent upon the diffusion coefficient of the solute in a gaseous mobile phase, while the term C_L depends upon the diffusion coefficient in a liquid stationary phase. Viscous stationary phases have larger C terms

In practice, the values for the coefficients of A, B and C in Figure 1.11 can be accessed by making several measurements of efficiency for the same compound undergoing chromatography at different flow rates, since flow and average linear speed are related. Next the hyperbolic function that best satisfies the experimental values can be calculated using, by preference, the method of multiple linear regression.

1.10.2 Golay's equation

A few years after Van Deemter, Golay proposed a modified rela-

turbulent
['tɜ:bjələnt]
adj.湍流的，激流的

eddy diffusion
涡流扩散

longitudinal
[ˌlɒndʒə'tju:dnl]
adj.纵向的

chaos ['keɪɒs]
n.混乱

consequently
['kɑnsəkwəntlɪ]
adv.因此，结果，所以

turbulence
['tɜ:bjələns]
n.湍流

instantaneous
[ˌɪnstən'teɪnɪəs]
adj.瞬间的，即时的，猝发的

hyperbolic
[ˌhaɪpə'bɒlɪk]
adj.双曲线的，夸张的

multiple['mʌltəpl]
adj.多重的，多样的，许多的；
n.倍数，[电]并联

modified
['mɒdɪfaɪd]
*adj*改进的，修改的，改良的

tionship reserved to capillary columns used in gas phase chromatography. There is no A term because there is no packing in a capillary column

$$H=\frac{B}{\overline{u}}+C_{L}\overline{u}+C_{G}\overline{u} \tag{1.32}$$

Expression 1.33 leads to the minimum value for the $HETP$ for a column of radius r, if the retention factor of the particular compound under examination is known.

The coating efficiency can then be calculated being equal to 100 times the ratio between the value found using expression (1.33) and that deduced from the efficiency ($H=L/N$) obtained from the chromatogram.

$$H_{\text{theo. min}}=r\sqrt{\frac{1+6k+11k^{2}}{3+(1+k)^{2}}} \tag{1.33}$$

1.11 Optimization of a chromatographic analysis

Analytical chromatography is used essentially in quantitative analysis. In order to achieve this effectively, the areas under the peaks must be determined with precision, which in turn necessitates well-separated analytes to be analyzed. A certain experience in chromatography is required when the analysis has to be optimized, employing all available resources in terms of apparatus and software that can simulate the results of temperature modifications, phases and other physical parameters.

modification
[ˌmɒdɪfɪˈkeɪʃn]
n.修改，修饰

In gas phase chromatography, the separations can be so complex that it can be difficult to determine in advance whether the temperature should be increased or decreased. The choice of column, its length, its diameter, the stationary phase composition and the phase ration (V_M/V_S) as well as the parameters of separation (temperature and flow rate), are among the factors which interact with each other.

composition
[ˌkɒmpəˈzɪʃən]
n.成分

The resolution and the elution time are the two most important dependent variables to consider. In all optimizations, the goal is to achieve a sufficiently complete separation of the compounds of interest in the minimum time, though it should not be forgotten that time will be required to readjust the column to the initial conditions to be ready for the next analysis. Chromatography corresponds, in fact, to a slow type of analysis. If the resolutions are very good then optimization consists to save time in the analysis. This can be done by the choice of a shorter column-recalling that the resolution varies with the square root of the column length (cf. the parameter N of formula 1.28 and Figure 1.10).

readjust
[ˌriəˈdʒʌst]
vt.再调整

formula
[ˈfɔːmjələ]
n.[数] 公式，准则

Figure 1.12 shows the optimization of a separation, by liquid

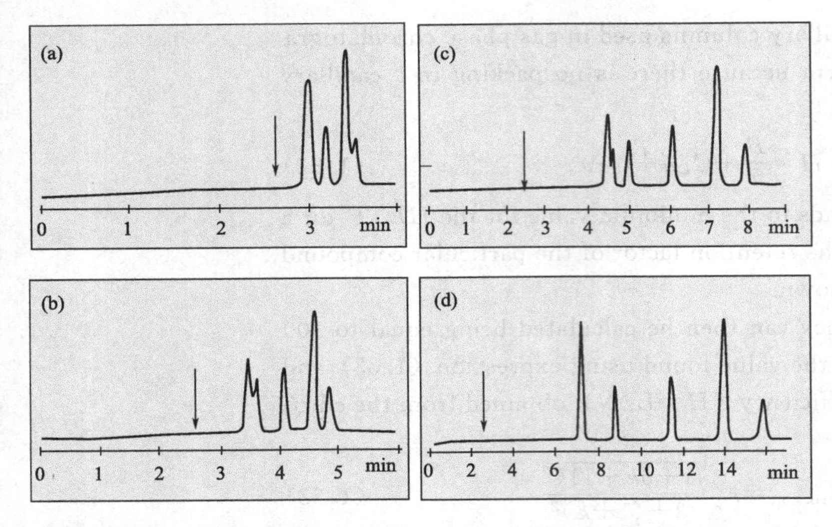

Figure 1.12 Chromatograms of a separation. The mobile phase in each trace is a binary mixture water/acetonitrile: (a) 50/50; (b) 55/45; (c) 60/40; (d) 65/35. The arrow indicates the dead time t_M (min)

chromatography, of a mixture of aromatic hydrocarbons. In this case, optimization of the separation has been carried out by successive modifications of the composition of the mobile phases. Note that by optimizing the sequence in this manner, the cycle time of analysis increases.

If only certain compounds present in a mixture are of interest, then a selective detector can be used which would detect only the desired components. Alternately, at the other extreme, attempts might be made to separate the largest number of compounds possible within the mixture.

Depending upon the different forms of chromatography, optimization can be more or less rapid. In gas phase chromatography optimization is easier to achieve than in liquid chromatography in which the composition of mobile phase must be considered: software now exists that can help in the choice of mobile phase composition. Based upon certain hypotheses (Gaussian peaks), the areas of poorly defined peaks can be found. The chromatographer must work within the limits bound by a triangle whose vertices correspond to three parameters which are in opposition: the resolution, the speed and the capacity (Figure 1.13). An optimized analytical separation uses the full potential of the selectivity which is the most efficient parameter. In the chromatographer's triangle shown, the optimized conditions are close to the vertex of resolution.

sequence [sɪkwəns] *n*.序列，顺序，续发事件

alternately [ˈɔltə:nətli] *adv*.交替地，轮流地，隔一个地

vertex [vɜ:teks] *n*.顶点

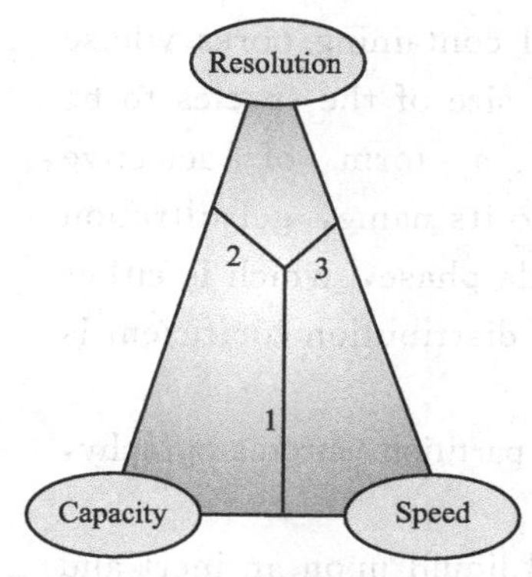

Figure 1.13 The chromatographer's triangle. The shaded areas indicate the domain corresponding to analytical chromatography based principally upon the five parameters K, N, k, α and R

1.12 Classification of chromatographic techniques

Chromatographic techniques can be classified according to various criteria: as a function of the physical nature of the phases; of the process used; or by the physico-chemical phenomena giving rise to the Nernst distribution coefficient K. The following classification has been established by consideration of the physical nature of the two phases involved (Figure 1.14).

classification
[ˌklæsɪfɪ'keɪʃn]
n.分类,类别,等级

1.12.1 Liquid phase chromatography (LC)

This type of chromatography, in which the mobile phase is a liquid belongs to the oldest known form of the preparative methods of separation. This very broad category can be sub-divided depending on the retention phenomenon.

belongs to
属于

(1) Liquid/solid chromatography (or adsorption chromatography)

The stationary phase is a solid medium to which the species adhere through the dual effect of physisorption and chemisorption. The physico-chemical parameter involved here is the adsorption coefficient. Stationary phases have made much progress since the time of Tswett, who used calcium carbonate or inulin (a very finely powdered polymer of ordinary sugar).

physisorption
[ˌfɪzɪ'sɔːpʃən]
n.物理吸附

carbonate
['kɑːbənət]
n. 碳酸盐;
vt. 使变成碳酸盐

(2) Ion chromatography (IC)

In this technique, the mobile phase is a buffered solution while the solid stationary phase has a surface composed of ionic sites. These phases allow the exchange of their mobile counter ion with ions of the same charge present in the sample. This type of separation relies on ionic distribution coefficients.

inulin['ɪnjəlɪn]
n.[生化] 菊粉

distribution
[ˈdɪstrəˈbjuʃən]
n.分布,分配

(3) Size exclusion chromatography (SEC)

The stationary phase here is a material containing pores whose dimensions are selected as a function of the size of the species to be separated. This method therefore uses a form of selective permeability at the molecular level leading to its name, gel filtration or gel permeation on the nature of the mobile phase, which is either aqueous or organic. For this technique, the distribution coefficient is called the diffusion coefficient.

(4) Liquid/liquid chromatography (or partition chromatography, LLC)

The stationary phase in an immobilized liquid upon an inert and porous material, which has only a mechanical role of support. Impregnation, the oldest procedure for immobilizing a liquid on a porous material, is a method now abandoned because of the elevated risk of washing out the column, which is called bleeding.

impregnation [ˌɪmpreg'neɪʃn] n.浸渍

(5) Liquid/bound phase chromatography

In order to immobilize the stationary phase (generally a liquid polymer), it is preferable to fix it by covalent bonding to a mechanical support. The quality of separation depends upon the partition coefficient K of the solute between an aqueous and organic phase in a separating funnel.

aqueous [ˈekwɪəs] adj.水的，水般的

1.12.2 Gas phase chromatography (GC)

The mobile phase in inert gas and as above this form of chromatography can be subdivided according to the nature of the phase components:

(1) Gas/liquid chromatography (GLC)

As indicated above the mobile phase here is a gas and the stationary phase is an immobilized liquid, either by impregnation or by bonding to an inert support which could be, quite simply, the inner surface of the column. This is the technique commonly called gas phase chromatography (GC). The gaseous sample must be brought to its vapour state. It was Martin and Synge who, in 1941, suggested the replacement of the liquid mobile phase by a gas in order to improve the separations. From this era comes the true beginnings of the development of analytical chromatography. Here once again it is the partition coefficient K that is involved.

replacement [rɪ'pleɪsmənt] n.更换

analytical [ˌænə'lɪtɪkl] adj.分析的，解析的

(2) Gas/solid chromatography (GSC)

The stationary phase is a porous solid (such as graphite, silica gel or alumina) while the mobile phase is a gas. This type of gas chromatography is very effective for analyses of gas mixtures or of compounds that have a low boiling point. The parameter concerned is the adsorption coefficient.

adsorption [æd'sɔːrpʃən] n.吸附(作用)

1.12.3 Supercritical fluid chromatography (SFC)

Here the mobile phase is fluid in its supercritical state, as carbon dioxide at about 50℃ and at more than 150bar (15MPa). The stationary phase can be a liquid or a solid. This technique combines the advantages of those discussed above: liquid/liquid and gas/liquid chromatography.

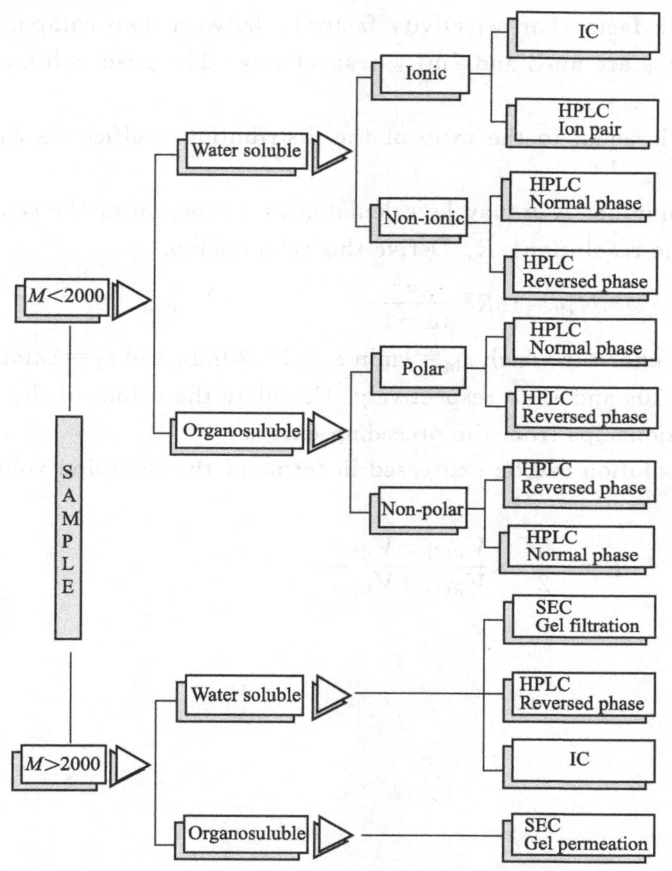

Figure 1.14 Selection guide for all of the different chromatographic techniques with liquid mobile phases. The choice of technique is chosen as a function of the molar mass, solubility and the polarity of the compounds to the separated

solubility [ˌsɒlju'bɪlətɪ] *n*.溶解度

Problems

1.1 A mixture placed in an Erlenmeyer flask comprises 6mL of silica gel and 40mL of a solvent containing, in solution, 100mg of a non-volatile compound. After stirring, the mixture was left to stand before a 10mL aliquot of the solution was extracted and evaporated to dryness. The residue weighed 12mg. Calculate the adsorption coefficient, K, of the compound in this experiment.

1.2 The retention factor (or capacity factor), k of a compound is defined as $k = m_S/$

m_M, that is by the ratio of the compound in equilibrium in the two phases. Show, from the information given in the corresponding chromatogram, that the expression used—$k=(t_R-t_M)/t_M$—is equivalent to this. Remember that for a given compound that relation between the retention time the time spent in the mobile phase t_M (hold-up or dead time) and the time spent in the stationary phase t_S, is as follows:

$$t_R=t_M+t_S$$

1.3

1. Calculate the separation factor (or selectivity factor), between two compounds, 1 and 2, whose retention volumes are 6mL and 7mL, respectively. The dead volume of the column used is 1mL.

2. Show that this factor is equal to the ratio of the distribution coefficients k_1/k_2 of these compounds ($t_{R(1)}<t_{R(2)}$).

1.4 The effective plate number N_{eff} may be calculated as a function of the separation factor α for a given value of the resolution, R. Derive this relationship.

$$N_{eff}=16R^2\frac{\alpha^2}{(\alpha-1)^2}$$

1.5 Consider two compounds for which $t_M=1$min $t_1=11.30$min and $t_2=12$min. The peak widths at half-height are 10s and 12s, respectively. Calculate the values of the respective resolutions using the relationships from the preceding exercise.

1.6 Explain how the resolution can be expressed in terms of the retention volumes in the following equation:

$$R=\frac{\sqrt{N}}{2}\times\frac{V_{R(2)}-V_{R(1)}}{V_{R(1)}+V_{R(2)}}$$

第1章 色谱法的发明

谁发明了色谱法这种在实验室中应用最广泛的技术？这个问题的答案目前还存在争议。19世纪50年代，Schonbein用滤纸部分分离了溶液中的物质。他发现并不是所有的溶液都会到达滤纸的同一高度。Goppelsroder（瑞士）发现溶液中的化学成分与其爬上试纸的高度之间存在联系。1861年，他写道："我相信，这个方法将会在快速确定混合染料的性质中非常实用。"

虽然这两个人对纸色谱法的进展做了有价值的工作，但是普遍认为是Michael S. Tswett于1900年后不久发明了现代色谱分离方法。通过他连续发表的文章，人们确实可以重建他的思维过程，这使得他即使不能被认为是这种重要分离方法的发明者，也可以被认为是其研究的先驱。他的研究领域涉及植物生物化学。当时人们使用乙醇可以很容易地从室内植物（house plants）的叶子中提取叶绿素和其他色素。通过蒸发溶剂得到黑色的提取物，它可以在其他溶剂（极性或非极性溶剂），特别是石油醚中溶解。然而，当时很难理解为什么使用石油醚无法直接从树叶中提取叶绿素。Tswett认为是某种分子力将植物叶绿素保留在叶子中，从而阻止了石油醚的提取。他此时也预见了吸附的原理。得到这一结论后，为了验证这种假设，他将这些色素溶解在石油醚中，用滤纸（纤维素）代替叶组织。他认识到滤纸收集了颜色，通过向混合物中添加乙醇，就可以重新萃取与之前相同的色素。

作为前期工作的延续，他决定对有机或无机的各类粉末进行试验。为了节省时间，他设计了一套系统，使他能够同时做几个实验。他将包装好的粉末放置在细管中，并将溶解在石油醚中的染料加入其中进行测试。他观察到在某些管子中的粉末变成了不同颜色相叠加的环，这表明颜料的性质不同其保留力也不同。通过使用合适的溶剂冲洗细管，他分离并获得了其中一些组分。现代色谱法就这样产生了。其后，在1906年，他发表了此实验结果，其中他写道：就像光谱中的光线一样，颜料混合物中有不同的组分，遵循着一定的规律，可以被碳酸钙柱子分离，并可进行定性和定量分析。他将这个过程称为色谱，对应的方法称为色谱法。

通过色谱法可以将混合物中的组分分开，这已经成为识别和鉴定气相或液相混合物的主要分析方法之一。其基本原理是在不互溶两相之间待分离组分的浓度分配平衡。其中一相在色谱柱中是不流动或者被固定在载体上的，所以称为固定相；而第二相被强迫流过第一个相，称为流动相。这两相的选择原则是样品在两相中的溶解度不同。不同化合物的迁移速度不同使其得以分离。在所有的仪器分析方法中，这种基于流体力学的方法应用最广泛。色谱法在涉及分子分析的所有实验室有着重要的地位。

1.1 分析色谱的一般概念

色谱法是与蒸馏、结晶或者分离萃取等类似的分离气相或液相混合物组分的物理化学方法。因为实验室中经常需要处理很多非均相液态混合物，或者固体，它们可以通过合适的溶剂溶解（因此也可以认为是混合物），因此这种方法的应用非常广泛。

基本的色谱过程可描述如下（图 1.1）：

1）在一个竖直中空玻璃管（圆柱形）中填充适当的细粉状固体，即固定相。

2）在此柱的顶部加有少量待分离为单独组分的样品混合物。

3）混合物样品被连续加入的流动相所携带，在重力作用下，携带着混合物中的各种成分一起流过柱子，这个过程叫作洗脱。如果组分以不同的速度迁移，它们将彼此分离开，并可以与流动相一起回收。

这个在柱子中进行的基本分离过程，从其被发现以来，已经用于许多化合物的大规模分离或纯化上（制备色谱法）。尤其是不需要收集纯组分进行鉴别，而是使用不同化合物的保留时间作为一种识别手段以后，色谱法已进化成为一个独立的分析技术。色谱分析时，需把一种光学装置接在柱子的出口，用于测量洗脱相的组分随时间的变化。这种形式的色谱，其目标不是简单地回收组分，而是控制它们的迁移，它首次出现在 1940 年左右，但其后发展一直相对缓慢。

色谱法对化合物的鉴定是通过比较来实现的：要确定此化合物可能是 A 还是 B，需要将这种未知的溶液在柱子中进行分离。然后，在相同的仪器和实验条件下，将未知化合物的保留时间与两个参照化合物 A 和 B 进行比较。通过保留时间的比较来判断未知化合物是 A 还是 B。

在这个实验中，对真实的分离没有影响（A 和 B 分别为纯产品），只是将它们的保留时间进行了对比。但是，这一实验中存在三个缺点：过程相当缓慢；绝对的识别是不可能实现的；样品和固定相之间的接触可能会改变其性质，进而改变其保留时间，甚至影响最后的结论。

这种使用两种相互接触但不混溶相的分离方法于 20 世纪首次提出，其成功也同样归功于植物学家 Tswett 所提出的色谱法和色谱图的概念。

这项技术从出现到现在已经有了很大的进步。现在，色谱技术可以通过计算机软件控制不同的分离参数。这项技术包含很多附件，以保证可以在一系列实验中获得重复性的结果。因此可以在数小时内连续分析样品，获取具有重复性的记录（图 1.2）。

对每个分离过程的记录称为色谱图。它对应于图纸或屏幕上的一个二维图，呈现了洗脱流动相组分在流过色谱柱时的变化。要获得色谱图，需要在柱子出口处放置一个传感器，这种传感器可以有很多类型。色谱图纵坐标为检测器信号强度，横坐标为保留时间或者洗脱体积。

仅仅通过保留时间鉴定化合物不够严谨。最好是将两种互补的方法结合起来，例如色谱和二次在线仪器（质谱或红外光谱）的结合。这些联用技术收集两种不同类型的信息（迁移时间和波谱）进行分析比对，可以确定纳克级别的复杂混合物的组成和含量。

1.2 色谱图

色谱图就是流出色谱柱的流动相中分析物含量随时间（偶尔用体积）变化的图例表示。色谱图中的基线是在没有洗脱物的情况下得到的。当色谱图的色谱峰数与混合物中的样品数一样多时，分析结束（图 1.3）。

组分用其保留时间 t_R 表征，t_R 代表了从样品开始检测到色谱图上出现最大峰处所需要的时间。在理想情况下，t_R 与注入量无关。

不被保留的组分，其洗脱时间 t_M 称为保持时间或死时间（曾经用 t'_0 表示），它是流动相通过柱子所需要的时间。

样品的保留时间和死时间之差是指定化合物的调整保留时间 t'_R。

在色谱中，传感器检测到的峰信号随化合物浓度呈线性变化，其对应的色谱峰面积会发生相同的变化。这是色谱图能进行定量分析的基础。

1.3 洗脱峰的高斯谱形

在色谱分析中，一个具有完美洗脱峰形的色谱图与随机误差的正态分布图有相同的形状，即高斯曲线。与经典符号对应，用 μ 来表示洗脱峰的保留时间，而 σ 则表示峰的标准偏差（σ^2 代表方差）。Y 代表位于探测器出口的信号对时间 X 的函数（图 1.3）。

这就是理想洗脱峰通常被描述为概率密度函数（式 1.2）的原因。

$$Y=\frac{1}{\sigma\sqrt{2\pi}}\cdot\exp\left[-\frac{(X-\mu)^2}{2\sigma^2}\right] \tag{1.1}$$

$$Y=\frac{1}{\sqrt{2\pi}}\cdot\exp\left[-\frac{x^2}{2}\right] \tag{1.2}$$

$$a=\frac{b}{f} \tag{1.3}$$

$$\mathrm{TF}=\frac{b+f}{2f} \tag{1.4}$$

式(1.2) 的特征是在 $x=\pm1$ 处有两个拐点（此时纵坐标的值为 0.242，最大值的 60.6%）的对称曲线（最大值为 $x=0$，$y=0.3999$）（图 1.3）。曲线拐点的宽度等于 2σ（$\sigma=1$）。

在色谱图中，$W_{1/2}$ 表示半峰高处的峰宽（$W_{1/2}=3.35\sigma$），而 σ^2 表示峰的方差。峰底的宽度用 W 表示，在峰高的 13.5%处测得。依据定义，对于高斯曲线，有 $W=4\sigma$。

真实的色谱峰往往明显偏离理想的高斯曲线。这由许多原因引起，特别是在柱头注入区浓度不均匀时，此外，流动相的速度在柱子边缘处是零而在柱子中心位置最大。

峰的不对称峰性通过两个参数进行表征，即在 10%的峰高处测量的倾斜因子，以及在 5%的峰高处测量的拖尾因子 TF（图 1.4）。

1.4 塔板理论

半个世纪以来，针对现代色谱技术已经提出了不同的理论来解释分析物在色谱柱中的迁移和分离。最有名的是应用统计方法（随机理论）的那些理论，如理论塔板模型或分子动力学方法。

解释色谱中化合物分离机理最古老的模型为Craig′s理论塔板模型，这个模型是一个静态方法，虽然现在已过时了，但是它却提供了一个描述组分分离过程的简单方法。

尽管色谱法是一种动态过程，但Craig模型认为是各溶质一起逐步移动的一系列不同的静态步骤。在液-固色谱中，这个基本过程用吸附/解吸周期来表示。这些步骤重现了柱上化合物的迁移图像，这与让一系列固定图像运动起来做动画的方式类似。每个步骤都对应于整体柱中一个新的平衡状态。

这些连续的平衡是塔板理论的基础，该理论将长度为L的柱子水平分成N个假想的从1到N的高度均为H的板。对每一块板来说，流动相中溶质的浓度与固定相中溶质的浓度处于平衡状态。在每个新的平衡中，由于溶质要通过塔板一个板或盘的距离，因此得名“塔板理论”。

理论塔板高度（$HETP$或H）由式(1.5)表示：

$$H=\frac{L}{N} \tag{1.5}$$

对于给定的板，物质在两相之间的分布采用多项式的方法进行计算。即在I时刻，J塔板含有质量为m_T的分析物，m_T由m_M和m_S两部分组成，m_M是$I-1$时刻从$J-1$层板上通过流动相带过来的分析物，m_S是在$I-1$时刻J层板上已有的分析物（图1.5）。

$$m_T(I,J)=m_M(I-1,J-1)+m_S(I-1,J)$$

假定对于每个理论塔板都有：$m_S=Km_M$和$m_T=m_M+m_S$，然后使用递推的方法，就可以计算m_T（以及m_M和m_S）。鉴于每个板的分析物浓度在两相之间处于平衡，只要分析物还没有达到柱出口，色谱柱中流动相的体积V_M中分析物的总质量是不变的。所以，色谱图对应于在连续平衡中的第（$N+1$）块板的运送质量（图1.6）。这个理论有一个很大的缺陷，在色谱柱中没有考虑由于化合物扩散而造成的分散作用。塔板理论最早来源于马丁和辛格（1952年诺贝尔化学奖获得者）提出的方法，他们用与蒸馏和逆流萃取类似的方法来描述色谱。

色谱柱中的塔板理论是由于历史原因而产生的，没有物理意义，这是它区别于蒸馏柱中的塔板理论之处。

溶质在色谱柱上的保留时间t_R可以分为两项：t_M（保持时间），即溶质在流动相中溶解以及以相同速度随流动相流动的时间；t_S，即在固定相中消耗的时间，此期间溶质不流动。一般认为组分从一个相到另一个相的连续转移中有足够的时间达到平衡。

在色谱相系统中，至少有三套平衡：溶质和流动相之间的平衡、溶质和固定相之间的平衡、流动相和固定相之间的平衡。在最近提出的一个色谱理论中，没有考虑固定相对分子的固定作用，而仅仅认为流动相流过固定相附近时流速变慢。

1.5 能斯特分配系数（K）

色谱的基本物理化学参数是平衡常数K，称为分配系数，即每个化合物在两相中的浓度之比。

$$K=\frac{C_S}{C_M}=\frac{\text{溶质在固定相中的摩尔浓度}}{\text{溶质在流动相中的摩尔浓度}} \tag{1.6}$$

K 值的变化很大，它在流动相为气相时可以很大（如 1000），而当流动相是凝聚相时则很小（如 2）。每种化合物只占用色谱柱中有限的空间，因此每处的浓度不同，所以色谱柱中的 C_M 和 C_S 时刻处于变化中，但其比例恒定。

可以将热力学关系应用到上述分配平衡中。利用在流动相（M）和固定相（S）中化合物的浓度 C，平衡常数 K（C_M/C_S）可根据色谱实验计算得到，因此，如果知道实验温度，这一过程的标准吉布斯自由能就可以推导出来：

$$C_M \Longleftrightarrow C_S \qquad \Delta G=-RT\ln K$$

在气相色谱中，可以很容易地确定两个不同温度下的 K，因此就有可能获得标准焓变 $\Delta H^{\ominus}$ 和标准熵变 $\Delta S^{\ominus}$（假设过程中熵和焓不变）：

$$\Delta G^{\ominus}=\Delta H^{\ominus}-T\Delta S^{\ominus}$$

这三个参数都是负值，表明变化是自发进行的。可以预料的是，当化合物从流动相向静止固定相移动时，熵会减小。同样，范特霍夫方程可以非常严格的方式来预测温度对化合物保留时间的影响。

显而易见，经典热力学是适用于色谱理论的详细研究的。

$$\frac{\mathrm{d}\ln K}{\mathrm{d}T}=\frac{\Delta H}{RT^2}$$

1.6 柱效

1.6.1 理论效率（理论塔板数）

分析物通过色谱柱时形成了持续扩展的区域（图 1.6）。通过方差 σ_1^2 测定的线性扩散 σ_1 会随迁移距离的增加而增加。当该距离达到总柱长 L 时，方差为：

$$\sigma_L^2=HL \tag{1.7}$$

依据塔板理论模型，该方法也能得出一个理论塔板高度 H 的数值和理论塔板数 N（$N=L/H$）。

因此任一表示洗脱峰及其方差的色谱图（图 1.7），均可用来测定该化合物的理论塔板数 N，见式(1.8)。

$$N=\frac{L^2}{\sigma_L^2} \quad \text{或} \quad N=\frac{t_R^2}{\sigma^2} \tag{1.8}$$

进而依据 $H=L/N$ 推导得出 H。

在色谱图上，σ 表示在 60.6% 色谱峰高度处的半宽峰，t_R 为化合物的保留时间。t_R 和 σ 应使用相同的单位来测量（如果流量恒定，可以使用时间、距离或洗脱体积）。如果 σ 用体积单位（流量）表示，4σ 对应于“峰的体积”，其包含了大约 95% 的注入化合物。这是由高斯曲线的性质（$W=4\sigma$ 和 $W_{1/2}=2.35\sigma$）即公式(1.9) 决定的。然而，由于大多数峰在它们的基线处会发生变形，式(1.9) 很少使用，而更倾向于使用式(1.10)。

$$N=16\frac{t_R^2}{W^2} \tag{1.9}$$

$$N=5.54\frac{t_R^2}{W_{1/2}^2} \tag{1.10}$$

N 是相对值，它取决于所选择的溶质以及所采用的操作条件。当事先不确定色谱柱能否成功实现分离目标时，通常会选择一个出现在色谱末端的组分作为参考值。

1.6.2 有效塔板数（真实效率）

在气相色谱中，为了比较不同设计的色谱柱对于给定化合物的性能，或者为了比较毛细管柱和填料柱的不同表现，通过使用扣除了任一化合物在流动相中的死时间得到的调整保留时间（$t'_R=t_R-t_M$）替代式(1.8)～式(1.10)中的总保留时间来获得更真实的值。之前的三个公式变为：

$$N_{eff}=\frac{{t'_R}^2}{\sigma^2} \tag{1.11}$$

$$N_{eff}=16\frac{{t'_R}^2}{W^2} \tag{1.12}$$

$$N_{eff}=5.54\frac{{t'_R}^2}{W_{1/2}^2} \tag{1.13}$$

但目前认为这三个公式用处并不是很大。

1.6.3 理论塔板高度

正如式(1.5)定义的，通过计算参比化合物的理论塔板等效高度 H，可以对不同长度的色谱柱进行比较。理论塔板高度 H 不是常数，其值取决于所选择的化合物以及实验条件。

长期以来，在气相色谱的计算中，使用真实的效率来计算被称为理论塔板的有效高度的调整值。其对应的方程是式(1.14)：

$$H_{eff}=\frac{L}{N_{eff}} \tag{1.14}$$

在色谱中，流动相经常是液体，色谱柱由球形颗粒填充，此时塔板调整高度 h 需要考虑粒子的平均直径 d_m，这就消除了粒子大小的影响。色谱柱长度和粒子直径的比值相同的色谱柱会得到相似的效率。

$$h=\frac{H}{d_m}=\frac{L}{Nd_m} \tag{1.15}$$

1.7 保留参数

在色谱中使用保留时间或保留体积有各种目的，特别是为了获得保留因子 k 和热力学参数时。下面只给出基本术语。

1.7.1 保留时间

保留时间（t_R）、死时间（t_M）和调整保留时间（t'_R）的定义之前已经给出（见 1.2 节）。

1.7.2 保留体积（或洗脱体积）

分析物的保留体积 V_R 表示流动相从入口流动到出口所需要的总体积。要预测该体积，可以使用不同的方法（直接或间接），这取决于流动相的物理状态。在以时间为横坐标的标准色谱图中，如果流速 F 是恒定的，V_R 可以使用式(1.16)进行计算：

$$V_R = t_R F \tag{1.16}$$

峰体积 V_{peak} 为该化合物流出色谱柱时洗脱该化合物所使用的流动相的体积。定义为：

$$V_{peak} = WF \tag{1.17}$$

1.7.3 死体积

色谱柱中流动相的体积（也称为死体积）V_M，为流动相所能到达的固定相间隙的体积。只要保证当前溶质不在固定相中保留，保留体积通常可以从色谱图中计算得到。死体积 V_M 可由 t_M 和流速 F 用下式得到：

$$V_M = t_M F \tag{1.18}$$

有时，在最简单的情况下，固定相的体积 V_S 可以通过空色谱柱的总内部体积减去死体积 V_M 来计算。

1.7.4 保留因子（或容量因子）

当一个总质量 m_T 的化合物加进色谱柱后，其分离成两个量：流动相中的质量 m_M 和固定相中的质量 m_S。在溶质顺着色谱柱迁移的过程中，这两个量保持恒定。它们的比率是恒定的且与 m_T 无关，称为保留因子 k：

$$k = \frac{m_S}{m_M} = \frac{C_S}{C_M} \times \frac{V_S}{V_M} = K\ \frac{V_S}{V_M} \tag{1.19}$$

保留因子，也称为容量因子，是体现色谱分析性能的一个重要参数。尽管 k 不随流量或柱长度而改变，但因为该值取决于实验条件，它也不是一个常数，因此有时使用 k' 而不是 k。

这个参数将色谱柱保留每个化合物的能力，不管是大还是小，都考虑了进来。理想情况下，k 应该大于 1 而小于 5，否则分析时间会太长。

得到 k 的实验方法如下：假设一种化合物在色谱柱中移动，根据 Craig 模型，每个分子交替地从流动相（随其顺柱流动）到固定相（分子被固定），其在固定相中的时间越长，则通过柱子的平均速度就会越小。现在想象同一化合物的 n 个分子的情况（质量为 m_T 的样品）。如果我们假设，在每个瞬间，固定在固定相中的分子数 n_S（质量 m_S）和存在于流动相中的分子数 n_M（质量 m_M）的比率与它们在每一个相中所停留的时间（t_S 和 t_M）的比率相同，那么这三个比率有相同的数值，即：

$$\frac{n_S}{n_M} = \frac{m_S}{m_M} = \frac{t_S}{t_M} = k$$

以一个分子为例，若其在固定相中的时间占其总时间的 75%，则其平均速度将比它一直在流动相中慢四倍。因此，如果 4μg 这样的化合物加入色谱柱中，平均会有 1μg 在流动相中，3μg 在固定相中。

如果知道一种化合物的保留时间 t_R，即 $t_R = t_M + t_S$，那么 k 的值可以从色谱图（$t_S = t'_R$）中获得（见图 1.7）：

$$k = \frac{t'_R}{t_M} - \frac{t_R - t_M}{t_M} \tag{1.20}$$

这个重要的关系也可以写作：

$$t_R = t_M(1 + k) \tag{1.21}$$

考虑到关系式(1.16) 和式(1.18)，溶质的保留体积可以写作：

$$V_R = V_M(1 + k) \tag{1.22}$$

或

$$V_R = V_M + KV_S \tag{1.23}$$

最后的表达式将实验参数与热力学分布系数 K 联系起来，对理想的色谱过程是有效的。

1.8 分离因子（选择性因子）

分离因子 α（式 1.24）是对同一色谱图中两个相邻峰 1 和 2 进行比较（图 1.8）。使用式(1.19) 和式(1.20) 会发现，分离因子可以表示为式(1.25)。

通过定义可知，α 大于 1（组分 1 洗脱速度比组分 2 快）：

$$\alpha = \frac{t'_{R(2)}}{t'_{R(1)}} \tag{1.24}$$

$$\alpha = \frac{k_2}{k_1} = \frac{K_2}{K_1} \tag{1.25}$$

1.9 分辨率

量化两个化合物之间分离情况的另一个指标是分辨率 R（又称分离度）。选择性因子没有考虑峰值宽度，相反，使用下式计算两种化合物 1 和 2 之间的分辨率 R（图 1.9）：

$$R = 2\,\frac{t'_{R(2)} - t'_{R(1)}}{W_1 + W_2} \tag{1.26}$$

考虑到一个参数可以用其他参数替换或为了简化，也可以使用式(1.27)。

将分辨率和效率、保留因子、分离因子关联起来也是有用的［当 $W_1 = W_2$ 时从式(1.26) 可得式(1.28)］。进行实验验证可得图 1.10 所示色谱图。

$$R = 1.77\,\frac{t_{R(2)} - t_{R(1)}}{\delta_1 + \delta_2} \tag{1.27}$$

$$R = \frac{1}{4}\sqrt{N} \times \frac{\alpha - 1}{\alpha} \times \frac{k_2}{1 + k_2} \tag{1.28}$$

$$R = \frac{\sqrt{N}}{2} \times \frac{\alpha - 1}{\alpha} \times \frac{k_1 - k_2}{k_1 + k_2 + 2} \tag{1.29}$$

1.10 色谱法的速率理论

在之前的所有讨论特别是塔板理论中，你可能会感到奇怪，色谱柱中的流动相速度和溶质的扩散速度并没有被考虑进来。在色谱中，速度本应该对分析物流过色谱柱的过程产生影响，进而影响其扩散，以及整个分析的质量。

速率理论考虑了溶质在两相之间的平衡，因此能够更真实地描述色谱柱的工作过程。其所关心的是分离过程的动力学，由此提出了气相色谱填充柱的第一动力学公式，即范第姆特方程。

1.10.1 范第姆特方程

这个方程基于高斯分布，类似于塔板理论。范第姆特在 1956 年提出了其众所周知的简化形式［式(1.30)］。该式将塔板高度 H 和色谱柱中流动相的平均线速度 $\overline{u}$ 联系起来（图 1.11）。

$$H=A+\frac{B}{\overline{u}}+C\overline{u} \tag{1.30}$$

三个实验基本参数 A、B 和 C 都与色谱柱的不同理化参数和实验条件有关。如果 H 以厘米为单位，A、B、C 的单位分别是 cm、cm^2、s（速度单位为 cm/s）。

这个方程表明，每一个色谱柱都存在一个最佳流速，此时对应着最小的 H，其预测了式(1.30) 所描述的曲线。

随着流速增加，效率的损失很明显，在通过增加流动相压力来加快色谱分离时，这种情况也会表现出来。

然而直观上很难预测流速太慢时对效率带来的损失。为了解释这种现象，必须考虑 A、B 和 C 三个参数的来源。从图中可以看出，每个参数代表一个因素的影响（图 1.11）。

范第姆特方程是一个有最小值（H_{min}）的双曲线，此时：

$$\overline{u}_{opt}=\sqrt{\frac{B}{C}} \tag{1.31}$$

(1) 涡流扩散项 $A=2\lambda d_p$

A 项对应于流动相流过固定相时的流动形态。粒子的尺寸（直径 d_p）、空间分配、装填的均匀性（装填因子 λ）的变化都可能使流通路径不同，从而引起溶质带被展宽，以及两相之间的不正常交换，最终导致湍流或涡流扩散，这两个现象在液相色谱中不太重要，而在气相色谱中没有这种现象［Golay 方程没有 A，参看（1.10.2）］。对于一个给定的色谱柱，无法减小 A 项。

(2) 分子扩散项 $B=2\lambda D_G$

B 项，可以用气相中分析物的扩散系数 D_G 和上述填料因子 λ 表示，与色谱柱中的纵向分子扩散有关。当流动相是气体时，此项很重要。此项是熵的结果，它提醒我们，系统将自发向最大自由度（即混乱度增大）方向进行，类似于一滴墨水在一杯水中扩散的现象。因此，如果流速太慢，待分离的化合物的混合速度会大于迁移速度。这就是色谱一旦开始进行，千万不能中断的原因，即使是短暂的中断，也会对实验效率造成影响。

(3) 传质项 $C=C_G+C_L$

C 项，与溶质在两相之间质量传递的阻力有关，当所需要的流速过大而无法达到平衡时，此项很重要。流动相中的局部湍流和浓度梯度减慢了平衡过程（$C_S \rightleftharpoons C_M$）。溶质在两相之间的扩散不是瞬时完成的，因此将导致不平衡的存在。流动相速度越快，展宽效果越明显。C 项中的不同因素之间不存在简单的公式。参数 C_G 取决于溶质在气相流动相中的扩散系数，而 C_L 项取决于液相固定相中的扩散系数。黏度越大的固定相，C 越大。

实际操作中，图 1.11 中的系数 A、B、C 可以通过对进行色谱分离的相同物质在不同流速下测定几次效率而获得，因为流速和平均线速度是相关的。最好使用多元线性回归来计算与实验值符合最好的双曲线方程。

1.10.2 Golay 方程

范第姆特提出其理论几年后，Golay 提出了针对气相色谱中毛细管柱的修正关系。因为毛细管柱无填料，因此没有 A 项。

$$H=\frac{B}{\overline{u}}+C_L\overline{u}+C_G\overline{u} \tag{1.32}$$

如果检测条件下被检测的特定化合物的保留因子已知，式(1.33) 可以得到半径为 r 的色谱柱的 $HETP$ 的最小值。

涂层效率可以通过式(1.33) 中的值与通过色谱图获得的效率（$H=L/N$）比值的 100 倍进行计算。

$$H_{\text{theo. min}}=\mathrm{r}\sqrt{\frac{1+6k+11k^2}{3+(1+k)^2}} \tag{1.33}$$

1.11 色谱分析法的优化

分析色谱主要用于定量分析。为了达到此目的，必须精确计算峰面积，因此需要分析物能很好地被分离。色谱分析需要应用所有可利用的资源，例如能模拟不同温度、相和其他物理参数的仪器和软件时，经验变得很重要。

在气相色谱法中，分离过程很复杂，以至于很难事先判断温度升高好还是降低好。色谱柱的长度、直径、固定相组成和相比率（V_M/V_S）以及分离的参数（温度、流量）等因素都相互影响。

分辨率和洗脱时间是需要考虑的两个最重要的变量。尽管需要考虑使用后调整柱子到原始状态所需要的时间，在所有的优化中，最终目标一般都是在最短的时间内对感兴趣的化合物进行高效的分离。实际上，色谱法是一种较慢的分析方法。如果分辨率很好，则通过优化参数来节约分析时间，可以选择较短的柱子，因为分辨率随着柱子长度的平方根而变化［参见式(1.28) 的参数 N 和图 1.10］。

图 1.12 为液相色谱中芳香烃的优化分离。在这个例子中通过连续改变流动相的组成使分离得到最优化。注意，这种方式的优化会增加分析循环的时间。

如果待检测的混合物中化合物类型已知，则可以选择只检测目标化合物的检测器，或者是另一个极端，可以尝试分离混合物中可能存在的所有化合物。

取决于色谱分析的不同形式，优化或快或慢。气相色谱的优化比液相色谱要容易，因为液相色谱中必须考虑流动相组成，目前有软件可以帮助选择流动相的组成；基于一定的假设（高斯峰），可以找出不明确的峰形。

色谱法必须在顶点对应于三个参数的三角形内工作，这三个参数是：分辨率、速度和峰容量（图 1.13）。优化后分析分离因为使用了最高效的参数而具有最佳的选择性。如色谱法的三角形所示，优化条件接近最佳分辨率。

1.12 色谱技术的分类

色谱技术可以根据不同的标准进行分类：相的物理性质；使用过程；或依据给出能斯特分布系数 K 的物理化学现象。下面的分类依据是两相的物理性质（图 1.14）。

1.12.1 液相色谱（LC）

这种流动相是液体的色谱法属于最古老的分离制备方法。这类色谱范围非常宽泛，可以根据保留现象细分。

（1）液/固色谱法（或吸附色谱法）

固定相是固体，组分通过物理吸附和化学吸附的双重作用而附着在固定相上。此处所涉及的物理-化学参数是吸附因子。固定相从 Tswett 时期到现在已经取得很大进步。Tswett 当时曾使用碳酸钙或菊粉（inulin）(普通糖的粉末状聚合物）作为固定相。

（2）离子色谱法（IC）

这里的流动相是缓冲溶液，而固定相是由离子位点组成的表面。这些相允许流动反粒子与样品中相同电荷离子之间的交换。这一类型的分离依赖于离子分配系数。

（3）体积排阻色谱法（SEC）

这里的固定相是依据待分离组分的大小而选择的不同孔径的多孔材料。这一方法因在分子水平上的选择渗透而获名。依据流动相的性质是水相或有机相，可以分为凝胶过滤色谱或凝胶渗透色谱。这项技术中，分配系数被称为扩散系数。

（4）液/液色谱（或分配色谱，LLC）

其固定相是在只起到固定力学支撑的惰性或多孔材料中分布的固定液体。浸透是在多孔材料中固定液体的古老手段，因为其增加了冲刷色谱柱的风险，又称为“流失”，现在已经不再使用。

（5）液相/束缚相色谱

为了固定固定相（通常为液体聚合物），最好的解决方法是使其与固相载体之间形成共价键。分离效果与溶质在分液漏斗中水相和有机相之间的分配系数 K 有关。

1.12.2 气相色谱（GC）

气相色谱即流动相为惰性气体，如上所述，这种色谱可以依据相的性质细分如下。

（1）气/液色谱法（GLC）

如上所述，这里的流动相是气体，固定相是通过浸渍或共价键键合到色谱柱表面上的不流动的液体。这项技术通常简称为气相色谱法（GC）。气态样品必须是蒸气状态。马丁和辛格曾在 1941 年建议为了提高分离效率可以用液相代替气相。从出现气相色谱开始，分析色谱才真正开始发展。分配系数 K 也因此出现了。

（2）气/固色谱法（GSC）

固定相是多孔固体（如石墨、硅胶或氧化铝），而流动相是气体。这种类型的气相色谱在气体混合物和低沸点化合物的分离中效率很高。关键参数是吸附系数。

1.12.3 超临界流体色谱（SFC）

这种色谱技术的流动相是处于超临界状态的流体，如在 50℃和超过 150bar（15MPa）下的二氧化碳，固定相可以是液体或固体。这种技术结合了上述两种方法（液/液色谱法和气/液色谱法）的优势。

Chapter 2

Thin Layer Chromatography

Thin layer chromatography (TLC), also known as planar chromatography, is an invaluable method used in chemistry and biochemistry, complementary to HPLC while having its own specificity. Although these two methods are applied differently, the principle of separation and the nature of the phases remain the same. Cheap and sensitive, this technique that is simple to use, can be automated. It has become essential principally since it is possible to undertake several separations in parallel. The development of automatic applicators and densitometers have led to nano-TLC, also called HPTLC, a highly sensitive technique which can be hyphenated with mass spectrometry.

complementary [kɑːmpli'mentərɪ] *adj*.补充的，互补的

principally ['prɪnsəplɪ] *adv*.主要地，大部分

parallel ['pærəlel] *n*.平行线，对比；*vt*.使…与…平行；*adj*.平行的，相同的

densitometer [densə'tɒmətə] *n*.比重计，浓度计，光密度计

2.1 Principle of TLC

The principle of the separation between phases of the sample components is similar to that of HPLC, though the migration of the constituents through the stationary phase is different. Separation is conducted on a thin layer (100-200μm) of stationary phase, usually based upon silica gel and deposited on a rectangular plate made out of glass, plastic or aluminum of a few centimeters in dimensions. To maintain the stationary phase on the support and to assure the cohesion of the particles, an inert binder like gypsum (or organic linker) is mixed into the stationary phase during the manufacture of the plate. The constituents can be identified by simultaneously running standards with the unknown.

There are three steps to conduct a separation with this technique.

2.1.1 Deposition of the sample

A small volume of sample (between a few nanolitres to a few microlitres), dissolved in a volatile solvent, is deposited close to the bottom of the plate as a small spot of about 1-2mm in diameter. This

migration [maɪ'greʃən] *n*.迁移，移民，移动

rectangular [rek'tæŋgjələ] *adj*.矩形的；成直角的

gypsum [dʒɪpsəm] *n*.石膏，石膏肥料

identify [aɪ'dentɪfaɪ] *v*.鉴定，辨认

simultaneously [saɪməl'teniəslɪ] *adv*.同时地

deposit is made either manually, or automatically, with a flat ended capillary (Figure 2.1). The spot can also have the form of a horizontal band of a few millimeters which is obtained by automatic spraying of the sample. This last method has the advantage of having a high reproducibility, indispensable of course, for quantitative analysis. The prepared plate is then placed in a glass developing chamber that contains a small amount of the appropriate developing solvent. The chamber is then covered (Figure 2.2). The position at which the sample has been deposited must be above the level of the solvent.

indispensable
[ˈɪndɪˈspensəbl]
adj.不可缺少的，责无旁贷的；
n.不可缺少的人或物

chamber
[ˈtʃeɪmbər]
n.室

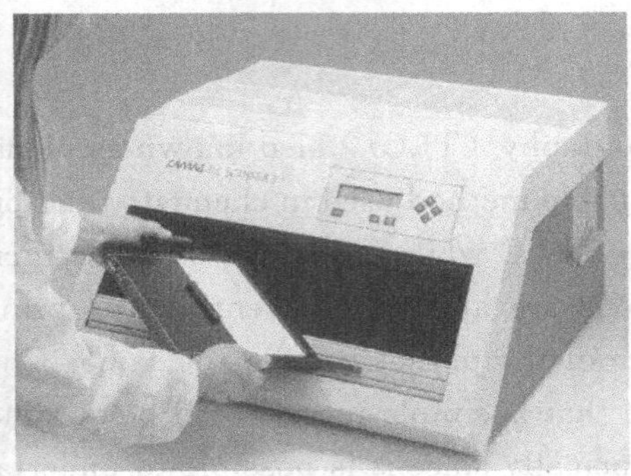

Figure 2.1 Automatic sample deposition device for TLC and a system to "read" the plate. Left, programmable applicator Linomat Ⅳ; Right, densitometer measuring the light either reflected or transmitted by the plate. The optical set up is similar to that of a UV/visible spectrometer (model Scanner 3, reproduced courtesy of Camag)

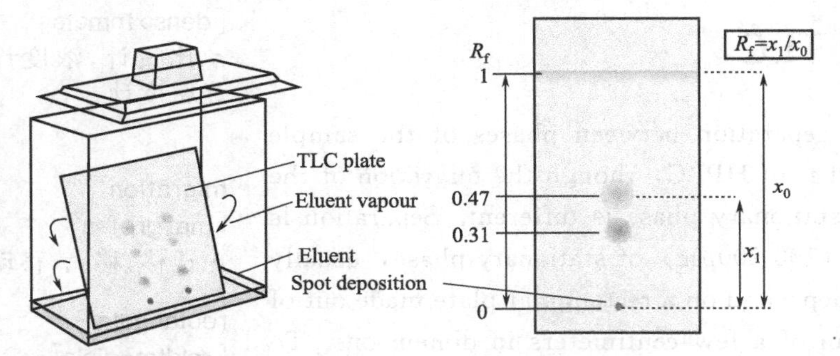

Figure 2.2 Developing chamber and TLC plate. Left, available in a variety of dimensions according to the size of the plates (of 5cm×5cm to 20cm×20cm), the chambers are made of glass and equipped with a tight fitting cover. Right, typical appearance after the TLC plate is partially dry. A faint line can be observed at the location of the solvent upper limit. This line is called the solvent front Calculation of R_f (cf. paragraph 2.4). A substance that does not migrate from the sample origin has a $R_f = 0$

2.1.2 Developing the plate

The mobile phase rises up the stationary phase by capillarity, moving the components of the sample at various rates because of their different degrees of interaction with the matrix and solubility in the solvent. Their separation may be complete in a few minutes. When the solvent front has travelled a sufficient distance (several centimeters), the plate is withdrawn from the chamber, the position attained by the mobile phase is immediately noted then it is evaporated.

When using a plate of reverse polarity ("RP-TLC"), the mobile phase will generally contain water. In this case it can be useful to add a salt such as lithium chloride which limits diffusion phenomena and thereby increases the resolution.

2.1.3 Identifying the spots

The localization of each component on the plate, which has now lost all of the eluent, consists to measure their migration distance from the original deposition. To locate colorless compounds the plate must be developed (Figure 2.2). In order to facilitate the visualization of the spots, manufacturers sell plates that contain a fluorescent salt of zinc which emits a bright green fluorescence when the plate is irradiated with a UV mercury vapour lamp ($\lambda=254$nm) and observed in a viewing cabinet. All compounds absorbing at this wavelength appear as a dark spot (or sometimes colored) against a bright green blue background.

Another method to make compounds visible, almost universally used, consists of heating the plate after spraying it with sulfuric acid which leaves charred blots behind. This approach is, however, not

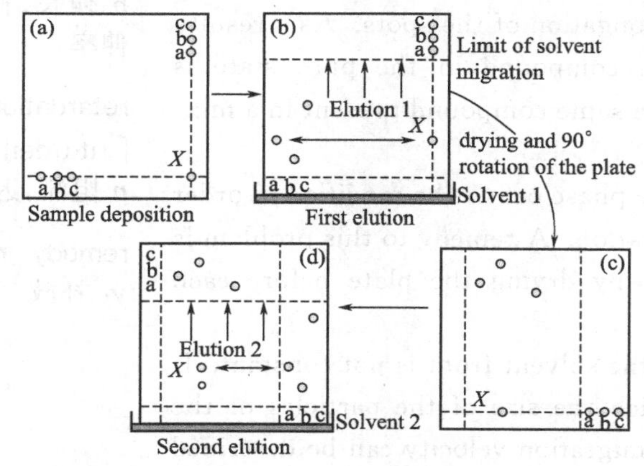

Figure 2.3 Two-dimensional TLC. An experiment with separation processes using two different solvents performed in two perpendicular directions. (a) Deposition of an unknown X and three standards; (b) migration in the first direction with the first solvent; (c) drying and rotation of the plate; (d) migration in the second direction with the second solvent. In conclusion, unknown compound X is a mixture of at least two compounds amongst which is the reference compound a (the same R_f in the two solvents) and one other compound which is not b. A typical application of this approach is the separation of amino acids

matrix [metrɪks]
n.基质

reverse[ri'vɜːs]
adj.相反的

polarity[pə'lærətɪ]
n.极性

lithium['lɪθɪəm]
n.锂(符号 Li)

chloride['klɔːraɪd]
n.氯化物

localization
[ləʊkəlaɪ'zeɪʃn]
n.[计] 定位

visualization
[vɪʒʊəli'zeɪʃən]
n.形象化，清楚地呈现

charred
[tʃɑːrd]
adj.烧黑了的，炭化的

adapted to quantitative TLC; in this case, development is effected by immersion of the plate, using a general (phosphomolybdic acid, vanillin), or specific (e.g. ninhydrin in alcoholic solution for amino acids) reagent. Hundreds of reagents have been described that serve to introduce chromophores or fluorophores groups into the analytes molecules after separation.

The use of square TLC plates allows two-dimensional chromatography to be carried out using two successive elutions with two different mobile phases (Figure 2. 3). A high degree of separation can be achieved.

2.2 Characteristics of TLC

TLC applies physico-chemical phenomena more complex than HPLC:

- TLC corresponds to a three-phase system between which equilibriums are established: solid (stationary), liquid (mobile) and vapour phases.
- The stationary phase is only partially equilibrated with the liquid phase before the migration of the compounds. Depending upon the manner in which the separation is obtained, the mobile phase may or may not be in equilibrium with the vapour phase.
- The adsorption phenomenon of the stationary phase is substantially reduced once a large part of the adsorption sites are occupied. This creates an effect of elongation of the spots. As a result, the R_f (retardation factor) of a compound in the pure state is slightly different from the R_f of the same compound present in a mixture.
- The flow rate of the mobile phase cannot be modified in order to improve the efficiency of a separation. A remedy to this problem is the multi-development technique, by drying the plate before each new cycle of migration.
- The speed of migration of the solvent front is not constant. It follows a complex function in which the size of the particles of the stationary phase play a part. The migration velocity can be described by a quadratic law: $x^2 = kt$. where x represents the distance of the migration front, t the time and k is a constant (Figure 2. 4). As a result the resolution between two spots depends greatly on the R_f values of the compounds. Resolution attains a maximum for an R_f value generally around 0. 3.

To sum up, the efficiency N of a TLC plate is very variable. The height equivalent of a theoretical plate has, as in HPLC, an optimal value.

chromophore [ˈkrəuməfɔː] n.[化学] 发色团

fluorophore [ˈfluərəfɔː] n.[化学] 荧光团

two-dimensional [ˌtuːdɪˈmenʃənəl] adj.二维的

substantially [səbˈstænʃəlɪ] adv.实质上

elongation [ˌiːlɒŋˈgeɪʃən] n.伸长，伸长率，延伸率

retardation [ˌriːtɑːˈdeɪʃən] n.阻滞，迟延，妨碍

remedy [ˈreməunder] vt.补救

quadratic [kwɑːˈdrætɪk] adj.[数] 二次的；n.二次方程式

resolution [rezəˈluʃən] n.分离度

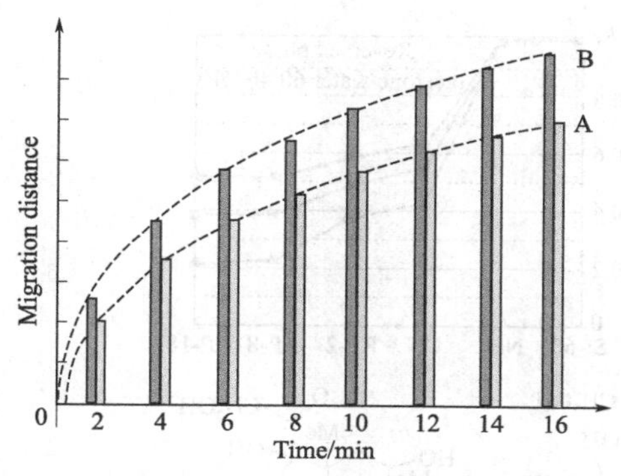

Figure 2.4 Migration distances of the mobile phase on a TLC plate. Over a sequence of regular time intervals, the quadratic progression of the eluent front can be seen. Curve A was obtained in a chamber unsaturated with eluent vapour. Curve B was obtained by saturating the chamber with eluent vapour

2.3 Stationary phases

Many physico-chemical parameters and several factors must be taken into account when choosing a good stationary phase. The size of the particles, their specific surface area, and the volume of the pores and the distribution of particle diameters are all factors that define the properties of the stationary phase. For nano-TLC the size of the particles is of the order of 4μm and the pores are 6nm.

The ratio between the silanols and siloxane groups determines the more or less pronounced hydrophilic character of the phase. As with HPLC, it is possible to use bonded silica in which various chains are bound by covalent bonds to silanol groups on the surface. Some phases incorporate alkyl chains (RP-2, RP-8, RP-18), while others contain organic functional groups (nitrile, amine, or alcohol), allowing these phases to be used with numerous mobile phases having the appropriate pH and salt concentrations (Figure 2.5). TLC plates can also be prepared to contain chemical groups with net positive or negative charges on the surface. This type of plate is used for ion-exchange TLC.

Modified cellulose supports are also used in TLC, either as fibres or chemically modified micro-crystalline powders. The most widely known is DEAE-cellulose, a phase fairly basic containing diethylaminoethyl groups. Other polar phases, with ion exchange properties can be employed for the separation of ampholytes.

Mineral binders such as gypsum make the stationary phase fragile, yet serve to advantage to recover the compounds after their separation. This can be done by scraping the zones of interest from the support and extracting the present compounds with a solvent.

properties [ˈprɔpətiz] *n*.性能，内容（property 的复数形式）

silanol [ˈsɪleɪnɒl] *n*.硅醇（甲烷）

siloxane [sɪˈlɒkseɪn] *n*.硅氧烷

hydrophilic [ˌhaɪdrəˈfɪlɪk] *adj*.亲水的

incorporate [ɪnˈkɔːpəreɪt] *vt*.包含，吸收，把……合并；*vi*.合并，混合；*adj*.一体化的

alkyl [ælkɪl] *adj*.烷基的，烃基的；*n*.烷基，烃基

concentration [ˌkɒnsnˈtreɪʃn] *n*.浓度，集中，浓缩

micro-crystalline [ˈmaɪkrəʊˈkrɪst(ə)laɪn] *adj*.微透明的；*n*.微晶体

diethylaminoethyl *n*.二乙氨乙基

ampholyte [ˈæmfəlait] *n*.两性电解质

fragile [ˈfrædʒəl] *adj*.脆的，易碎的

zone [zəʊn] *n*.区域

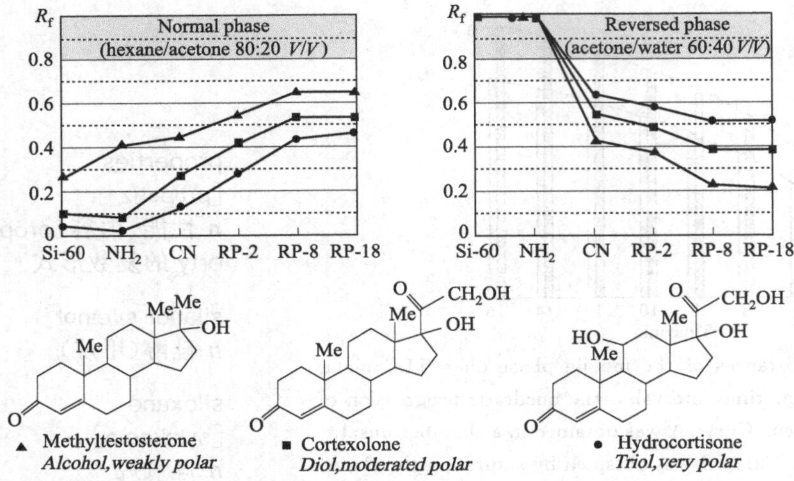

Figure 2. 5 Study of the separation of three steroids of different polarities upon six stationary phases with two binary solvent systems. Evolution of the R_f values and inversed migration caused by the change of the eluting mobile phase, one polar and the other non-polar (reproduced courtesy of Merck)

2.4 Separation and retention parameters

Each compound is defined by its retardation factor R_f (unitless) that corresponds to its relative migration compared to the solvent. R_f values lie between 0 and 1 (2. 1).

$$R_f = \frac{\text{Distance run by the solute}}{\text{Distance run by the solvent front}} = \frac{x}{x_0} \quad (2.1)$$

The efficiency N of the plate for a compound whose migration distance is x and spot diameter w is given by Equation (2. 2) and H (HETP) through Equation (2. 3):

$$N = 16\,\frac{x^2}{w^2} \quad (2.2)$$

and

$$H = \frac{x}{N} \quad (2.3)$$

In order to calculate the retention factor k of a compound or the selectivity coefficient between two compounds, the distances migrated along the plate are compared, for matching, with the migration times read on the chromatogram. Assuming that the ratio of the migration velocities u/u_0 is the same on the plate as on the column (which is really only an approximation), then R_f and k can be linked:

approximation [əˌprɑːksəˈmeɪʃn] n.近似法；近似值

such that

$$R_f = \frac{x}{x_0} = \frac{u}{u_0} = \frac{t_0}{t} = \frac{1}{k+1}$$

$$k=\frac{1}{R_f}-1 \tag{2.4}$$

R_f values often depend on the solvent use in the TLC experiment and temperature. Thus, the most effective way to identify a compound is to spot known substances on the same plate, next to the unknown.

substance
['sʌbstəns]
n.物质,基本内容

Finally, the resolution can be given by the relation:

$$R=2\frac{x_2-x_1}{w_1+w_2} \tag{2.5}$$

2.5 Quantitative TLC

In order to use TLC as a quantitative method of analysis, it is essential to quantify the spots (Figure 2.6), along with definitions for all of the usual parameters (specificity, range of the domain of linearity, precision, etc.). This is done by placing the plate under the lens of a densitometer (or scanner) that can measure either absorption or fluorescence at one or several wavelengths. This instrument produces a pseudo-chromatogram that contains peaks whose areas can be measured. In fact, it is actually an isochronic image of the separation at the final instant. In TLC, a spot is usually detectable if it corresponds at least to a few of a compound UV absorbent.

essential
[ɪ'senʃl]
adj.基本的,必要的

pseudo-chromatogram
['sju:dəʊ-krɒ'mætəgræm]
n.伪色谱分离

isochronic
[ˌaɪsəʊ'krɒnɪk]
adj.等时的,同时的

Plate type RP-18
UV detection at 254nm
Eluting phase:acetone/water 50:50 *V/V*
Climb:20min for 8cm.

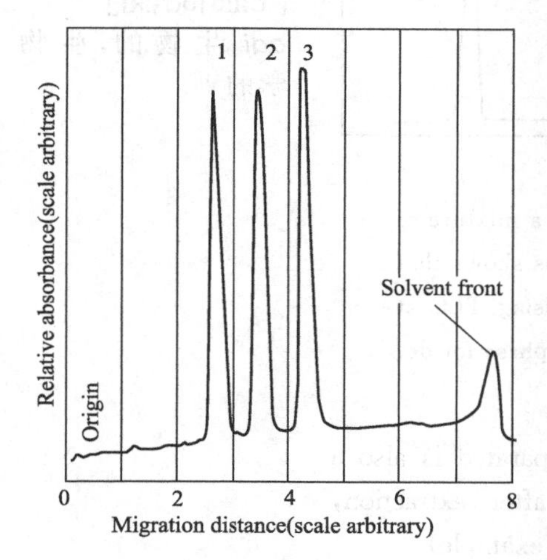

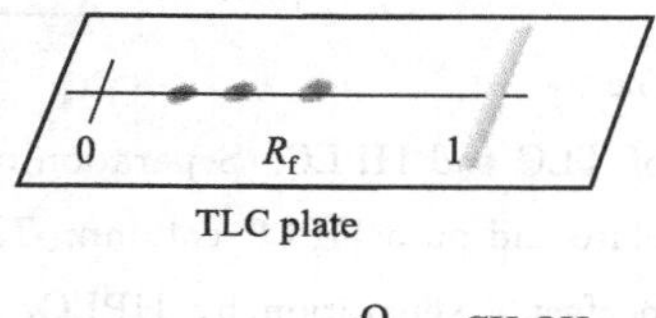

1— Cortexone, R^1=H and R^2=H
2— Corticosterone, R^1=OH and R^2=H
3— Cortisone, R^1=O and R^2=OH

Figure 2.6 Separation of three steroids by TLC on a phase of reversed polarity. The migration distance increases with the polarity of the compound. This pseudo-chromatogram has been obtained by scanning the TLC plate. The same separation effected by HPLC would lead to a chromatogram in which the order of the peaks would be reversed, a compound strongly retained having the longest elution time

In order to reveal radioactively labelled compounds (β-emission), there exist densitometers that are equipped with a video camera giving an image of the radioactive distribution on the plate.

radioactively
[ˌreɪdiəʊ'æktɪvlɪ]
adv.放射性地,有辐射地

The former procedure of autoradiography, obtained by putting a photoplate in contact with the TLC plate is rather insensitive (exposures could take up to 48h). The new densitometers, known as Charpak machines, which are also used in gel electrophoresis, have sufficient sensitivity for detecting activities in the order of a few Becquerels per mm^2. In many applications, TLC serves as an alternative tool of HPLC (Figure 2.7). Although this technique requires more manual manipulations than HPLC, new improved tools for spotting, migrating, gradient elution, development and recording, confer the necessary reproducibility. Compared with HPLC, TLC is able to treat more samples in the same time period by the setting up of analyses in parallel on the same plate. The plate, only used once and disposable, allows for rapid sample preparation with less risk of loss or contamination. It is very useful for biological samples.

autoradiography [ɔːtɔːreɪdɪ'ɒgrəfɪ] *n*.射线自显迹法，自动射线照相术

Becquerel [bekərel] *n*.[核] 贝克(勒尔)(放射性强度的国际制单位)

gradient ['grediənt] *n*.梯度

elution [ɪ'ljuːʃən] *n*.洗脱淋洗，洗提

contamination [kənˌtæmɪ'neɪʃən] *n*.污染，沾污，污染物

biological [ˌbaɪə'lɒdʒɪkl] *adj*.生物的，生物学的

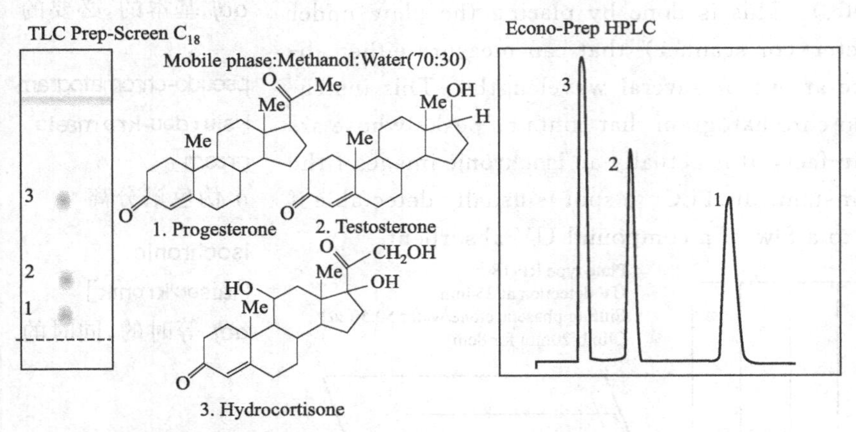

Figure 2.7 Comparison of TLC and HPLC. Separation of a mixture of three steroids on a TLC plate and on a HPLC column. This shows that it is possible to quickly perfect a separation by HPLC, using TLC at first, with the same stationary phase and the same mobile phase for development or elution

The TLC plate on which the products have been separated is also a means, provisionally, to preserve very small samples after extraction, which can serve for other analyses (mass spectrometry for example).

A recent technological advance allows the eluent to migrate at a constant speed through the application of positive gas pressure in the migration chamber, an effect especially developed which gains in both time and quality of separation.

High-performance thin layer chromatography (HPTLC) is an improvement of the technique where the sorbent material (eg. silica gel 60) has a finer particle size and a narrower particle size distribution than conventional TLC. HPTLC plates have an improved surface homogeneity and are thinner. The resolution is improved, analysis

homogeneity [ˌhɑːmədʒə'niːəti] *n*.同质，同种，同次性(等于 homogeneousness)

times are shorter and it is sufficient to apply nanolitres or nanograms of sample (Nano-TLC).

Problems

2.1 A mixture of two compounds A and B migrates from the origin to leave two spots with the following characteristics (migration distance x and spot diameter w).

$x_A = 27\text{mm}$ $w_A = 20\text{mm}$

$x_B = 33\text{ mm}$ $w_B = 25\text{mm}$

The mobile phase front was 60 mm from the starting line.

1. Calculate the retardation factor R_f, the efficiency N and the HETP H for each compound.
2. Calculate the resolution factor between the two compounds A and B.
3. Establish the relation between the selectivity factor and the R_f of the two compounds. Calculate its numerical value.

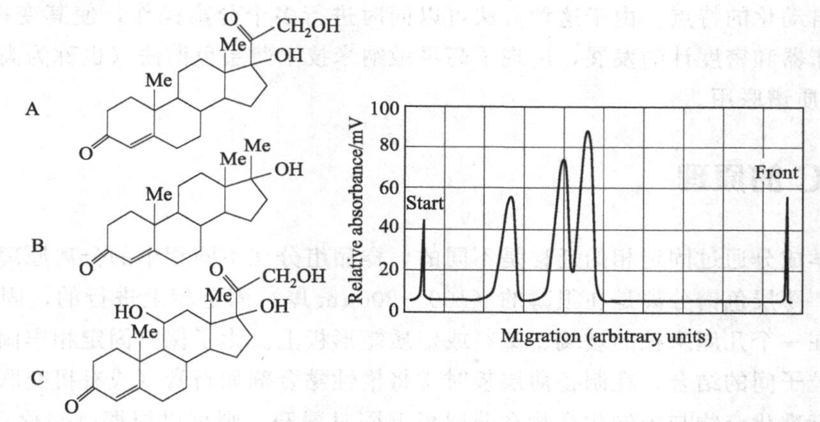

2.2 The following figure represents the results of scanning a TLC plate in normal phase (mobile phase: hexane/acetone 80/20). The three compounds have the structures A, B and C.

1. Indicate compounds A, B and C, from the identification of the three principal peaks of the recording.
2. What would have been the order of elution of these compounds if examined by an HPLC column containing the same types of stationary and mobile phases?
3. What would have been the order of elution of these compounds if examined by an HPLC column containing a phase of type RP-18 with a binary mixture of acetonitrile/methanol (80/20) as eluent?
4. Calculate the R_f and the HETP for the compound which migrates fastest upon the plate (use the transposed formulae of column chromatography, in particular that giving efficiency, with x as distance of migration).

第2章

薄层色谱法

薄层色谱法（TLC），也称平面色谱法，在化学和生物化学研究中非常实用的一种方法，由于其自身特性可与高效液相色谱相互补。尽管薄层色谱法和高效液相色谱法在应用上存在差异，但是其分离原理和两相性质是相同的。薄层色谱法具有价格低廉，灵敏度高，操作简单，易自动化的特点。由于这种方法可以同时进行多个分离操作，使其变得非常重要。随着自动涂布器和密度计的发展，出现了高灵敏纳米技术薄层色谱法（也称为高效薄层色谱法），并可与质谱联用。

2.1 TLC的原理

尽管化学成分通过固定相的迁移是不同的，样品组分在不同相中的分离原理与高效液相色谱法类似。薄层色谱分离是在很薄的（100～200μm 厚）固定相上进行的，固定相通常是将硅胶沉积在一个几厘米长的玻璃、塑料或铝质矩形板上。为了保持固定相牢固地结合在板子上及保证粒子间的结合，在制备薄层板时常将惰性黏合剂如石膏（或有机交联剂）加入固定相中。将标准化合物与未知化合物在薄层板上同时展开，则可以根据已知化合物的 R_f 值对组分进行定性。

采用TLC技术的分离分三步进行。

2.1.1 点样

将少量体积的样品（介于纳升到微升之间）溶解在挥发性溶剂中，在接近薄层板底端处点样，点样原点直径控制在1～2mm以内。可用一端平整的毛细管采用手动或自动的方式点样（图2.1），也可以采用自动喷涂技术将样品喷涂成几毫米高的水平带。后者的优点是重现性高，是进行定量分析不可或缺的方法。然后将点样的薄层板放入有适量展开溶剂的玻璃展开室内，盖上盖子进行展开（图2.2）。点样位置必须高于展开剂液面。

2.1.2 薄层板的显影

流动相（展开剂）因毛细作用沿着薄层板上的固定相缓慢上升，由于样品各组分与固定相的相互作用及在展开剂中的溶解性不同而表现出不同的迁移速率。样品成分的分离可以在几分钟内完成。当溶剂前沿移动足够的距离（几厘米）后，将薄层板从展开室内取出，立即标出展开剂到达的位置，然后挥发溶剂。

当使用反相板（RP-TLC）时，流动相通常含有水。在这种情况下，可以添加氯化锂一类的盐以抑制样品扩散现象，从而提高分辨率。

2.1.3 斑点的识别

薄层板上展开溶剂挥发后，通过测量各组分自原点移动的距离进行定位，无色组分需要使用显色剂显色后定位。为了便于斑点的可视化，薄层板制造商通常会在出售的板中添加含有锌的荧光盐，在紫外汞灯（$\lambda=254\text{nm}$）照射下会发出明亮的绿色荧光，可在观察柜中查看。所有在该波长下有吸收的化合物会在蓝绿色背景中出现一个暗点（或其他颜色）。

另一种普遍使用的让化合物显色的方法是喷洒硫酸后加热薄层板，薄板上会留下烧黑的斑点印迹。然而，这种方法不适于定量薄层色谱。这种方法一般采用通用（磷钼酸、香草醛）或特定（如用于氨基酸显色的茚三酮醇溶液）的显色剂作浸板处理。分离后有上百种试剂可将发色基团或荧光基团引入到分析物分子中。

方形薄层板也可以应用到二维色谱中，即使用两种不同的展开剂进行连续展开（图 2.3）。该方法可以获得较高的分离度。

2.2 TLC 的特性

TLC 应用的物理化学现象比 HPLC 更加复杂。

• TLC 相当于建立一个固相（固定相）、液相（流动相）和气相三相之间的平衡系统。

• 在化合物发生迁移前固定相与流动相只是部分达到相平衡。根据分离的方式，流动相可以与气相达到平衡或不平衡。

• 一旦吸附位点被大量占用，固定相的吸附作用会大大降低。这会导致斑点拖尾。其结果是，纯化合物的 R_f（比移值）与混合物中存在的同一种化合物的 R_f 略有不同。

• 不能为了提高分离效率而改变流动相的流速。对此问题的补救办法是采用多级展开技术，即在每次展开前对薄层板进行干燥。

• 溶剂前沿的迁移速度不是恒定的。它遵循一个复杂的函数，其中固定相的颗粒尺寸发挥部分作用。流动速率可用二次方程 $x^2=kt$ 描述 x 代表移动距离，t 代表时间，k 是一个常数（图 2.4）。因此两个斑点之间的分辨率很大程度上取决于两个化合物的 R_f 值。一般 R_f 值在 0.3 左右分辨率达到最大。

综上所述，TLC 板的分离效率 N 可变化性大。其理论塔板高度与 HPLC 一样有一个最优值。

2.3 固定相

选择一个好的固定相，需要考虑许多物化参数和其他一些因素。粒径、颗粒的比表面积、孔容、粒径分布等都是影响固定相性质的因素。对于纳米薄层色谱而言，固定相的粒径是 $4\mu\text{m}$，孔径是 6nm。

硅烷醇与硅氧烷基团之间的比例决定了固定相亲水性的强弱。与 HPLC 的固定相一样，TLC 可以使用键合硅胶，各种官能团通过共价键与其表面上硅烷醇基团结合。一些固定相含有烷基链（如 RP-2、RP-8、RP-18），而另一些固定相中含有有机官能团（腈、胺或醇），这些固定相可以在适当的 pH 值和盐浓度的流动相中使用（图 2.5）。也可以制备表面带有净正电荷或净负电荷基团的 TLC 板，这种类型的板用于离子交换型 TLC。

除了纤维、化学修饰的微晶粉之外，修饰的纤维素也可用于 TLC。最为大家所熟知的是 DEAE-纤维素固定相，其含二乙氨乙基基团。其他具有离子交换性质的极性固定相可用于两性电解质物质的分离。

虽然矿物黏合剂（如石膏）的加入会让固定相易碎，但有助于分离后获得所需的有机物。可以在分离之后刮取感兴趣的固定相区域，并结合溶剂提取来获得相应的化合物。

2.4 分离和保留参数

化合物的定性通过比移值 R_f（无量纲）来确定，该值与样品相对于溶剂的移动距离有关。R_f 值位于 0 和 1 之间。

$$R_f=\frac{\text{溶质移动的距离}}{\text{溶剂前沿移动的距离}}=\frac{x}{x_0} \tag{2.1}$$

对于一种化合物，其分离效率 N 通过式(2.2) 计算。其迁移距离为 x，斑点直径是 w，H（理论塔板数）可通过式(2.3) 给出：

$$N=16\,\frac{x^2}{w^2} \tag{2.2}$$

$$H=\frac{x}{N} \tag{2.3}$$

为了计算一种化合物的保留因子 k 或两种化合物间的选择性系数，可以比较它们在薄板上的迁移距离，相应的迁移时间也可以在色谱图中获得。假设在硅胶板和色谱柱中的迁移速率 u/u_0 是相同的（其实也接近相同），R_f 和 k 就可以建立联系（式 2.4）：

$$R_f=\frac{x}{x_0}=\frac{u}{u_0}=\frac{t_0}{t}=\frac{1}{k+1}$$

$$k=\frac{1}{R_f}-1 \tag{2.4}$$

R_f 值通常取决于 TLC 实验中的展开剂和温度。因此，确定一个化合物最有效的方法是在一薄板上点取已知化合物，然后在相同的板上点取未知化合物。

最后，分辨率可根据下列关系式求出：

$$R=2\,\frac{x_2-x_1}{w_1+w_2} \tag{2.5}$$

2.5 TLC 的定量分析

为了利用薄层色谱进行定量分析，斑点的定量（图 2.6）及所有常规参数（特异性、线性范围、精确度等）的确定都是至关重要的。用光密度计（或扫描仪）测定薄层板对一个或多个紫外线的吸收程度或产生荧光的强度可确定这些参数。仪器会产生一个可以测量峰面积的伪色谱图。事实上，它是最终分离的实时图像。在 TLC 上，如果一个斑点有紫外吸收，那么这个斑点通常可以检测。

为了检测放射性标记的化合物（β-放射），需要使用装备有视频相机的光密度计，用于拍摄薄板上的放射性分布图像。放射自显影之前须把感光片与薄层色谱板接触，其敏感性较低（曝光时间最长可达 48h）。新型光密度计 Charpak 仪，除了以上应用，还可用于凝胶电泳，对于检测活性物有足够的灵敏度，其检测限可达到每平方毫米几贝克级。

在很多应用上，TLC 可以代替 HPLC（图 2.7）。与 HPLC 相比，TLC 需要更多的手动操作。但新开发的工具为点样、展开、梯度洗脱、显影和记录提供了较好的重现性。与 HPLC 相比，TLC 可在同一薄层板上建立平行分析，在相同时间内处理更多样品。为了快速制备样品并减少分析物的损失和避免污染，薄层板通常只使用一次就弃掉。这对于生物样品非常有用。

薄层色谱板可作为样品分离的一种方法，暂时保存少量提取后的样品以用于其他分析(如质谱法)。

最新的技术可通过对迁移室施加恒定正相气压使洗脱液以等速率移动，它对分离时间和分离性能具有特定的效果。

与传统的 TLC 相比，高效薄层色谱法（HPTLC）在许多方面进行了改进，如吸附剂材料（如硅胶 60）粒径更小，粒径分布更窄。改良后的 HPTLC 薄层板表面更均匀也更薄。分辨率提高，分析时间更短，足够胜任纳升或纳克级的样品分析（如纳米 TLC）。

Chapter 3

Gas Chromatography

In gas chromatography, the components of a vaporized sample are separated by being distributed between a mobile gaseous phase and a liquid or a solid stationary phase held in a column. In performing a gas chromatographic separation, the sample is vaporized and injected onto the head of a chromatographic column. Elution is brought about by the flow of an inert gaseous mobile phase. In contrast to most other types of chromatography, the mobile phase does not interact with molecules of the analytes. The only function of the mobile phase is to transport the analytes through the column.

Two types of gas chromatography are encountered: gas-liquid chromatography (GLC) and gas-solid chromatography (GSC). Gas-liquid chromatography finds widespread use in all fields of science where its name is usually shortened to gas chromatography (GC). Gas-solid chromatography is based on a solid stationary phase in which retention of analytes occurs because of physical adsorption. Gas-solid chromatography has limited application because of semipermanent retention of active or polar molecules and severe tailing of elution peaks. The tailing is due to the nonlinear character of adsorption process. Thus, this technique has not found wide application except for the separation of certain low-molecular-mass gaseous species.

In gas-liquid chromatography, the mobile phase is a gas, and the stationary phase is a liquid that is retained on the surface of an inert solid by adsorption or chemical bonding.

In gas-solid chromatography, the mobile phase is a gas, and the stationary phase is a solid that retains the analytes by physical adsorption. Gas-solid chromatography permits the separation and determination of low-molecular-mass gases, such as air components, hydrogen sulfide, carbon monoxide, and nitrogen oxides.

Gas-liquid chromatography is based on partitioning of the analyte between a gaseous mobile phase and a liquid phase immobilized on the surface of an inert solid packing or on the walls of capillary tubing.

vaporize
[ˈveɪpəraɪz]
vi.蒸发

widespread
[ˈwaɪdspred]
adj.广泛应用，普及的

analyte
[ænəˈlɪt]
n.（被）分析物，分解物

semipermanent
[ˌsemɪˈpɜːmənənt]
adj.非永久(性)的，半固定的，半永久的

adsorption
[ædˈsɔːpʃən]
n.吸附(作用)

stationary
[ˈsteɪʃənrɪ]
adj.不动的，固定的，静止的，不变的

hydrogen
[ˈhaɪdrədʒən]
n.[化]氢

partition[pɑːˈtɪʃn]
vt.分配

capillary[kəˈpɪlərɪ]
n.毛细管

The concept of gas-liquid chromatography was first enunciated in 1941 by Martin and Synge, who were also responsible for the development of liquid-liquid partition chromatography. More than a decade was to elapse, however, before the value of gas-liquid chromatography was demonstrated experimentally and this technique began to be used as a routine laboratory tool. In 1955, the first commercial apparatus for gas-liquid chromatography appeared on the market. Since that time, the growth in applications of this technique has been phenomenal. Currently, several hundred thousand gas chromatographs are in use throughout the world.

enunciate [ɪ'nʌnsɪeɪt] *vt* 确切地说明

3.1 Instruments for gas-liquid chromatography

Many changes and improvements in gas chromatographic instruments have appeared in the marketplace since their commercial introduction. In the 1970s, electronic integrators and computer-based data-processing equipment became common. The 1980s saw computers being used for automatic control of such instrument parameters as column temperature, flow rates, and sample injection. This same decade also saw the development of very high-performance instruments at moderate costs and, perhaps most important, the introduction of open tubular columns that are capable of separating components of complex mixtures in relatively short times. Today, some 50 instrument manufacturers offer about 150 different models of gas chromatographic equipment at costs that vary from \$1000 to over \$50000. The basic components of a typical instrument for performing gas chromatography are shown in Figure 3.1.

parameter [pə'ræmɪtə] *n*.参量，参数

tubular ['tjuːbjələ(r)] *adj*. 管子形的，管状的

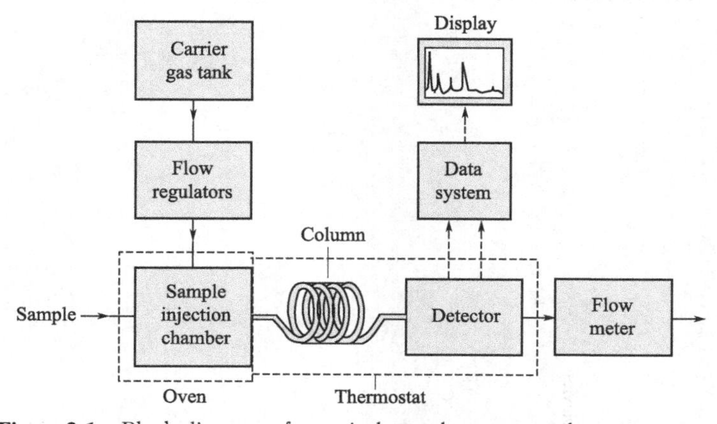

Figure 3.1 Block diagram of a typical gas chromatograph

3.1.1 Carrier gas system

The mobile phase gas in gas chromatography is called the carrier gas and must be chemically inert. Helium is the most common mobile phase, although argon, nitrogen, and hydrogen are also used. These gases are available in pressurized tanks. Pressure regulators, gauges, and flow meters are required to control the flow rate of the gas. Classically, flow rates in gas chromatographs were regulated by controlling the gas inlet pressure. A two-stage pressure regulator at the gas cylinder and some sort of pressure regulator or flow regulator mounted in the chromatograph were used. Inlet pressures usually range from 10to 50 psi (lb/in^2) above room pressure, yielding flow rates of 25 to 150mL/min with packed columns and 1 to 25mL/min for open tubular capillary columns. With pressure-controlled devices, it is assumed that flow rates are constant if the inlet pressure remains constant. Newer chromatographs use electronic pressure controllers both for packed and for capillary columns.

With any chromatograph, it is desirable to measure the flow through the column. The classical soap-bubble meter shown in Figure 3. 2 is still widely used. A soap film is formed in the path of the gas when a rubber bulb containing an aqueous solution of soap or detergent is squeezed; the time required for this film to move between two graduations on the buret is measured and converted to volumetric flow rate (see Figure 3. 2). Note that volumetric flow rates and linear flow velocities are related. Bubble flow meters are now available with digital readouts that eliminate some human reading errors. Usually, the flow meter is located at the end of the column, as shown. The use of electronic flow meters has become increasingly common. Digital flow meters are available that measure mass flow, volume flow, or both. Volumetric flow measurements are independent of the gas composition. Mass flow meters are cali-

helium
[ˈhiːlɪəm]
n.〈化〉氦

argon
[ˈɑːrgɒn]
n.〈化〉氩

inlet
[ˈɪnlet]
n.进口

detergent
[dɪˈtəːdʒənt]
n.洗涤剂，去垢剂

volumetric
[ˌvɒljʊˈmetrɪk]
adj.体积的

Figure 3.2 A soap-bubble flow meter

brated for specific gas compositions, but, unlike volumetric meters, they are independent of temperature and pressure.

3.1.2 Sample injection system

For high column efficiency, a suitably sized sample should be introduced as a "plug" of vapor. Slow injection or oversized samples cause band spreading and poor resolution. Calibrated microsyringes, such as those shown in Figure 3.3, are used to inject liquid samples through a rubber or silicone diaphragm, or septum, into a heated sample port located at the head of the column. The sample port (see Figure 3.4) is usually kept at about 50℃ greater than the boiling point of the least volatile component of the sample. For ordinary packed analytical columns, sample sizes range from a few tenths of a microliter to 20μL. Capillary columns require samples that are smaller by a factor of 100 or more. For these columns, a sample splitter is often needed to deliver a small known fraction (1 : 100 to 1 : 500) of the injected sample, with the remainder going to waste. Commercial gas chromatographs intended for use with capillary/columns incorporate such splitters, and they also allow for splitless injection when packed columns are used.

silicone
['sɪlɪkəʊn]
n.硅树脂

splitter
['splɪtə]
n.分离器,分流器

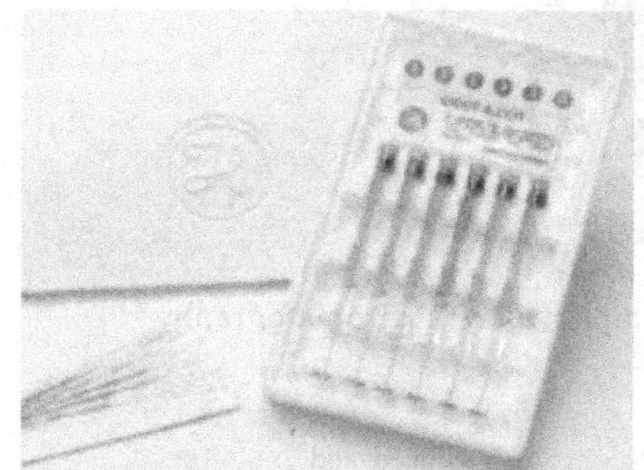

Figure 3.3 A set of microsyringes for sample injection

For the most reproducible sample injection, newer gas chromatographs use autoinjectors and autosamplers, such as the system shown in Figure 3.5. With such autoinjectors, syringes are filled, and the sample is injected into the chromatograph automatically. In the autosampler, samples are contained in vials on a sample turntable. The autoinjector syringe picks up the sample through a septum on the vial and injects the sample through a septum on the chromatograph. With the unit shown, up to 150 sample vials can be placed on the turntable. Injection volumes can vary from 0.1μL with a 10μL syringe to 200μL with a 200μL syringe. Standard deviations as low as 0.3% are common with autoinjection systems.

autoinjector
[ɔːtɔɪndʒektər]
n.自动进样器

syringe
[sɪ'rɪndʒiː]
n.进样器

septum
['septəm]
n.隔膜

For introducing gases, a sample valve, such as that shown in Figure 3.6, is often used instead of a syringe. With such devices, sample sizes can be reproduced to better than 0.5% relative. Liquid

valve
[vælv]
n.阀，真空管

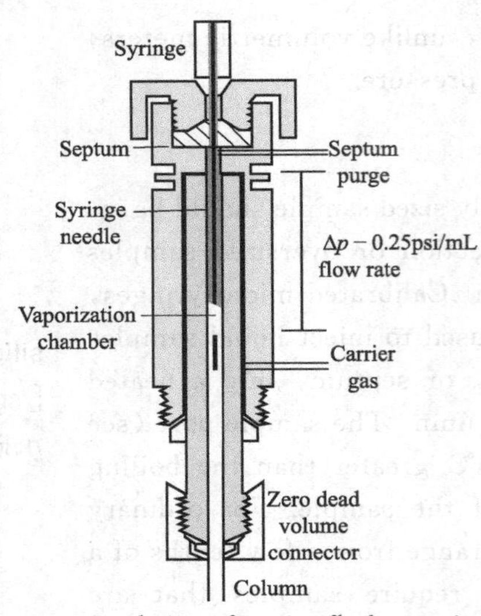

Figure 3.4 Cross-sectional view of a microflash vaporizer direct injection

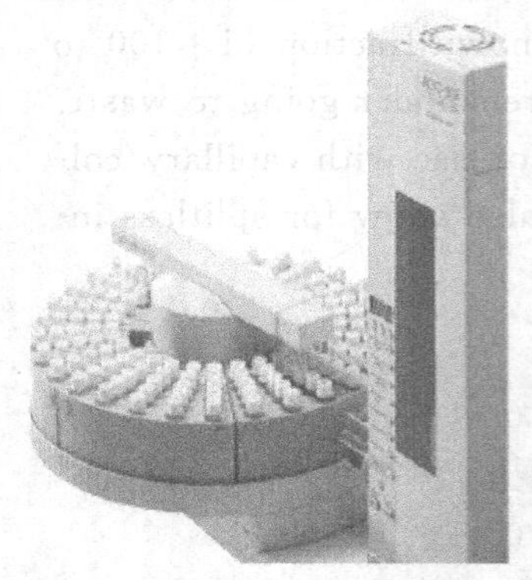

Figure 3.5 An autojection system with autosampler for gas chromatography

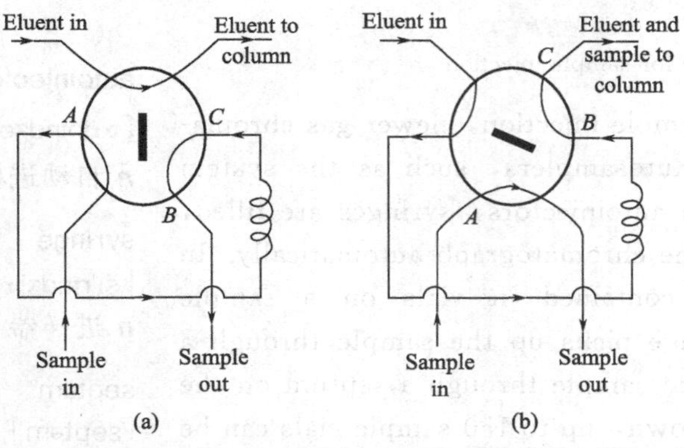

Figure 3.6 A rotary sample valve. Valve postion (a) is for filling the sample loop ACB; postion (b) is for introduction of sample into the column

samples can also be introduced through a sampling valve. Solid samples are introduced as solutions or alternatively are sealed into thin-walled vials that can be inserted at the head of the column and punctured or crushed from the outside.

puncture [ˈpʌŋktʃə] *v*. 在（某物）上穿孔，刺穿（某物）

3.1.3 Column configurations and column ovens

The columns in gas chromatography are of two general types: packed columns or capillary columns. In the past, the vast majority of gas chromatographic analyses used packed columns. For most current applications, packed columns have been replaced by more efficient and faster capillary columns. Chromatographic columns vary in length from less than 2m to 60m or more. They are constructed of stainless steel, glass, fused silica, or Teflon. In order to fit into an oven for thermostating, they are usually formed as coils having diameters of 10 to 30cm (see Figure 3.7).

majority [mə'dʒɔːrəti] n.多数(获胜的)

silica ['sɪlɪkə] n.二氧化硅

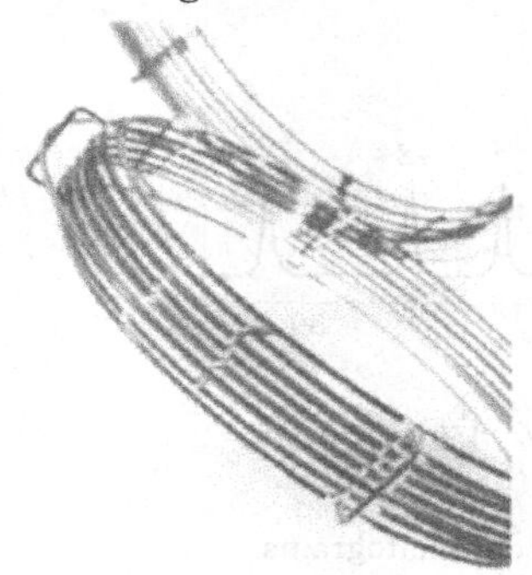

Figure 3.7 Fused-silica capillary columns

Column temperature is an important variable that must be controlled to a few tenths of a degree for precise work. Thus, the column is normally housed in a thermostated oven. The optimum column temperature depends on the boiling point of the sample and the degree of separation required. Roughly, a temperature equal to or slightly above the average boiling point of a sample results in a reasonable elution time (2 to 30min). For samples with a broad boiling range, it is often desirable to use temperature programming whereby the column temperature is increased either continuously or in steps as the separation proceeds. Figure 3.8 shows the improvement in a chromatogram brought about by temperature programming.

variable ['veərɪəbl] n.变量

temperature programming 程序升温

In general, optimum resolution is associated with minimal temperature. The cost of lowered temperature, however, is an increase in elution time and, therefore, the time required to complete an analysis. Figures 3.8(a) and 3.8(b) illustrate this principle.

Analytes of limited volatility can sometimes be determined by forming derivatives that are more volatile. Likewise, derivatization is used at times to enhance detection or improve chromatographic performance.

volatility [ˌvɒlə'tɪlətɪ] n.挥发性

derivative [dɪ'rɪvətɪv] n.衍生物

3.1.4 Chromatographic detectors

detector [dɪ'tektəː] n.检测器

Some detectors have been investigated and used with gas chrom-

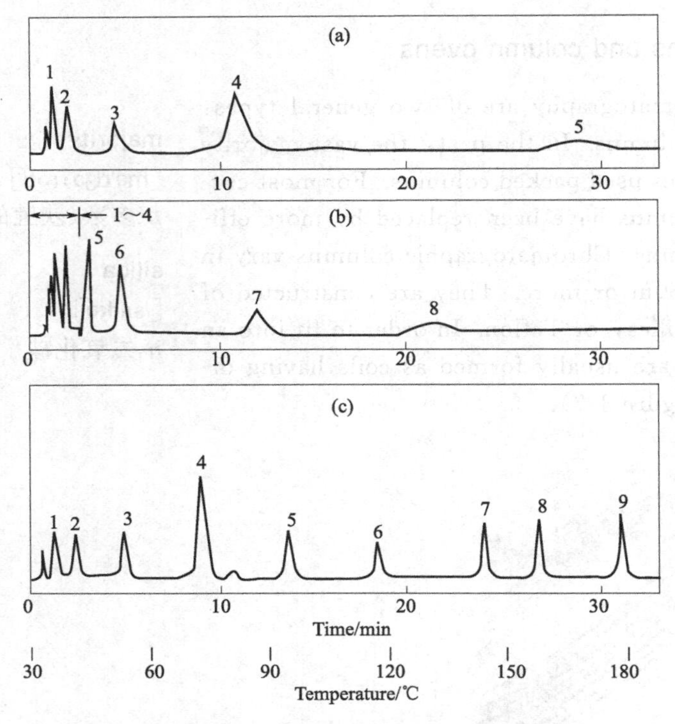

Figure 3.8 Effect of temperature on gas chromatograms.

(a) Isothermal at 45℃; (b) Isothermal at 145℃;

(c) Programmed at 30℃ to 180℃

atographic separation. We first describe the characteristics that are most desirable in a gas chromatographic detector and then discussed the most widely used devices.

Characteristics of the ideal detector

The ideal detector for gas chromatography has the following characteristics:

- Adequate sensitivity. In general, the sensitivities of present-day detectors lie in the range of 10^{-8} to 10^{-15} g solute/s.
- Good stability and reproducibility.
- A linear response to solutes that extends over several orders of magnitude.
- A temperature range from room temperature to at least 400℃.
- A short response time that is independent of flow rate.
- High reliability and ease of use. To the greatest extent possible, the detector should be foolproof in the hands of inexperienced operators.
- Similarity in response toward all solutes or, alternatively, a highly predictable and detective response toward one or more classes of solutes.
- Nondestructive of sample.

characteristic
[ˌkærəktəˈrɪstɪk]
n.特性,特征

stability
[stəˈbɪlətɪ]
n.稳定(性),稳固

reproducibility
[rɪprədju:səˈbɪlɪtɪ]
n.再现性,可重现性

magnitude
[mægnɪtju:d]
n.广大,重大

reliability
[rɪˌlaɪəˈbɪlətɪ]
n.可靠,可信赖

solute
[sɒˈlju:t]
n.溶解物,溶质

3.2 GC instrument component design

The third and final major component analytes encounter after the injector and the column is a detector. This produces an electrical signal (usually analog, but often converted to digital) which is proportional to either the concentration or the mass flow rate of the analyte molecules in the effluent stream. The signal is displayed as a chromatogram on a chart recorder, or more often these days, on the screen of desktop computer data system. Retention times are automatically calculated, heights of peaks are measured, or they are automatically integrated to obtain their areas, and peaks can be identified by their elution within a retention time window, and quantitated by comparison to the areas or heights of a quantitative standard. We will not discuss the details of the operation of this signal processing equipment, but will describe only the operation and characteristics of the most useful GC detectors.

(1) Universality vs. selectivity

If a detector responds with similar sensitivity to a very wide variety of analytes in the effluent, it is said to be universal (or at least almost so—no GC detector is absolutely universal). Such detectors are valuable when one needs to be sure that no components in the separated sample are overlooked. In the other extreme, a selective detector may give a significant response to only a limited class of compounds: those containing only certain atoms, (e.g., atoms other than the ubiquitous C, H, and O atoms of the majority of organic compounds), or possessing certain types of functional groups or substituents which possesses certain affinities or reactivities. Selective detectors can be valuable if they respond to the compounds of interest while not being subject to interference by much larger amounts of coeluting compounds for which the detector is insensitive.

(2) Destructive vs. nondestructive

Some detectors destroy the analyte as part of the process of their operation (e.g., by burning it in a flame, fragmenting it in the vacuum of a mass spectrometer, or by reacting it with a reagent). Others leave it intact and in a state where it may be passed on to another type of detector for additional characterization.

(3) Mass flow vs. concentration response

In general, destructive detectors are mass flow detectors. If the flow of analyte in effluent gas stops, the detector quickly destroys whatever is in its cell, and the signal drops to zero. A nondestructive detector does not affect the analyte, and the concentration measurement can continue as long as the analyte continues to reside in the

component
[kəmpəunənt]
n.成分，零件

proportional
[prə'pɔːʃənl]
adj.比例的，成比例的

quantitate
[kwɔntiteit]
v.测定(估计)的数量，用数量来表示

describe
[dɪ'skraɪb]
vt.描写，形容，叙述，描绘，作图

universality
[ˌjuːnɪvəː'sælətɪ]
n.普遍性，通用性，普适性，一般性

selectivity
[səˌlek'tɪvətɪ]
n.选择性，专一性

sensitivity
[ˌsensə'tɪvətɪ]
n.敏感，感受性，(仪器等的)灵敏性

ubiquitous
[juː'bɪkwɪtəs]
adj.无所不在的

coelute
[kuː'luːt]
共洗脱，
一起洗脱出来

destructive
[dɪ'strʌktɪv]
adj.破坏性的，毁灭性的，有害的

effluent
['efluənt]
adj.发出的，流出的

detector cell, without decline in the signal. Some types of nondestructive detectors (e. g., the ECD) measure the capture of an added substance (e. g., electrons). The "saturation" of this process causes a signal loss, so they are mass-flow detectors.

(4) Requirement for auxiliary gases

Some detectors do not function well with the carrier gas or flow rates from a capillary column effluent. Makeup gas, sometimes the same as the carrier gas, may be required to increase flowrates through the detector to levels at which it responds better and/or to suppress detector dead volume degradation of resolution achieved on the column. Some detectors require a gas composition different from that used for the GC separation. Some detectors require both air and hydrogen supplied at different flowrates than the carrier to support an optimized flame for their operation. Makeup flow dilutes the effluent but does not change the detection mechanism from concentration to mass-flow detection.

(5) Sensitivity and linear dynamic range

Detectors (both universal, and of course selective ones) vary in their sensitivity to analytes. Sensitivity refers to the lowest concentration of a particular analyte that can be measured with a specified signal-to-noise ratio. The more sensitive the detector, the lower this concentration. The range over which the detector's signal response is linearly proportional to the analyte's concentration is called the linear dynamic range. Some exquisitely sensitive detectors have limited linear dynamic ranges, so higher concentrations of analytes must be diluted to fall within this range. Another less than satisfactory solution to a limited linear dynamic range is to calibrate against a multilevel nonlinear standard curve. This requires injections of more standards and is more prone to introduce quantitative error. Dilution will not work satisfactorily if there is a wide range of concentrations in the sample. A very insensitive detector will perforce a more limited dynamic range, and multilevel standard curves or dilution will be of no avail with it.

Figure 3.9 compares the sensitivities and dynamic ranges of several of the most common types of GC detectors. The further to the left the range bar extends, the more sensitive the detector. The wider the bar, the greater the dynamic range. Each vertical dotted line denotes a span of three orders of magnitude (1000), so the overall ranges covered are very large. The values along the x-axis assume an injected volume of $1\mu L$ of solution. The ranges are approximate, their exact end points will depend on the design and model of the detector and optimization of its operating conditions.

detector cell
检测池

decline
[dɪ'klaɪn]
n.下降

saturation
[ˌsætʃə'reɪʃn]
n.(达到)饱和状态,饱和度

dilute
[daɪ'lju:t]
v.稀释

error[erə(r)]
n.[数]误差

multilevel
[mʌltɪ'levəl]
n.多级,多层次

solution
[sə'lu:ʃn]
n.溶液

optimization
[ˌɒptɪmaɪ'zeɪʃən]
n.最佳化,最优化

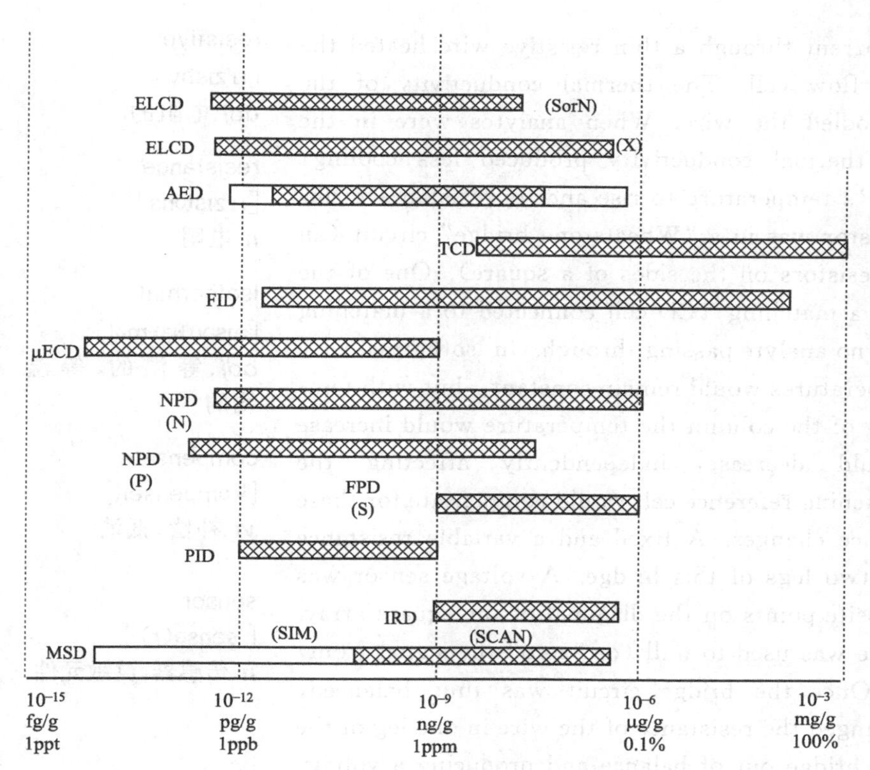

Figure 3.9 Approximate limits of detection (left end of bar) and dynamic ranges for 12 GC detectors from 1 μL sample injected

They provide a good general comparison. The IRD and MSD at the bottom are hyphenated method spectroscopic detectors, operable in several modes. Note that the "original" GC detector, the TCD, is the least sensitive, but also the only one suitable for handling neat (i.e., 100% pure single component) samples. The FID has the greatest single-mode dynamic range, while the "micro-ECD" is more sensitive, but with a more limited dynamic range. Let us proceed to describe the operation, characteristics, and applications for each of these nonhyphenated method detectors.

spectroscopic [ˌspektrəˈskɒpɪk] *adj*.分光镜的，借助分光镜的

3.2.1 Thermal conductivity detector (TCD)

TCD characteristics is as follows: universal (except for H_2 and He); non-destructive; concentration detector; no auxiliary (aux.) gas; works better with a parallel column; insensitive; limited dynamic range.

auxiliary [ɔːgˈzɪlɪərɪ] *adj*.辅助的

The TCD was the first widely commercially available GC detector, in the era when all the columns were packed, and samples were neat (i.e., not diluted solutions) mixtures to be separated. It measured differences in the thermal conductivity and/or specific heat of highly thermally conductive (either H_2 or He) carrier gas when diluted by small concentrations of much less conductive analyte vapor

conductivity [ˌkɒndʌkˈtɪvətɪ] *n*.传导率，电导率

(anything else). A current through a thin resistive wire heated the wire in the detector flow cell. The thermal conductivity of the flowing carrier gas cooled the wire. When analytes were in the stream, their lower thermal conductivity produced less cooling, which caused the wire's temperature to rise and its resistance to increase. This wire resistor was in a "Wheatstone bridge" circuit (an arrangement of four resistors on the sides of a square). One of the other resistors was in a matching TCD cell connected to a matching column and flow with no analyte passing through. In isothermal GC, carrier flows and temperatures would remain constant, but with temperature programming of the column the temperature would increase and the flow would decrease, independently affecting the conductivity. The matching reference cell would compensate for these effects on the resistance changes. A fixed and a variable resistance constituted the other two legs of this bridge. A voltage sensor was connected across opposite points on the diagonal of the square array. The variable resistance was used to null (or "zero" the signal from) the voltage sensor. Once the bridge circuit was thus balanced, passage of analyte changed the resistance of the wire in one leg of the bridge, throwing the bridge out of balance and producing a voltage signal proportional to the resistance change. The designs of two TCD cells are illustrated in Figure 3.10. This design of TCD cell was also referred to as a katharometer cell, a name for a gas thermal conductivity measurement device.

resistive
[rɪ'zɪstɪv]
adj.电阻的

resistance
[rɪ'zɪstəns]
n.电阻

isothermal
['aɪsəʊ'θɜːməl]
adj.等温的，等温线的

compensate
['kɒmpenseɪt]
vt.补偿，抵消

sensor
['sensə(r)]
n.传感器，灵敏元件

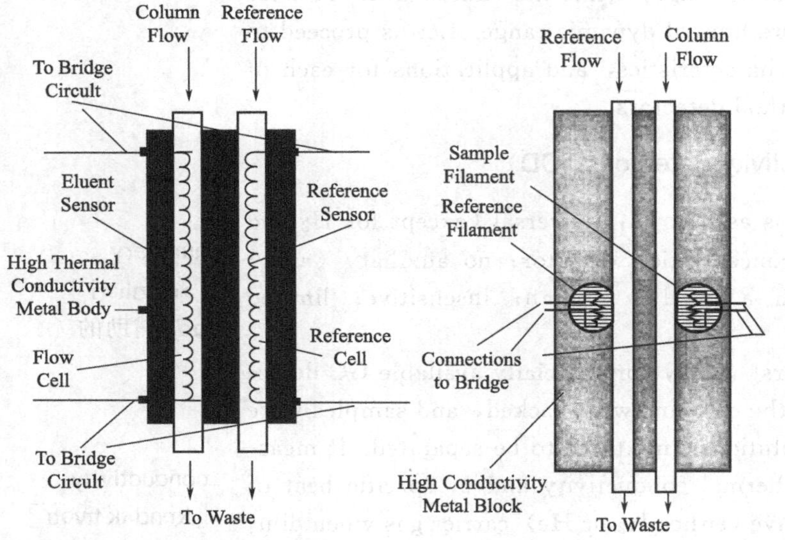

Note the application of the TCD for the detection of "fixed gases" in the chromatogram of Figure 3.10. These are generally not seen by other detectors (except hyphenated GC-MS). Note the low

Column Flow
Reference Flow
To Bridge Circuit
Eluent Sensor
Reference Sensor
High Thermal Conductivity Metal Body
Reference Cell
Flow Cell
To Bridge Circuit
To Waste
Reference Flow
Column Flow
Sample Filament
Reference Filament
Connections to Bridge
High Conductivity Metal Block
To Waste

Figure 3.10 Diagrams of two types of thermal conductivity detectors (TCD)

sensitivity signal for H_2, whose thermal conductivity is the only one to closely match that of the He carrier gas used. If only H_2 were being measured, it would be better to use N_2 as the carrier. This would yield a peak signal in the negative direction. In Figure 3.10 note that the detector is described as a u-TCD. This must employ miniaturized cells to be compatible with the low carrier flows of the open-tubular PLOT column and the need for small detector cell volumes to avoid extra-column band broadening.

3.2.2 Flame ionization detector (FID)

FID characteristics is as follows: nearly universal (all carbon compounds except CO, CO_2, HCN, but not many inorganic gases); destructive; mass flow detector; needs H_2 and air or O_2 aux. gas; sensitive with wide dynamic range.

The FID is the most commonly employed detector, as it gives a response to almost all organic compounds. On a molecular basis, the signal is roughly proportional to the number of carbon atoms in the molecule. Hydrogen and oxygen (or air) must be separately provided to fuel a flame in the detector cell. As illustrated in the diagram of Figure 3.11, the H_2 is introduced and mixed with the carrier effluent from the GC column. Even if H_2 is used as capillary carrier gas, an additional separately controlled H_2 supply is necessary to adjust the appropriate fuel supply for the flame. The mixed gas enters the cell through a jet, where air or O_2 flows past to serve as the flame oxidizer supply. An electrical glow plugin the cell can be pulsed to ignite the flame. The fuel and oxidizer flows are adjusted with needle valves to achieve a stable flame with optimal FID response, often by bleeding a volatile unretained hydrocarbon into the carrier stream to provide a reference signal. The jet tip is charged by several hundred volts positive relative to several "collector electrodes" or a "collector ring" surrounding the flame. In the absence of eluting organic analytes, no current flows in the jet-collector circuit. When a carbon-containing analyte elutes into the flame, the molecule breaks up into smaller fragments during the cascade of oxidation reactions. Some of these are positively charged ions, and they can carry current across the flame in the circuit. Although the ionization efficiency of the FID is low, its base current is also very low. Hence its signal-to-noise ratio is very high. Against such a low background, even very small ionization current scan be accurately measured using modern electronics which draw very low currents (high input impedance, voltage measurement circuits). Hence the good sensitivity and extraordinary dynamic range of this detector, often exceeding six orders of magnitude.

inorganic
[ˌɪnɔːˈgænɪk]
adj.无机的

oxygen
[ˈɒksɪdʒən]
n.氧，氧气

appropriate
[əˈprəuprɪət]
adj.恰当的，合适的

fragment
[ˈfrægmənt]
n.碎片，片段

current
[ˈkʌrənt]
n.电流

signal-to-noise ratio
信噪比

impedance
[ɪmˈpiːdns]
n.阻抗，全电阻，电阻抗

voltage measurement circuits
电压测量电路

extraordinary
[ɪkˈstrɔːdnrɪ]
adj.非凡的，特别的

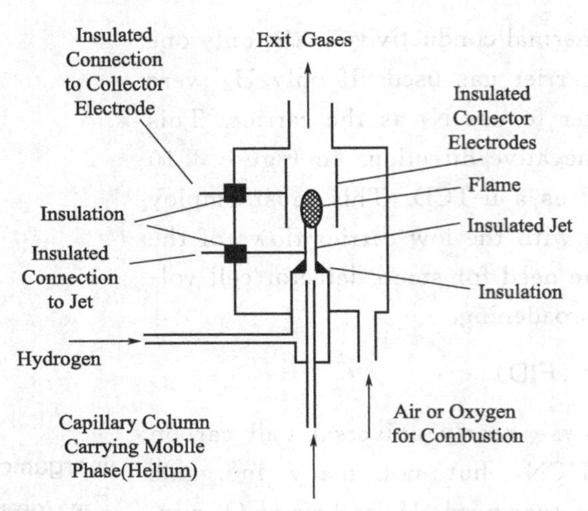

Figure 3.11 Diagram of the flame ionization detector (FID)

One can detect essentially any organic compound that can survive passage through the GC to the detector. A very wide range of analyte concentrations can be measured in a single run. The FID is the forerunner of a series of more selective and sometimes even more sensitive ionization detectors. It led the way to GC as it is currently practiced, with analytes being present in dilute concentrations in a preparation or extraction solvent solution instead of being mixtures of the neat compounds. Precise quantitation requires calibration against reference standards of the analytes being measured. The responses on a compound weight basis are sufficiently similar that for approximate quantitation, calibration for many components can be made against a single reference compound peak area. The wide dynamic range minimizes the number of calibration levels required for an adequate standard curve. The flame gases can be supplied from compressed gas cylinders or generated on site by gas generators employing hydrolysis of water, as well as compression and filtration (to avoid organic contaminants) of ambient air. If H_2 is used as carrier instead of He, all compressed gas cylinders could be dispensed with.

detect[dɪ'tekt]
vt.查明，发现，洞察，侦察，侦查

forerunner
[ˈfɔːrʌnə(r)]
n.先驱，先驱者

approximate
[ə'prɒksɪmət]
adj.大概的，极相似的

hydrolysis
[haɪ'drɒlɪsɪs]
n.水解

3.2.3 Electron capture detector (ECD)

ECD characteristics is as follows: very selective (for organic compounds with halogen substituents, nitro and some other oxygen-containing functional groups); non-destructive; concentration detector; needs to use N_2 or argon/CH_4 as carrier, or as makeup gas if used with H_2 or He capillary carrier flow; extreme but highly variable sensitivity but with limited dynamic range.

If the FID brought GC into the realm of characterizing dilute solutions of organics, it was the ECD that allowed it to spark a revolution in the understanding of the threat of bioaccumulation in tissue

spark [spɑːk]
vt.发动，触发

and bioconcentration up food chains, of lipophilic, persistent organic pollutants (POPs) in the environment. The classic example of this was the discovery of the threat posed by the organochlorine pesticide DDT, and its subsequent banning. J. E. Lovelock (subsequently famous as the founder of the Gaia theory of the whole earth as an organism) invented this deceptively simple but exquisitely sensitive and selective detector. It is not too much of a reach to claim that observations which were only made possible by the use of this GC detector set off the revolution in environmental consciousness in the 1960s.

bioconcentration [bi:əukɒnsent'reɪʃən] *n*.生物浓缩

organochlorine [ɔ:gənəu'klɔ:ri:n] *n*.有机氯

consciousness ['kɒnʃəsnəs] *n*.意识，观念

How then does this little marvel function? Two versions are illustrated in Figure 3.12. The right hand diagram displays features of the initial design (operation in DC mode). The column effluent enters from the right. A nickel foil doped with radioactive ^{63}Ni [α β particle (energetic electron)-emitter] constantly bombards the carrier gas, ionizing some of its molecules, creating an atmosphere of positive ions and the negative electrons which have been knocked loose. This radioactive source is chosen for its ability to withstand the high operating temperatures GC detectors may require to prevent condensation of high boiling analytes. A low voltage (several volts, instead of the hundreds of volts across the FID) between the negative inlet side of the detector and the positive outlet side is sufficient to set up a current between these two electrodes. Under this low voltage, it is the more mobile, lighter free electrons which carry the so-called "standing current" to the positive electrode.

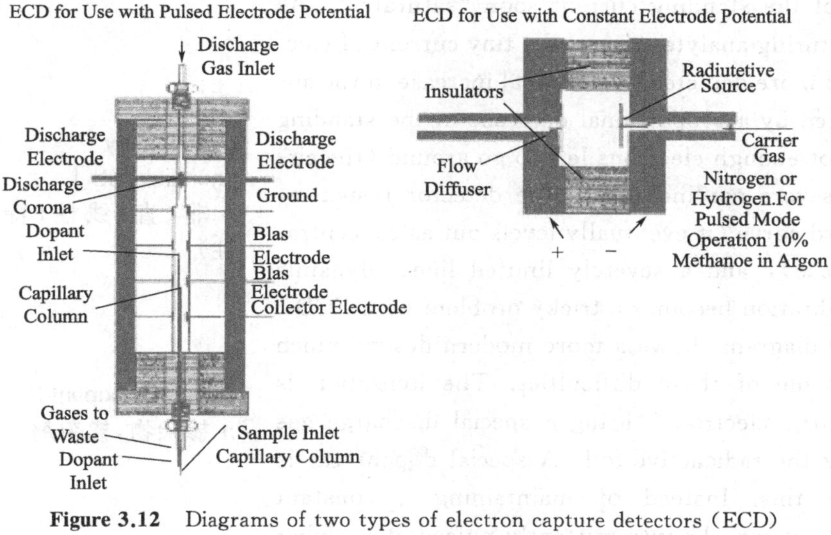

feature ['fi:tʃə(r)] *n*.特征，特点

nickel foil 镍箔片

bombard [bɒm'bɑ:d] *vt*.轰击

radioactive [ˌreɪdiəu'æktɪv] *adj*.放射性的

Figure 3.12 Diagrams of two types of electron capture detectors (ECD) (Cazes, used with permission)

If an analyte with highly electronegative, highly polarizable substituents enters the cell, the electrons of the standing current can be captured by these molecules. The mobility of the captured negative

polarizable [pəulәraɪzəbl] *adj*.可极化的

charge is drastically reduced, and some may be more easily neutralized by collision with the positive ions which are generated. It is the depression of this standing negatively charged electron current which is displayed reversed and appears as a peak signal. The electron capture sensitivity varies dramatically (one to two orders of magnitude) with the nature and number of the substituents: I>Br>Cl≫F among the halogens, and 5-8 halogen atoms>4>3≫2≫1. Note that most electronegative F is surpassed by more polarizable Cl, Br, and I, increasing in that order, so it is the polarizability which dominates, although both are necessary for electron capturing effectiveness. Some other groups such as $—NO_2$, —NO, and so on fall somewhere on the low side of these responsiveness series. Hydrocarbons are essentially unresponsive, although in great excess they can depress the sensitivity of the ECD response of ECD active compounds if they coelute with them. So with the ECD, calibration standards for every analyte are necessary to perform quantitative work.

neutralize [ˈnjuːtrəlaɪz] *v*.使失效，抵消，中和

electronegative [ɪˌlektrəʊˈnegətɪv] *adj*.负电的，带负电的

polarizability [ˈpəʊləraɪzəˈbɪlətɪ] *n*.极化性，极化率，极化度，可极化性

Note that ECDs require often special carrier or makeup gases. Most early ECD cells were designed for high packed column flow rates, and will need makeup gas at higher flow rates and of different composition when He or H_2 are employed as capillary carrier gases. Use of a radioactive ionization source in the detector requires a license from the NRC in the US, and a program of regular wipe tests to detect radioactivity leaks in and around the detector. Another problem is that the depression of the standing current soon "saturates". As the level of electron capturing analytes rises, the tiny current of electrons becomes more and more depleted. Additional increase in the analyte level is not reflected by a proportional decrease in the standing current-there just are not enough electrons left to go around (the circuit). This results in severe nonlinearity of the detector response, curvature of the standard curve (it eventually levels out as concentrations continue to increase), and a severely limited linear dynamic range. Quantitative calibration becomes a tricky problem with ECDs.

ionization [aɪənaɪˈzeɪʃn] *n*.离子化

nonlinearity [nɒnlɪnɪˈærɪtɪ] *n*.非(直)线性(特性)

The left hand ECD diagram shows a more modern design which attempts to alleviate some of these difficulties. The ionization is achieved by the "discharge electrode" using a special discharge gas flow, thereby replacing the radioactive foil. A special dopant gas is introduced to enhance this. Instead of maintaining a constant potential across the cell, it may be intermittently pulsed at a higher voltage. Typically, a pulse at 30V of around1μs repeated at 1ms intervals (a duty cycle of 0.1%) collects all free electrons in the ECD cell, while during the "off period" the electrons re-establish equilibrium with the gas. The very fast collection time is enabled by

dopant [ˈdəʊpənt] *n*.掺杂物，掺杂剂

admixture of 5%-10% methane in the argon ECD gas, which serves to enhance the voltage-induced migration rate of the free electrons. The current during the pulsed collection is integrated to produce the standing current level. The benefit of this method of controlling the standing current is to extend the linear dynamic range, and most modern ECDs are designed to operate as pulsed-mode ECDs. Such an extended linear dynamic range is illustrated for the "μ-ECD" in Figure 3.9. The micro-(μ) designation indicates that this model is equipped with a much smaller cell volume adapted for the effluent flow rates of capillary columns. Dispensing with the need for dilution of the effluent stream with makeup gas also increases the sensitivity even more. All other things remaining equal, pushing the sensitivity even lower extends the dynamic range since it is mainly limited by the effects of saturation of the standing current at the upper end of the range. The earlier model, large volume, DC-mode ECDs often had a linear dynamic range of only 100 or less. This made measuring multiple analytes (especially if they covered a much larger range of concentration) within a valid calibration range very complicated. Many dilutions and reassays were often required to do good quantitative work— great contrast to operating within the huge dynamic range of the FID. In contrast the special selectivity and extraordinary sensitivity of the ECD for certain classes of compounds, such as multiply-chlorinated DDT, were indispensable for the discovery of the bioconcentration pathways of that compound and others such as PCBs, dioxins, and PBDEs (polybrominated diphenyl ether flame retardants). It is an important general principle that the detection limits for given analytes in chromatographic methods applied to real samples from environmental or biological systems are more often determined by the selectivity of the detector against background coeluting interferences than by its absolute instrumental signal to noise sensitivity limit.

admixture
[ədˈmɪkstʃə(r)]
n.混合,混合物

dynamic
[daɪˈnæmɪk]
adj.动态的

multiply-chlorinated
adj.多氯化的

3.2.4 Sulfur-phosphorous flame photometric detector (SP-FPD)

SP-FPD characteristics is as follows: heteroatom selective (for organic compounds with S or P atoms; separately); destructive; mass-flow detector; less sensitive and shorter dynamic range than FID.

When organic compounds containing S or P atoms are burned in an FID, the flame conditions can be adjusted to produce a lower temperature which enhances the emission from S_2 fragments at 394nm or HPO fragments at around 515nm. The FID electrodes are omitted, and a temperature resistant window or fiber-optic light guide in the side of the detector cell passes the emitted light to filters designed to

enhance
[ɪnˈhɑːns]
vt.加强,提高,增加

isolate these wavelengths and pass the characteristic emission to a sensitive photomultiplier tube. Operated in S-selective mode it is useful to characterize the organosulfur compounds in complex petroleum mixtures, as the much higher levels of coeluting hydrocarbon peaks give minimal response. Operated in P-selective mode it is a sensitive detector for organophosphorus pesticide trace residues in complex environmental mixtures. Only the sulfur mode response range is illustrated in Figure 3. 9.

photomultiplier [fəutəu'mʌltɪplaɪə] *n*.光电倍增器，光电倍增管

3.2.5 Nitrogen-phosphorous detector (NPD)

NPD characteristics is as follows: heteroatom selective (for organic compounds with N or P atoms; separately); more sensitive for N and P compounds than FID, as well as selective for them; destructive; mass-flow detector; more sensitive than FID for N, more sensitive than P-FPD for P, but with less dynamic range than the FID.

heteroatom ['hetərəuætəm] *n*.杂原子

The NPD is yet another variation on the workhorse of GC detectors, the FID. Comparison of Figure 3. 13 for an NPD to Figure 3. 11 for the FID highlights the similarities. Both are defined as "thermionic detectors", that is, the high temperature of a flame breaks the eluting analytes into fragments, some of them positive ions, which release electrons to carry a current under the influence of the voltage between two electrodes. The NPD is operated under fuel (i.e., H_2)-rich conditions. Under these conditions the normal carbon compound FID thermionic response is suppressed by orders of magnitude. The "new element" in the detector is rubidium in the form of a glass or ceramic bead doped with a rubidium salt, which is heated by immersion in the flame, but is also additionally and variably heated by passing current through thin wires which support the bead in the flame. What exactly happens near the surface of the bead while

rubidium [rʊbɪdiəm] *n*.铷

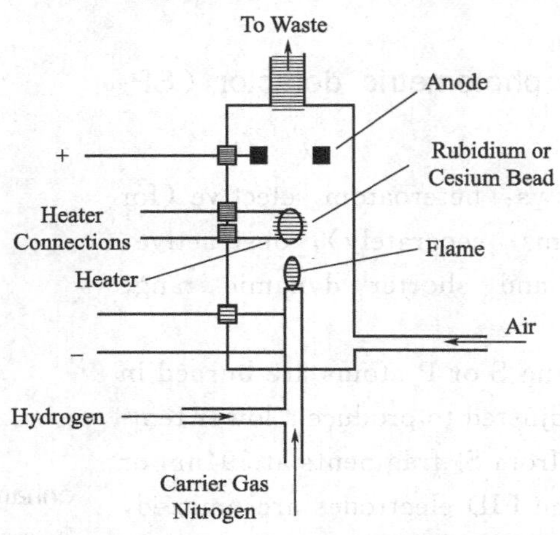

Figure 3.13 Diagram of the nitrogen-phosphorus detector (NPD) (Cazes, used with ermission)

organonitrogen or organophosphorus compounds are decomposing in this not-so-hot flame is complex and poorly understood. In some designs, the flame would not be self-supporting, were it not for the independently heated bead embedded in it. Somehow C-N or C-P containing fragments interact with easily ionizable rubidium atoms to produce ions and electrons which then produce a signal like that in the FID; but one which is exquisitely sensitive and selective to either N or P atoms (depending on how the detector parameters are set). This is illustrated in Figure 3. 9. The extreme sensitivity of this detector to some organophosphorus compounds explains why it has largely supplanted the P-FPD for this application, and no range is depicted for that mode of the FPD operation in the comparison figure. The NPD produces no response to N atoms not bound to carbon in organic molecules. This is a fortunate circumstance which renders it immune from interference from ubiquitous atmospheric N_2.

organophosphorus [ˌɔgənəu'fɒsfərəs] *adj*.有机磷的

immune [ɪ'mjuːn] *adj*.免疫的，有免疫力的，不受影响的

ubiquitous [juː'bɪkwɪtəs] *adj*.普遍存在的

3.3 Gas chromatography-mass spectrometry (GC-MS)

If we consider their wide availability and capability, GC-MS instruments could have been said to provide the largest total amount of analytical power available to the instrumental analysis world. With the more recent spread of LC-MS instrumentation to serve the biochemical research market, they now must share this status. If analytes are volatile and thermally stable, capillary GC-MS can identify and quantitate hundreds (even thousands in GC-GC-TOFMS runs) of them from a single injected mixture. It is the quadrupoles and ion-traps which best meet the earlier-mentioned scan speed criterion. Magnetic sector instruments, with their higher mass resolution but slower scan speeds, will require that the GC peaks be broadened and slowed down. This works against the goals of faster analyses and better chromatographic resolution. On the other hand, so-called "fast GC", employing short, narrow-bore, thin-film capillary columns, will require MS-scan acquisition speeds of 50-500s^{-1}, which are attainable only by TOF-MS instruments.

The major problem for GC-MS is interfacing. Mass spectrometers form and move their ions at highly reduced pressures, and the mass analyzer sections need to be operated at even lower pressures. The effluent of a GC consists primarily of carrier gas around atmospheric pressure (760torr). The MS vacuum pumps must remove this fast enough to maintain the necessary low pressure. With packed column flows of 10—40mL/min, this was impractical. With capillary GC flows around 1mL/min, modern diffusion pumps, or

volatile ['vɒlətaɪl] *adj*.易变的

quadrupole ['kwɒdrəpəul] *n*.四极

ion-trap 离子阱

acquisition [ˌækwɪ'zɪʃn] *n*.采集，获得，购置物

interfacing [ˌɪntə'feɪsɪŋ] *n*.接口

impractical [ɪm'præktɪkl] *adj*.不切实际的，不现实的

diffusion [dɪ'fjuʒən] *n*.扩散，扩散

even better, powerful turbo molecular pumps, could achieve this. Another reason to prefer capillary GC to packed column GC separations! Many modern GC-MS instruments simply introduce the end of the capillary column into the ion source (direct coupling). A separate pump evacuates the small, largely confined volume of the source, while the ions are extracted through a small orifice into the mass analyzer region, which is kept at a lower pressure by another pump. Such a MS design is called a differentially pumped system. To interface packed columns, a device called a jet separator (Figure 3.14) preferentially removed light He or H_2 carrier gas atoms or molecules from the effluent, allowing a smaller but enriched concentration of analyte molecules to enter the ion source at a lower total gas flow rate. The much higher speed of the lighter carrier gas particles causes most of them to spread more widely in the vacuum of the separator chamber, and not to pass on through the conical orifice leading to the ion source. Jet separators are not much used or needed for GC-MS now.

turbo molecular *adj*.涡轮分子的

orifice[ˈɒrɪfɪs] *n*.孔，洞口

vacuum [ˈvækjuəm] *n*.真空；*v*.用真空吸尘器清扫

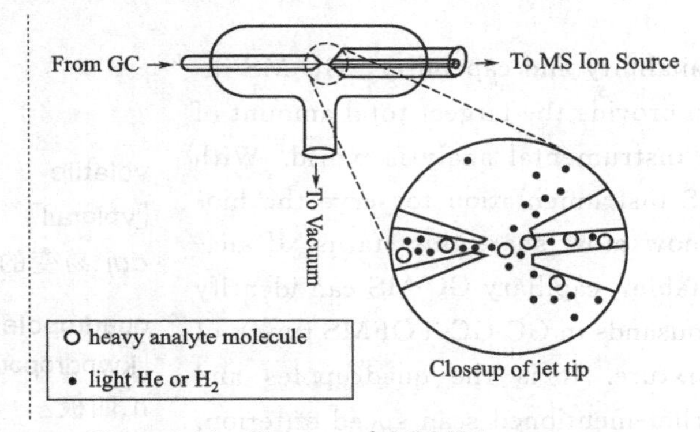

Figure 3.14 Diagram of GC-MS jet separator

In principle, one could monitor a GC effluent with a nondestructive detector like a TCD, pass it on through a suitable interface, and acquire and print out the mass spectra of peaks as they are eluting. With fast digital data converters and the speed and power of modern desktop PCs, it is better to simply acquire mass spectral data as a continuous sequence of spectra or selected mass fragment signals as the GC run proceeds. A data file in computer memory consists of a sequence of MS scans ordered by retention time. There are three main modes of acquiring and employing such data files:

(1) Full-scan

If the file consists of a sequence of full-scan spectra, the total counts from ions of all masses in each scan may be summed, and these totals plotted against successive scan numbers (or equivalently, GC retention time). This display is called a total ion chromatogram (TIC). Here the mass spectrometer acts like an ionization detector,

interface [ˈɪntəfeɪs] *n*.界面，接口，交界面

spectra [ˈspektrə] *n*.波谱，光谱

successive [səkˈsesɪv] *adj*.连续的

using whatever MS ion-source ionization mode was selected (EI, CI, etc.). The TIC chromatogram will appear similar to that obtained by a FID detector. One difference is that the FID output digitized by a chromatographic data system will likely define each peak at many more points across its width than the number of scans that the MS could make (unless it is the very fast scanning TOF-MS). For this reason, integration of FID peaks yields more accurate quantitation than TIC peaks.

output [ˈautput] n.输出信号

The advantage of the TIC data file is that by selecting any scan comprising a TIC peak one can display the mass spectrum of the compound in the peak. If the peaks are sharp and narrow, the level of the analyte may vary significantly over the time of the scan, thus distorting the proportions of mass fragments in the spectrum. It would be best therefore to select a scan at the top of the peak. Even better would be to sum scans across the whole peak to get an average spectrum. Even better than that would be to select and sum an equal number of scans of background mass spectra near the peak where nothing else is obviously eluting. Then subtract that sum from the peak average sum, and thereby obtain a cleaner mass spectrum, with spurious fragments from column bleed eliminated. This illustrates the power and flexibility of manipulating a data file of continuous, contiguous MS scans. We have combined the great separation power of capillary GC to the great characterization and identification power of MS. This is particularly feasible because in GC the samples are in the vapor state, which is mandatory for the MS ionization and fragment identification process to proceed. The vast libraries of EI-MS spectra are based on this mode of operation. In many cases it is possible to automate the process of peak detection, spectral selection, library spectra searching, and separated compound identification by matching to a library spectrum. A naive person might be forgiven for wondering whether GC-MS eliminates the need for an analyst. Not quite. But the interfacing of vapor-separation GC with vapor-requiring MS, when combined with the fluency of PC processing of digital data files is an excellent combination.

spurious [ˈspjuəriəs] adj.假的，伪造的

automate [ˈɔːtəmeɪt] vt & vi.(使)自动化，使自动操作

matching [ˈmætʃɪŋ] n.匹配

(2) Mass chromatograms (XIC)

One may seek to locate and measure only certain categories of analytes in the chromatogram of a much larger and more complex mixture. These might have very characteristic mass fragments. An example would be mono-, di-, and trim ethylnaphthalene isomers in a complex mixture of petroleum hydrocarbons such as a fuel oil. Their mass spectra consist mostly of a single intense molecular ion (M^+) at masses 142, 156, or 170, respectively. We can program a GC-MS data system to extract from the full-scan GC-MS files and

complex [kəm ˈpleks] adj.复杂的，复合的

plot only the ions of these masses in three separate chromatograms. Such plots are called extracted ion chromatograms (XIC), from their mode of production, or mass chromatograms, since they display peaks whose spectra contain the selected ion mass (ES). If in the mass chromatogram of a petroleum sample at $M^+ = 142$, we observed a pair of large peaks close together, we might well suspect them of being from the two possible monomethyl naphthalene isomers, and could confirm our suspicion by inspecting the complete spectrum of the scans at the center of each peak. Thus, we maybe able to determine the retention times of various compounds without injecting a standard. Unfortunately, mass spectra of isomers of such compounds are indistinguishable, so we will need some sort of retention information based on standards run on the particular GC stationary phase to say which isomer is which. Such a process could be repeated at many other characteristic masses for the different classes of hydrocarbons. The data for all masses in the range scanned, at every point in the TIC chromatogram, are present in the full-scan GC-MS data file.

suspicion
[səˈspɪʃn]
n.怀疑，疑心

(3) Selected ion monitoring (SIM)

In full-scan mode the mass spectrometer is acquiring counts at any particular ion mass for only a brief portion of the scan. For example, if scanning from mass 50 to 550 each second, with unit mass resolution, the detector spends less than 2ms at each mass during one scan. If we have only a few classes of analytes we wish to measure, and we know their characteristic major mass fragments, we can program the MS to acquire counts only at the selected masses, thereby increasing the dwell time at each mass, increasing the signal-to-noise ratio, and improving the sensitivity. If we wanted the three successive methyl naphthalene isomer distributions, we could monitor at only $M^+ = 142, 156$, and 170. In 1s each of these three ions would be monitored for a little less than 330ms instead of 2ms, greatly improving the sensitivity over the 50-550 full-scan acquisition. In fact, we could cut the cycle time from 1s to 0. 2s, still acquire for 66ms per cycle, but now be sampling and defining the peak shape 5 times/s instead of once per second. If we know that each class of isomers elutes over a unique range of retention times, we can set MS acquisition to monitor just their most characteristic and abundant ions during this period and achieve even greater sensitivity. This mode of acquisition is called SIM. SIM improves sensitivity by collecting more counts at the masses of interest, and it improves quantitative precision by enabling the GC-MS peak to be defined and integrated using more points. The extension of the linear range of measurement to lower values with SIM vs. full-scan is represented by the two sec-

dwell
[dwel]
vt.存在于

naphthalene
[ˈnæfθəliːn]
n.萘(球)

tions of the GC-MS sensitivity range. Ultimate SIM sensitivity is a complex function of MS ionization efficiency for the particular analyte compound, number of different masses monitored, dwell time, and cycle time. GC-MS-SIM can reach detection limits below that of the FID and in the range of the ECD. It is far more selective than any of the general GC detectors. This sort of analysis is characterized as "target compound analysis" since the system is tuned or programmed to select specific characteristic ions from specific target compounds expected to elute in specific retention time ranges.

To reiterate, the difference between XIC and SIM chromatograms is that in the former case the desired ion masses are extracted from a full-scan data file, while in the latter case the MS is directed to acquire the data only at those selected masses. The ability to improve sensitivity and quantitative precision by using SIM applies mainly to magnetic-sector (including use of high resolution MS) and quadrupole MS instruments. The mode of operation of ion-trap MS and TOF-MS instruments yields near-optimum sensitivity in full scan mode. Thus, there is generally no provision for SIM acquisition on these instruments, and quantitation is done on XIC files. The MS in GC-MS is a destructive, mass-flow, detector. Only with mass spectrometric detection can the analyst use the "perfect" internal standard; namely, the identical chemical species, labeled with stable isotopes of atoms with a higher mass (e.g., ^{2}H, ^{13}C, ^{15}N, ^{18}O, etc.). This procedure, isotope-dilution mass spectrometry (IDMS), can correct for differences in sample preparation recovery, derivatization efficiency, MS ionization efficiency, and so on, for which use of a different chemical species as an IS may not fully. However, IDMS IS materials are difficult to make, expensive to purchase, and commercial products are limited to only several thousand especially important target analytes.

reiterate
[rɪ'ɪtəreɪt]
vt.重申，反复地做

3.3.1 Gas chromatography-mass spectrometry

Operators generally fall into one of two schools: those who consider the mass spectrometer as the ultimate selective detector for their gas chromatograph and those who consider the gas chromatograph as an expensive inlet system for their mass spectrometer. Practitioners equally knowledgeable in both areas and dedicated to the overall process are less common. Settlage and Jaeger point out that regardless of the cause, GC-MS technology falls far short of its potential. Nonetheless, GC-MS has developed into one of the most fruitful techniques in analytical chemistry. GC-MS is capable of providing both quantitative and qualitative data by means of spectral interpretation procedures. Figure

practitioner
[præk'tɪʃənə]
n.从业者

knowledgeable
['nɒlɪdʒəbl]
adj.博学的，有见识的，精明的

interpretation
[ɪnˌtɜːprɪ'teɪʃn]
n.解释说明

3. 15 is a block diagram that shows how the components of a GC-MS-computer system are interrelated.

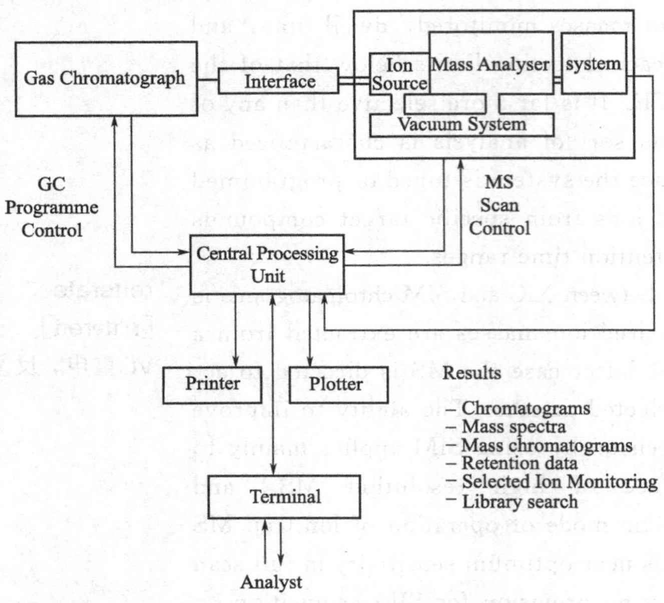

Figure 3.15 Block diagram of a GC-MS-computer system

The mass spectrometer receives the separated chromatographic zones which enter the ion source, maintained at a pressure of 10^{-2} Pa to ensure efficient production of ions from the neutral molecules. Ion production can be achieved by electron bombardment (electron ionization), chemical ionization (CI), field ionization or field desorption techniques. The various ions must then be separated in the mass analyser unit and recorded as a spectrum of ions according to their mass-to-charge (m/z) ratios and relative abundances. Various types of mass separation methods, including quadrupole and magnetic/electrostatic sector spectrometers with high scanning rates, have been used but, in general, the quadrupole instrument is preferred for GC-MS because of its relatively high sensitivity, resolution of about 1000 amu, and ability to operate at pressures up to 10^{-2} Pa, with scan rates of 10^{-1} s per mass decade. The only disadvantage in using quadrupole analysers is their relatively low resolving power. With applications requiring high resolution as, for example, in the case of suspected interference, high resolution magnetic sector or double-focusing instruments are needed. More recently, Fourier Transform mass spectrometers and ion trap detectors have been introduced. The mass spectrum is a line spectrum showing the peak of the molecular ion and the signals corresponding to the individual fragments. Figure 3. 16 depicts the mass spectrum of 1,1,1-trichloroethane. The molecular ion indicates the molecular mass of the parent compound. Lines grouped around this peak result from the natural abundances of vari-

efficient
[ɪ'fɪʃnt]
adj.有效率的

desorption
[dɪ'sɔːpʃən]
n.解吸附作用

interference
[ˌɪntə'fɪrəns]
n.干涉

abundance
[ə'bʌndəns]
n.丰度，丰富

ous isotopes. Further structural information can be obtained from a

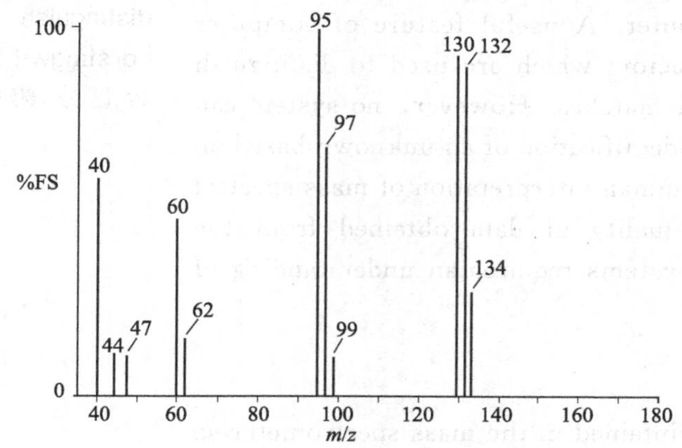

Figure 3.16 Electron-impact mass spectrum of trichloroethane at 70eV ionizing energy

consideration of the fragment ions and their magnitude observed in the mass spectrum.

The amount of useful data generated by a GC-MS is so overwhelming that it is neither practical nor possible to obtain and analyse it without a computer. This can be illustrated by the following example of a typical gas chromatographic analysis of an environmental sample performed on an open tubular column. If we suppose the chromatogram required a 60min programmed temperature run and a 2s mass spectral scan rate, at the completion of the chromatogram 1800 mass spectra will be stored. Each peak in a mass spectrum consists of a mass and abundance value, so a typical mass spectrum of about 100 peaks requires the storage of 200 numbers. The entire analysis will therefore require storage space for more than $200 \times 1800 = 360000$ numbers. Additional storage space is required for such data as retention times. Hence, modern instruments use a data system for the automatic acquisition, storage and efficient processing of the measured data, as well as for instrument calibration and control.

For a specific set of experimental conditions, the mass spectrum of a compound is like a fingerprint. The task of identifying an unknown peak is greatly simplified by comparing the mass spectrum for the peak with other spectra stored in a reference library. Many collections of mass spectra are available, although several of these are specialized and contain only a few thousand spectra. Reasons for this are readily apparent; extensive quality control measures are essential to ensure that only verified, accurate mass spectra free of interferences are included. The largest collection is the Wiley-NBS mass spectral data base, which currently contains less than 100000 compounds. This data base may seem large but it must be remembered that the number of known organic compounds exceeds 5 million. Manual

fingerprint [ˈfɪŋɡəprɪnt] *n*.指纹，指印；*vt*.采指纹

comparison with reference spectra is tedious and comparison is usually performed with a computer. A useful feature of computer matching is the calculation of factors which are used to distinguish between good, average and poor matches. However, no system can give a completely unambiguous identification of an unknown based on search data only. The need for human interpretation of mass spectral data remains. Evaluating the quality of data obtained from the various commercially available systems requires an understanding of different search strategies.

distinguish
[dɪˈstɪŋgwɪʃ]
vt.区分，辨别

3.3.2 Interface

A high vacuum must be maintained in the mass spectrometer so that molecular reactions can be avoided. However, if carrier gas from the gas chromatograph is admitted directly to the ion source, the pressure will rise. Many methods have been described for interfacing the gas chromatograph and mass spectrometer to ensure that transfer of sample between the two is maximized, while removing most of the carrier gas. A poorly designed interface can compromise the performance of the column, the mass spectrometer, or both. An ideal interface should be such that: band broadening is minimized by a minimum dead volume; the pressure is reduced at the column exit to approximately 10^{-2} Pa at the ion source; there are no discrimination effects against thermally or chemically labile compounds owing to active sites or heating of the interface; and there is efficient transfer of the entire sample to the ionization chamber.

transfer
[trænsˈfɜː(r)]
vt.使转移，使调动

minimize
[ˈmɪnəˌmaɪz]
v.把…减至最低数量

Ideally, the interface should isolate the column from the vacuum in order to minimize the effects of the vacuum on the column efficiency. Isolation is essential for packed columns, and in early instruments, an all-glass interface, known as a jet separator, was widely adopted as the most efficient interface. Isolation becomes increasingly less important for narrow bore and long (>15m) open tubular columns where low carrier gas flow-rates are usual. Hence, because of the popularity of open tubular columns and improvements in vacuum pump design, direct connection of the column to the ion source is now normal practice. The ion source is designed to cope with the low flow encountered with open tubular columns (<5mL/min) by using high capacity diffusion pumps. Direct connection can take several forms. The easiest method is direct connection of the column to the ion source, with no attempt to isolate the column from the vacuum. This approach works well for longer narrow-bore columns, but even here there is the disadvantage that the column cannot be removed without venting the mass spectrometer. Alternatively, some isolation can be effected by placing a small-bore capillary tube, either

isolate
[ˈaɪsəleɪt]
vt.使隔离，使孤立

platinum-iridium or fused silica, in the transfer line. Several reports indicate that platinum-iridium alloy is responsible for catalytic decomposition of certain substances. A third and most successful method of direct configuration is the open-split.

configuration [kən ˌfɪgə'reɪʃn] n.配置，布局，构造

3.3.3 Recording and analysis

Gas chromatography can be used for both qualitative and quantitative analyses. For qualitative analysis, retention time serves as a good index for identifying a component. Retention time is defined as time taken by a given component to elute from the column and is measured as time elapsed from the point of injection to the peak maximum. This property is characteristic of a sample and a liquid phase at a given temperature. It can be reproduced to an accuracy of better than 1% with appropriate control of temperature and flow rate.

maximum ['mæksɪməm] n.最大的量、体积、强度等

The signal from the eluted components is conventionally recorded on a strip chart recorder, where time is the abscissa and the signal, in millivolts, is the ordinate. At present, the signal is more likely to be collected digitally and then plotted electronically with a computer (Figure 3.17).

millivolt ['mɪlivəult] n.毫伏

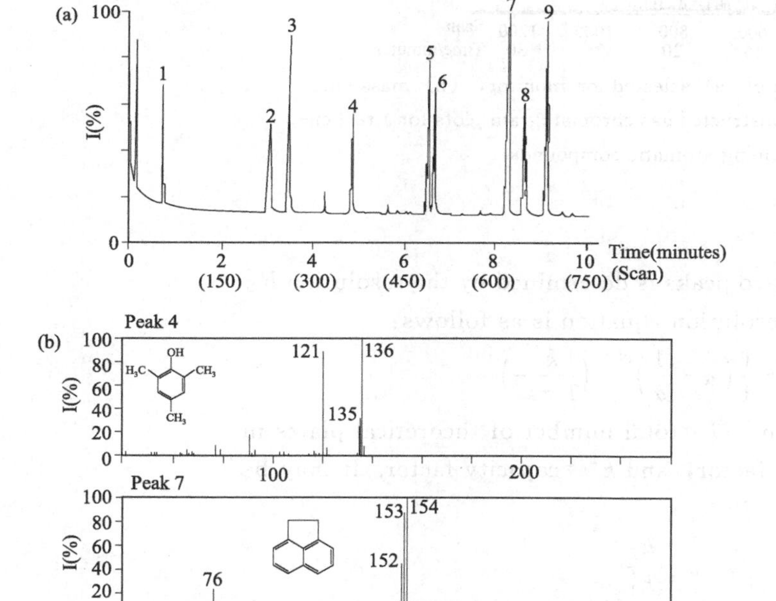

Figure 3.17 (a) Reconstructed gas chromatogram of a test mixture performed on a 25 m BP-1 open tubular column temperature programmed from 60℃ to 130℃ at 4℃/min with direct coupling to the mass spectrometer. Compounds are identified: 1—1,2-dimethylbenzene; 2—isooctanol; 3—2,6-dimethylphenol; 4—2,4,6-trimethylphenol; 5—2,6-dimethylundecane; 6—nicotine; 7—1,2-dihydroacenaphthylene; 8—dodecan-1-ol; 9—2-methyltetradecane; (b) The mass spectra obtained from peaks 4 and 7

The area of the peak for a given sample is proportional to its concentration. This allows exact determination of the content of a given component by calculation against a known standard at comparable concentration. The accuracy of quantification relates to sample concentration and preparation, injection, detection, and integration methods. An accuracy of better than 1% is possible with electronic digital integrators or computers (see Figure 3.18).

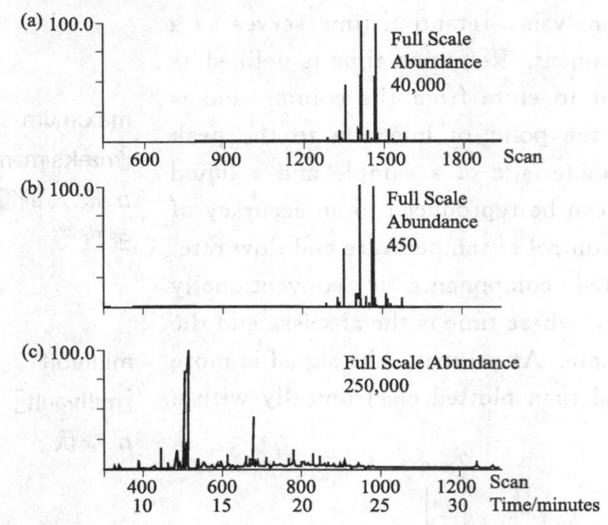

Figure 3.18 Comparison of (a) selected ion monitor, (b) mass chromatogram, and (c) reconstructed gas chromatogram plots for a real environmental sample containing aromatic compounds.

3.3.4 Resolution

Separation between two peaks is determined by the resolution R_S of a given system. The resolution equation is as follows:

$$R_S = \frac{1}{4}\left(\alpha - \frac{1}{\alpha}\right)^{N^{1/2}}\left(\frac{k'}{1+k'}\right)$$

where R_S = resolution, N = total number of theoretical plates in a column, α = separation factor, and k' = capacity factor. It may be recalled that

$$\alpha = \frac{k'_2}{k'_1}$$

$$k' = \frac{\text{retention time of a peak} - \text{retention time of unretained peak}}{\text{retention time of unretained peak}}$$

In its simple form, the resolution equation can be described as follows:

$$R = \frac{t_2 - t_1}{(W_1 + W_2)/2}$$

where t_2 and t_1 = retention times of peak 2 and peak 1, respectively and W_2 and W_1 = peak widths of peak 2 and peak 1,

respectively [rɪ'spektɪvlɪ] *adv*. 各自地，各个地，分别地

respectively. The number of plates required for a given separation may be calculated from the resolution equation as follows:

$$N_{req} = 16R_S^2 \left(\frac{\alpha}{\alpha - 1}\right)\left(k_2' + \frac{1}{k_2'}\right)^2$$

where N_{req} = number of plates required for a given separation and k_2' = the peak capacity of the second peak.

3.4 Selection of a stationary phase

The selection of a stationary phase is generally based on overall knowledge of the field, literature survey, and actual experimental work. A reliable stationary phase should meet the following criteria:

- Good solubility for sample components because components with poor solubility elute rapidly.
- Good differentiation for sample components-this relates to their partition coefficient.
- Good thermal stability-instability can result from catalytic influence of solid support at high temperatures.

Poor volatility; that is, the stationary phase should be nonvolatile and have a vapor pressure of 0.01 to 0.1mm at operating temperatures.

- Poor reactivity with components of the sample at operating temperatures.

It is important to have at least some idea about the chemical nature of sample component (s). If this information is not available, an intelligent guess is made as to the possible nature of the sample. Alternatively, it would be necessary to analyze a sample for all classes of compounds to assure that none of the components is missed. The basis plan is to try to classify the sample into one of five groups, on the basis of the polarity of its components.

criteria
[kraɪ'tɪərɪə]
n.标准，准则（criterion的名词复数）

elute
[ɪ'lju:t]
vt.洗提

volatility
[ˌvɒlə'tɪlətɪ]
n.挥发性，挥发度

3.4.1 Sample classification

Sample can be divided into various groups, based on their polarity.

(1) Group 1 (polar)

This group contains compounds with both a donor atom, such as N, O, or F, and an active hydrogen atom. They include:

* Alcohols
* Fatty acids
* Phenols
* Amines (primary and secondary)
* Nitro compounds with α-H atoms
* Nitriles with α-H atoms

polarity
[pə'lærəti]
n.极性

phenol
['fi:nɒl]
n.酚

amine
[ə'mi:n]
n.胺

(2) Group 2 (highly polar)

The compounds capable of forming a network of hydrogen bond are included in this group. They include:

* Water
* Glycerol
* Amino alcohols
* Hydroxy acids
* Dibasic acids
* polyphenols

glycerol
[glɪsərɒl]
n.甘油，丙三醇

(3) Group 3 (intermediate polarity)

The compounds containing a donor atom but no active H atoms are included in this group.

* Aldehydes
* Ethers
* Esters
* Ketones
* Tertiary amines
* Nitro compounds with no α-H atoms
* Nitriles with no α-H atoms

aldehyde
[ˈældɪhaɪd]
n.醛

(4) Group 4 (low polarity)

The compounds with active H atoms but no donor atoms belong to this group.

* Halogenated hydrocarbons
* Aromatic hydrocarbons
* Olefins

hydrocarbon
[ˌhaɪdrəˈkɑːbən]
n.碳氢化合物，烃

(5) Group 5 (nonpolar)

The compounds showing no hydrogen-bonding capacity are placed in this group.

* Saturated hydrocarbons
* Mercaptans
* Sulfides
* Halocarbons such as CCl_4

mercaptan
[məˈkæptæn]
n.硫醇

halocarbon
[ˈhæləˌkɑːbən]
n.卤代烃，卤碳

Suitable stationary phases for these samples are given in the next section.

3.4.2 Suitable stationary phases

Stationary phases can be similarly classified to assist selection for use with a given group of compounds.

(1) Stationary phases suitable for polar compounds (Group 1)

* β,β-Oxydipropionitrile
* XE-60

(2) Stationary phases suitable for highly polar compounds (Group 2)

* Carbowaxes (various molecular weights)
* Carbowax 20M-TPA
* FFAP
* Diglycerol

carbowax [kɑːbəuwæks] n.碳蜡，聚乙二醇，水溶性有机润滑剂

(3) Stationary phases suitable for compounds with intermediate polarity (Group 3)

* Dibutyl tetrachlorophthalate
* Tricresyl phosphate
* OV-17

(4) Stationary phases suitable for compounds with low polarity and for nonpolar compounds (Group 4 and 5)

* SE-30
* Squalane
* Apiezons
* Hexadecane

The structures of some of the commonly used liquid phases are given below

$HO—(CH_2—CH_2—O)_n—H$

Carbowax

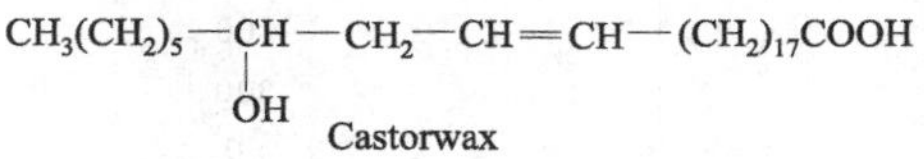

$CH_3(CH_2)_5—CH(OH)—CH_2—CH=CH—(CH_2)_{17}COOH$

Castorwax

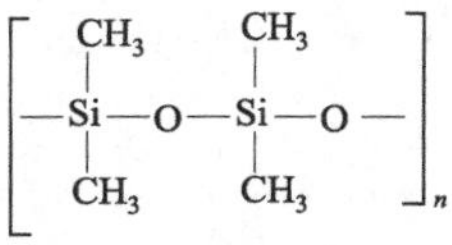

$[—Si(CH_3)_2—O—Si(CH_3)_2—O—]_n$

SE-30

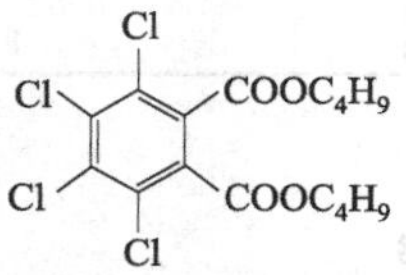

Cl, Cl, Cl, Cl, $COOC_4H_9$, $COOC_4H_9$

Dibutyl tetrachlorophthalate

CH_3, CH_3, CH_3, O, O—P=O, O

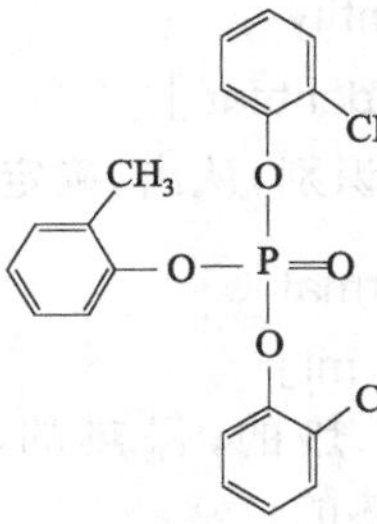

Tricresyl phosphate

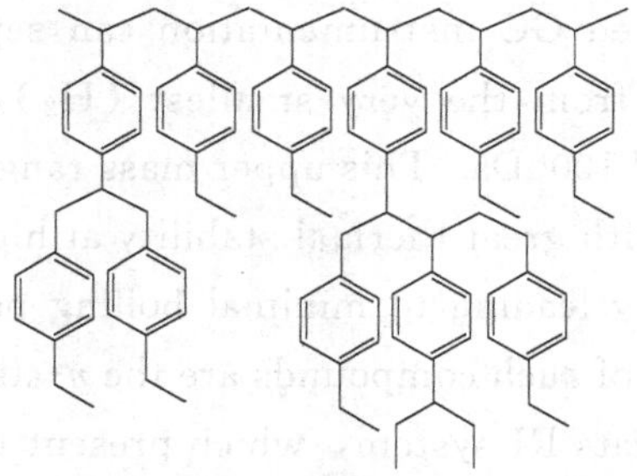

Porapak

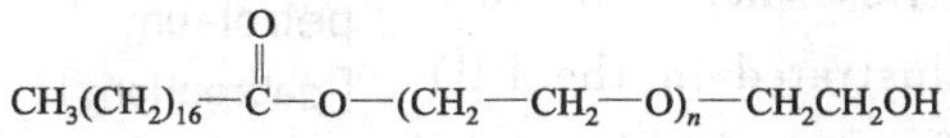

$CH_3(CH_2)_{16}—C(=O)—O—(CH_2—CH_2—O)_n—CH_2CH_2OH$

Ethofat

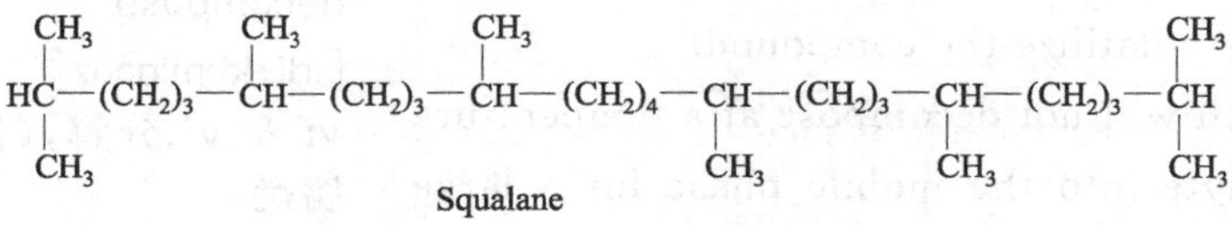

$(CH_3)_2CH—(CH_2)_3—CH(CH_3)—(CH_2)_3—CH(CH_3)—(CH_2)_4—CH(CH_3)—(CH_2)_3—CH(CH_3)—(CH_2)_3—CH(CH_3)_2$

Squalane

$—(CH_2—CH_2—O—CH_2—CH_2—O—C(=O)—CH_2—CH_2—C(=O)—O)_n—$

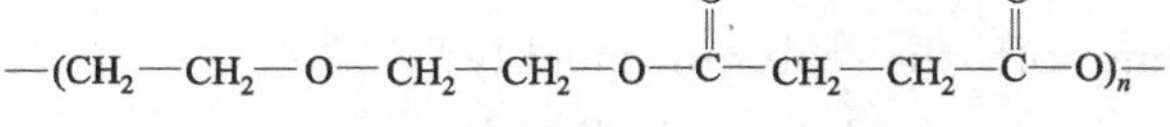

DEGS

The liquid phases should be operated under the recommended maximum temperatures (see Table 3.1)

Table 3.1 Recommended maximum temperature for operation of liquid phases

Liquid phase	Maximum temperature/℃
Carbowax 400	125
Carbowax 1500	200
Carbowax 4000	200
Carbowax 20M-TPA	250
FFAP	275
Diglycerol	120
Castorwax	200
β,β-oxydipropionitrile	100
Silicone XE-60	275
Tetracyanoethylated pentaerythritol	180
Ethofat	140
Dibutyl tetrachlorophthalate	150
Tricresyl phosphate	125
OV-17	300
Silicone SE-30	300
Squalane	100
Apiezon J	300
n-Hexadecane	50
OV-1	350

3.5 The scope of GC analysis

Appropriately configured GC instrumentation can separate and identify molecules ranging from the very smallest (H_2), to those with masses on the order of 1000Da. This upper mass range limit applies only for compounds with great thermal stability at high temperatures and very low polarity leading to minimal boiling points for a given mass. The prototype of such compounds are the *n*-alkanes, the reference series for the Kovats RI system, which present the characteristic "picket fence" profile of a homologous series when they elute under linear temperature programs. This is illustrated in the FID chromatogram of a light fuel oil. Heavier petroleum crude oils might extend this pattern out to *n*-C_{60} or beyond. The limiting factors are:

- The ability to efficiently volatilize the compound.
- A stationary phase which will not decompose at a temperature sufficient to partition the analyte into the mobile phase for a large enough percentage of the time. This will enable it to pass through the column in a reasonable time, and come off the column quickly enough to produce a peak well above the noise background. Recall

identify [aɪ'dentɪfaɪ] *vt*.识别，认出，确定

thermal ['θɜːml] *adj*.热的，保热的，温热的

petroleum [pə'trəulɪəm] *n*.石油

decompose [ˌdiːkəm'pəuz] *vt & vi*.分解，(使)腐烂

that very slowly eluting compounds with long retention times elute so gradually as they exit the column into the detector, that their peaks spread out and eventually become not much higher than and indistinguishable from background noise and baseline drift.

• The column tubing itself must be stable at the necessary temperature. The polyimide scratch-protective outer coating of fused-silica capillaries is itself stable only to about 380℃. Some exceptional stationary phases can exceed this limit, and may require use of metal capillaries, which for some analytes at those temperatures must be deactivated-perhaps with an inner coating of silica, which will not yield a flexible column coil.

exceptional [ɪk'sepʃənl] *adj*. 优越的，杰出的，例外的，独特的

3.6 Limitations of GC

GC serves well for analyzing many mixtures of compounds significant for environmental pollution monitoring, checking processes and contaminants in synthetic chemical manufacture, characterizing complex petroleum hydrocarbon mixtures, and of course it shines at the low end of the compound volatility range. The huge area of application of chromatographic instrumentation in the 21st century is in biochemical, medical, and pharmaceutical research. The chemistry of life sciences is mostly chemistry in aqueous media, and many of the compounds are correspondingly polar, to function in this very polar medium. The complexity of life processes demands large, complex molecules: enzymes, proteins, receptors, hormones, the double helix of DNA, RNA, nucleotides, and so on. These things are generally polar, full of acidic and basic functionality, and too large to be volatilized into a GC without suffering catastrophic thermal decomposition, even if resort to derivitization is attempted. Chromatography in the liquid state, at temperatures closer physiological values, will be required to separate, identify, and quantitate these complex mixtures. GC led the way in instrumentation for chromatography, but the torch is being passed to those methods which are carried out in liquids. One does not see many GC instruments in genomics or proteomics laboratories.

pharmaceutical [ˌfɑːmə'suːtɪkl] *n*. 药物；*adj*. 制药(学)的

genomics [dʒə'nəumɪks] 基因组学

proteomics [p'rəutiəumɪks] 蛋白质组学

Problems

3.1 How do gas-liquid and gas-solid chromatography differ?

3.2 What kind of mixtures are separated by gas-solid chromatography?

3.3 Describe a chromatogram and explain what type of information it contains.

3.4 What variables must be controlled if satisfactory quantitative data are to be obtained from chromatograms?

3.5 What are the principal advantages and the principal limitations of each of the detectors listed in Part 3.2?

3.6 Why are gas chromatographic stationary phases often bonded and cross-linked? What do these terms mean?

3.7 List the variables that lead to (a) band broadening and (b) band separation in gas-liquid chromatography.

第3章

气相色谱法

在气相色谱中，挥发性样品组分通过在流动的气相和液相或色谱柱内的固体固定相间进行分配得到分离。气相色谱分离时，样品先气化并注入色谱柱的柱头，用惰性气体流动相洗脱。与其他色谱技术相比，气相色谱的流动相不与分析物分子发生作用。流动相的功能只是携带分析物通过色谱柱。

气相色谱仪有两种类型：气-液色谱（GLC）和气-固色谱（GSC）。其中气-液色谱在所有科学领域应用广泛，通常缩写为气相色谱（GC）。气-固色谱法基于固定相物理吸附作用对分析物产生的保留。由于气-固色谱对活性或极性分子的半永久保留和严重的峰拖尾，使其应用受到限制。拖尾是因非线性吸附。因此，这种色谱技术除了应用于某些低分子量气体的分离外，没有广泛使用。

对于气-液色谱，其流动相是气体，固定相是液体，液体通过吸附作用或化学键合被固定在惰性载体表面。

对于气-固色谱，其流动相是气体，固定相是固体，对分析物的保留是通过物理吸附作用。气-固色谱能对低分子量气体进行分离和检测，如空气组分、硫化氢、一氧化碳、氮氧化物。

气-液色谱是基于分析物在气体流动相和固定在惰性载体表面液相间的或毛细管内壁间液相的分配。马丁和辛格于1941年首次确切地阐述了气液色谱的概念，他们也为液-液分配色谱的发展做出了贡献。通过十几年的努力，实验已经证明了气-液色谱的使用价值，该技术已开始作为实验室常规分析仪器使用。1955年，市场上出现了第一台商品化的气-液色谱。从那时起，该技术得到广泛的应用。目前，有数十万台气相色谱仪在世界各地使用。

3.1 气-液色谱仪器

自从气相色谱商品化后，市场上出现的气相色谱仪有了很大的改进和变化。在20世纪70年代，电子积分仪和计算机数据处理设备普及。80年代见证了电脑用于仪器参数的自动控制，如柱温、流速、进样等参数。同样在这十年里开发了中等成本的高效色谱仪，也许最重要的是，出现了能在相对较短时间内分离复杂组分的开管柱。今天，全世界大约有50个仪器制造商提供约150种不同型号的气相色谱仪设备，其成本在1000～50000美元之间，有的甚至高于50000美元。一个典型气相色谱仪器的基本部件如图3.1所示。

3.1.1 载气系统

气相色谱的流动相称为载气，它必须是化学惰性的。虽然氩气、氮气和氢气都可以作为载气，但氦气是使用最普遍的流动相。这些气体储存在高压钢瓶中。用减压阀、压力表和流量计控制气体的流速。气相色谱仪载气的流速通常通过控制进气压力来调节。钢瓶出来的气体经 2 级压力调节，和固定在色谱仪器上的压力调节器或者流量计调整。入口的压力通常为 10～50 psi（$1b/in^2$，$1 lb/in^2=1psi=6.894kPa$），高于室温大气压。此压力下填充柱内载气流速为 25～150mL/min，而开管毛细管柱的载气流速为 1～25mL/min。如果用压力控制装置保持入口压力恒定，柱内载气流速也是恒定的。较新的色谱仪使用电子压力控制器控制填充柱和毛细管柱内载气流速。

对于任何色谱仪，我们都希望测出载气在色谱柱内的真实流速。图 3.2 所示的经典的皂膜流量计至今仍被广泛使用。当含有肥皂或清洁剂的水溶液橡皮球被挤压时，就会在气路形成皂膜。测量皂膜在滴定管两刻度间移动所需要的时间，将其转化为体积流速（见图 3.2），体积流速和线性速度是相关的。数字化皂膜流量计消除了人为读数误差。流量计一般安装在色谱柱的末端。现在电子流量计使用越来越普遍。数字化流量计可测量质量流速、体积流速或者同时测定二者。体积流速的测量与气体组成无关。质量流量计能校准特定气体组分流速，但是，与体积流量计不同的是，质量流量计与温度和压力无关。

3.1.2 进样系统

为了获得高的柱效，适量的样品应以蒸汽塞子的形式注入色谱柱内。进样缓慢或进样超量都会引起色谱峰扩展和分离度变差。使用校准过的微量进样器（见图 3.3），通过橡胶或硅胶隔膜垫，将液体样品注入色谱柱前端的加热样品口。样品口（见图 3.4）温度通常比样品中最不易挥发性成分沸点高出 50℃以上。对于普通分析填充柱，进样量的范围是从零点几微升到 20μL。毛细管柱要求进样量是填充柱进样量的 1/100 或更小。对这类柱子，通常样品分流器只将需要的一小部分（1∶100 到 1∶500）注入色谱柱内，其余的以废液形式弃掉。商品化的气相色谱主要将这种分流器与毛细管柱配合使用，当使用填充柱时，可以不分流进样。

如图 3.5 所示，进样重现性最好的是使用较新的有自动进样装置的气相色谱仪。用自动进样器吸取样品，自动注入色谱仪。使用自动进样器前，先将样品装入小瓶中，放入样品转盘上，自动进样器穿过样品瓶的隔膜吸取样品，然后刺穿气相色谱隔膜将样品注入色谱柱内。如图所示转盘上可放置多至 150 个样品瓶。进样体积为 0.1μL（用 10μL 进样器）到 200μL（200μL 进样器）。自动进样系统的标准偏差通常低至 0.3%。

进气体样品时，通常使用图 3.6 所示的进样阀而不是进样器。使用这样的装置，进样的重现性可高于 0.5%。液体样品也可以通过进样阀进样。固体样品可以溶液的形式进样，也可将样品密封在薄壁瓶中，插入在色谱柱的头部，然后从外部刺穿或打碎小瓶。

3.1.3 色谱柱参数和柱温箱

气相色谱有两种类型色谱柱：填充柱和毛细管柱。在过去，绝大多数气相色谱分析使用填充柱。现在效率高和分析速度快的毛细管柱取代了填充柱。色谱柱长度一般是 2～60m 甚至更长。色谱柱通常由不锈钢、玻璃、熔融二氧化硅或特氟隆材料制成。为了能装入柱箱中使用，它们通常做成直径为 10～30cm（参见图 3.7）的线圈。

柱温是一个十分重要的参数，必须精密控制到小数点后一位。因此，色谱柱通常装在一个恒温的柱箱内。最佳柱温取决于样品的沸点和分离度的要求。一般来说，柱温等于或略高于样品的平均沸点会产生合理的洗脱时间（2～30min）。对于沸程宽的样品，通常采用连续或逐步程序升温的方式实现分离。图 3.8 显示了程序升温对色谱分离情况的改善。

总之，最佳分离度与最低温度有关。然而温度越低洗脱时间越长，因此，完成一次分析所需的时间较长。图 3.8(a) 和图 3.8(b) 说明了其原理。

挥发性差的分析物，有时需要衍生化以生成挥发性强的物质，然后用气相色谱测定。同样，衍生化有时用来提高检测的灵敏度或改善色谱分析性能。

3.1.4 色谱检测器

这里研究并考察了用于气相色谱分离的几种检测器。首先介绍用于气相色谱分析的最理想检测器的特点，然后探讨使用最多的检测器。

理想的气相色谱检测器应具有以下特点：

• 灵敏度高。目前的检测器对溶质的灵敏度范围为 10^{-15}～10^{-8}g/s。

• 具有良好的稳定性和重现性。

• 对溶质有几个数量级线性响应。

• 适应温度范围从室温到 400℃。

• 响应时间短，与流速无关。

• 操作简单，可靠。检测器应尽可能地简单，方便那些没有操作经验的人员使用。

对所有的溶质有相似的响应。或者说，对一种或更多种类的溶质有高度预测和检测的响应。

• 不破坏样品。

3.2 气相色谱检测器部件的设计

分析物进入进样口，流经色谱柱之后遇到的第三个，也就是最后的主要部件是检测器。分析物进入检测器会产生电信号（通常是模拟信号，但要转换为数字信号），信号的大小与气流中分析物分子的质量流速或浓度成正比。信号在图表记录器上以色谱图形式出现，最近信号也可以显示在计算机处理系统的显示屏上。自动记录保留时间、自动测量峰高或自动采集峰面积，根据保留时间确定色谱峰，通过与标准物质的峰面积或峰高比较进行定量分析。这里不详细介绍信号处理设备，仅描述最常用的气相色谱法检测器的操作方法和特点。

(1) 通用型与选择型

如果一个检测器对多种流出物具有相似的灵敏度，那么就可以把它看作是通用型检测器（或至少大多数通用，没有检测器是绝对通用的）。当需要知道分离样品中是否有组分忽略时，这种检测器是非常有使用价值的。另一种极端情况是，选择型检测器只对有限种类含有特定的原子的化合物具有较高的灵敏度，这些化合物含有特定的原子（如，除了绝大部分有机化合物含有的 C、H 和 O 以外的原子），或具有特定类型的官能团或具有一定亲和力和反应活性的取代基。假如选择型检测器对目标化合物有响应，而对浓度高得多的共流出组分不灵敏而使分析不受杂质的干扰，这样的检测器很有价值。

(2) 破坏性与非破坏性

一些检测器在操作过程中会破坏分析物的结构（如，在火焰中燃烧，质谱分析需要在真空条件将分析物打成碎片，或者与其他试剂反应）。另一种类型的检测器对物质不产生破坏作用。分析物结构不变，然后把分析物传递给另一种检测器进行进一步的表征。

（3）质量型与浓度型

通常来说，破坏型检测器是质量型检测器。如果载气流中没有分析物，检测器将迅速破坏进入它内部的物质，同时信号下降为零。非破坏性检测器不影响被测物，只要待测物继续驻留在检测器内，且在信号中不下降的条件下，浓度测量就可以连续进行。一些非破坏性检测器（如，ECD）能测量捕获的物质（如电子），这个过程如果“饱和”会导致信号丢失，所以它们是质量型检测器。

（4）对辅助气体的要求

纯载气或者从毛细管柱流出物的流速会使某些检测器不在最佳状态。尾吹气，有时与载气一样，可以用来提高通过检测器的气体流速，以达到使检测器产生良好的响应或者降低检测器衰减的死体积，从而提高色谱柱上的分离度。一些检测器使用的气体不同于色谱分离使用的载气。一些检测器在运行时需要提供不同流速的空气和氢气以产生最优的火焰，而不是载气产生的火焰。尾吹气会稀释色谱柱流出物，但不会改变检测机理而把浓度型变为质量型。

（5）灵敏度和线性范围

不同类型的检测器（通用型和选择型）对分析物的灵敏度不同。灵敏度是指根据一定的信噪比能够检测出一个特定分析物的最低浓度。检测器越灵敏，能够检测的浓度越低。检测器的信号反应与被测物的浓度呈线性关系的范围称为线性范围。一些高灵敏的检测器线性范围比较窄，因此，浓度大的分析物必须稀释到这个范围内。对线性范围窄的另一个解决方法是对多水平非线性标准曲线进行校准。这就要求进样更标准，但也更容易产生定量误差。如果样品具有很宽的浓度范围，稀释不会产生令人满意的效果。灵敏度低的检测器线性范围窄，做多水平标准曲线或稀释都不会起作用。

图 3.9 比较了几个常见气相色谱检测器的灵敏度和线性范围。条形图越靠近左侧，检测器就越灵敏。条形越宽，线性动态范围就越大。每条垂直虚线表示的跨度为三个数量级(1000)，所以整体覆盖范围是非常大的。沿 X 轴的值假设是溶液进样量为 1μL。这个范围是近似的，其确切的终点由检测器的设计、型号和操作条件优化决定。

此图对检测器的灵敏度和线性动态范围进行了综合比较。底部的 IRD 和 MSD 是与光谱检测器联用的，有几种操作模式。注意：“原始”的气相色谱检测器，即 TCD 灵敏度最低，也是唯一一个适合处理纯品（即 100%纯单组分）的检测器。FID 具有最大的单模式的动态范围，而“微型 ECD”更灵敏，但线性动态范围窄。下面将介绍非联用检测器的操作、特点和应用方法。

3.2.1 热导检测器（TCD）

热导检测器的特点：通用型检测器（除了 H_2 和 He）；非破坏性、浓度型检测；不需要辅助气；使用一个平行色谱柱性能更好；灵敏度低；线性动态范围窄。

TCD 是第一个商品化并广泛应用的气相色谱检测器，所用色谱柱都是填充柱，分离的是纯物质混合物样品（即不是稀释的溶液）。热导检测器测量热导率的差异和/或测量热导率低的分析物蒸汽稀释的高热导率（无论是氢气或氦气）载气的比热差异。电流通过检测器流通池中的细电阻丝并将其加热。流动的导热载气冷却电阻丝。当载气流中有分析物时，由于分析物热导率越低带走的热量就越少，导致电阻丝的温度上升，其电阻增加。这个电阻位于“惠斯登电桥”（方形的四条边有四个电阻）电路中。另一个电阻在与之匹配的热导池中，与一个匹配的色谱柱连接，没有分析物通过。等温气相色谱的载气流速和温度保持不变，但使用程序升温时，随着温度升高，载气的流速会减小，以致影响载气的热导率。匹配的参比池抵消产生电阻变化影响。一个固定电阻和一个可变电阻构成了惠斯登电桥的另两个边。电压传感器连接在正方形对角线两端对应的点上，可变电阻使电压传感器（或“零”）归零。桥路平衡后，分析物通过会改变惠斯登电桥的一端的电阻，使惠斯登电桥失去平衡，产生一个与电阻变化成正比的电压信号。两个热导池的设计如图 3.10 所示。热导检测器池体的设计是参考用于气体热导率测量的装置——导热计池体。

图 3.10 是热导检测器对“特定的气体”的检测，这些气体通常不能用其他检测器检测（除气相色谱-质谱联用外）。热导检测器对氢气灵敏度低，其热导率是唯一一个与所使用的载气氦气的热导率相匹配的。如果只测量氢气，氮气作为载气较好，会产生一个负的信号峰。在图 3.10 中，可以看到检测器是一个微型 TCD。这就要求必须采用小型化的池体，与低载气流速的开管柱配合使用，同时需要小检测池体积来避免超柱谱带增宽。

3.2.2 火焰离子化检测器（FID）

FID 的特点如下：

近乎通用型检测器（除了无机气体如 CO、CO_2、HCN 外所有含碳的化合物）；破坏性；质量型检测器；需要氢气、空气或氧气作为辅助气体；灵敏度高；线性动态范围宽。

FID 是使用最普遍的检测器，对几乎所有的有机化合物都会产生响应。在分子水平上，该信号大致与分子中的碳原子数成正比。检测池中作为燃料的氢气和氧气（或空气）必须分开供给，这些气体的作用是产生火焰。如图 3.11 所示，氢气与从色谱柱流出的载气流出物混合进入检测器。即使氢气作为毛细管色谱柱的载气，也要附加一个独立控制的氢气供应，为火焰调节提供合适的流量。混合气体通过喷嘴进入检测池，其中空气或氧气流是产生火焰的氧化剂。在检测池中的电子发光塞以脉冲形式点燃火焰。用针形阀调节燃气和氧化剂流量产生响应稳定最佳的 FID 火焰，在火焰中经常产生不被保留的挥发性烃类物质，这个信号为参比信号。喷嘴的尖端与正对的火焰周围的收集极或收集环之间的电压在几百伏以上。在载气流中不存在有机分析物时，在喷嘴-收集极电路中没有电流产生。当含碳的洗脱分析物进入火焰，分子通过氧化反应分解成较小的碎片。这些碎片一些是带正电荷的离子，它们可以在电路中移动产生电流。尽管 FID 离子化效率低，其基流也很低。但它的信噪比是非常高的。在这样低的背景下，即使是非常小的电离电流，使用现代电子技术也可以准确地测量和绘制出非常低的电流（高输入阻抗，电压测量电路）。因此，这种检测器具有很高的灵敏度和极大的动态范围，往往超过六个数量级。

从气相色谱柱进入检测器的任何有机化合物都可以用 FID 检测。浓度相差较大的分析物，可一次测量完成。FID 是一系列更具选择性甚至更灵敏的离子检测器的先驱。它引导 GC 发展到当前应用，分析物在制剂或萃取溶剂中以较低浓度存在，而不是以纯物质的混合物形式存在。精确定量需要以测量分析物的标准物质作为参考标准。化合物的重量反应与定量反应非常相似，许多组分的标准曲线是以对应的参考化合物的峰面积来校准建立的。宽动态范围最大限度地减少标准曲线所需的校准水平数。燃气由气体压缩钢瓶提供，或由气体发生器通过水解产生，如同压缩和过滤（以避免有机污染物）环境空气一样。如果氢气代替氦气作为载气，则无需配备压缩气体钢瓶。

3.2.3 电子捕获检测器（ECD）

ECD 的特点如下：选择性强（对含有卤素、硝基和其他含氧官能团的有机化合物）；非破坏性；浓度型检测器；需要使用氮气或氩气/甲烷为载气，或者如果氢气或氦气为毛细管载气，需要使用尾吹气；灵敏度极高但可变，动态范围窄。

如果说 FID 的使用使气相色谱能够测定低浓度的有机物，那么 ECD 让人们充分了解了组织和环境中食物链中亲脂、持久性有机污染物（POPs）生物富集的危险性，经典的例子是有机氯农药 DDT 危害的发现，以及随后的禁用。J. E. 洛夫洛克（因创立整个地球作为一个有机体的“盖亚理论”而著名）发明了这个看似简单但极其灵敏兼具选择性的检测器。尽管人们还没有达成一致性意见，但调查表明气相色谱检测器的使用可能激起了 20 世纪 60 年代环境意识上的革命。

那么这个小小的奇迹是怎么产生的呢？图 3.12 列出了 2 种类型的 ECD。右边的图显示的是 ECD 初始设计的特点（在 DC 模式下操作）。色谱柱流出物从右侧进入。掺杂放射性 ^{63}Ni [α、β 粒子（高能电子）发射源] 镍箔片不断轰击载气，载气部分分子电离，产生正离子和负电子的气体氛围。选择此放射源是它能承受 GC 检测器的高温，以防止高沸点分析物在检测器内冷凝。在检测器的负入口侧和正出口侧两个电极之间的低电压（几伏，而不像 FID 中几百伏）足以使两个电极之间产生电流。在这种低电压下，更易移动，更轻的自由电子带着所谓的“稳定电流”向正极移动。

如果电负性高，含有容易被极化取代基的分析物进入检测器，稳定电流中的电子被这些分子捕获。被捕获的负电子流动性急剧降低，其中一些可能更容易与产生的正离子碰撞而被中和。这种稳定负电荷电子流的下降被反向显示从而出现一个峰信号。电子捕获的灵敏度根据取代基的本性和数量发生显著变化（1～2 个数量级）：卤族元素的顺序为 I＞Br＞Cl≫F，5～8 卤代原子＞4＞ 3≫2≫1。需要注意的是：电负性最大的 F 原子被更易极化的 Cl、Br、I 超过，按顺序依次增加，虽然电负性和极化对于有效捕获电子是必要的，但极化是占主导的。其他一些基团如—NO_2、—NO 等会落在反应偏低的一侧。如果碳氢化合物和 ECD 反应灵敏的活性化合物一起进入检测器，虽然它们不参与反应，但它们会大大降低 ECD 对活性化合物的反应灵敏度。如果使用 ECD，对于每一种分析物的定量分析校准必须使用标准物质。

ECD 通常需要特殊的载气或尾吹气。大多数早期 ECD 池体设计是为了满足填充柱高流速的需要。当氢气或氦气用于毛细管载气时，需要高流速和不同成分的尾吹气。放射性离子源的使用在美国需要 NRC 颁发执照并按照一定的程序定期监测检测器内部和周围辐射泄漏的情况。另一个是直流电流衰减至饱和的问题。随着电子捕获分析物浓度增加，越来越多的微小电流被忽略不计。分析物浓度的增大并没有使电流按比例减小，仅表现为电路中没有足够的电子。这导致检测器产生严重的非线性响应，标准曲线弯曲（当浓度继续增加，曲线变平坦），线性动态范围剧烈变窄。用 ECD 定量校准成为一个棘手的问题。

图 3.12 左边的 ECD 图为一个更现代的设计，它试图解决右图出现的一些难题。通过使用一种特殊流动的放电气体作为“放电电极”实现电离，取代了放射性箔片。通过引入一种特殊掺杂气体以提高放电性能。使用较高的间歇脉冲电压，而不是恒定电压通过检测池，通常以 1ms 的间隔在 1μs 内重复 30V 脉冲电压（0.1%的占空比），这个电压能收集 ECD 池体内的所有自由电子，而在关闭期，电子与气体重新建立平衡。快速的收集时间是由氩气中的

5%～10%甲烷实现的，甲烷提高了自由电子的电压诱导迁移率。脉冲采集期间产生的电流积分是产生稳定电流的基础。控制稳定电流的好处是可以扩展线性动态范围，而最先进的ECD设计就是以脉冲模式操作的。图 3.9 微型电子捕获检测器扩展了线性动态范围。微（型）设计表明，该模型是配备了一个较小池体积以匹配毛细管柱流出物的流量。省掉用尾吹气来稀释柱流出气流，提高了灵敏度，甚至改善了其他性能。其他参数保持不变，会使灵敏度更高，动态范围更宽，这是因为ECD主要受稳定电流饱和度上限的影响，早期的模型，大容量，直流模式，线性动态范围只有100或更少。这就使得在一个有效的校准范围测量多个分析物（特别更大的浓度范围）变得非常复杂。与具有宽动态范围的FID对比，做好定量分析ECD必须要进行多次稀释和重测。相比之下，对某些化合物具有特殊选择性和超高灵敏度的ECD，对多氯滴滴涕和其他的如多氯联苯（PCBs）、二噁英和多溴联苯醚（PBDEs，多溴联苯醚阻燃剂）的生物富集途径的发现是必不可少的。重要的通用原则：应用色谱分析法，环境或生物体系实际样品中具体分析物的检测限，往往取决于选择的检测器的背景重叠干扰信号，而不是受仪表信噪比的绝对灵敏度限制。

3.2.4 硫-磷火焰光度检测器（SP-FPD）

硫-磷火焰光度检测器的特点如下：对杂原子具有选择性（分别对含S或P原子的有机化合物有响应）；破坏性；质量型检测器；灵敏度低，比FID动态范围窄。

当含S或P原子的有机化合物在FID中燃烧时，调整火焰产生较低温提高S_2碎片在394nm和HPO碎片在515nm处的放射强度，产生的光经耐高温窗口或检测池一侧的光纤光导纤维，再通过过滤器分离出394nm或515nm的光。这些特征波长的光传递到敏感的光电倍增管上。在S选择模式下操作，测定复杂油气混合物中有机硫化合物是非常有用的，因为共存烃的峰响应信号最小。在P选择模式操作，能检测复杂环境混合物中有机磷农药的痕量残留分析，使用硫磷火焰光度检测器灵敏度高。硫的模式响应范围如图 3.9 所示。

3.2.5 氮磷检测器（NPD）

NPD的特点如下：

对杂原子具有选择性（分别对含有N或P原子的有机化合物响应）；对含氮、磷化合物比氢火焰离子化检测器更灵敏更具有选择性，对化合物具有破坏性；质量型检测器；对氮比氢火焰离子化检测器更灵敏，对磷比P-FPD更灵敏，但比FID动态范围窄。

NPD是气相色谱的重要检测器的另一种变形。图 3.13 中的NPD与图 3.11 中的FID突出了相似之处。两者都为“热离子检测器”，也就是说高温火焰将洗脱分析物打成碎片，其中一些是正离子，形成碎片时产生的电子在电压的作用下，在两个电极间形成电流。氮磷检测器在富氢火焰条件下工作。在此条件下，正常的碳化合物在FID热离子响应缩小数个量级。这是因为在检测器增加了一个新的部件，铷盐掺杂在玻璃或陶瓷形成铷珠。铷珠用火焰加热和连接铷珠的细导线通过的电流加热。但当有机氮和有机磷化合物在不太热的火焰中分解，铷珠表面发生什么样的变化？这个机理是很复杂，难于理解。在某些设计中，如果没有镶嵌在检测器内独立加热的铷珠，火焰不会自支持。无论如何，C-N或C-P的碎片都会与易电离铷原子作用产生离子和电子，产生类似于FID的信号。而且对氮原子或磷原子极其灵敏且具有选择性（取决于检测器参数如何设定）。如图 3.9 所示，对有机磷化合物的高灵敏解释了NPD为什么取代了P-FPD。但没有描述FPD操作模式的比较数据。NPD对有机分子中不与碳原子键合的氮原子没响应，这样免除了空气中无处不在的氮气的干扰。

3.3 气相色谱-质谱法（GC-MS）

从气相色谱-质谱的广泛应用性和性能看，可以说 GC-MS 在仪器分析界中分析能力最强。随着 LC-MS 仪器最近也致力于服务生化研究市场，它们的应用也越来越广泛。如果分析物是挥发性和热稳定的，毛细管 GC-MS 可以对单次进样的混合物中几百个物质进行定性和定量分析（用 GC-GC-TOFMS 运行甚至达上千）。四极杆和离子阱是最符合早期提出的扫描速度的标准。磁扇仪器，具有更高的质量分辨率，但扫描速度较慢，会使气相色谱峰变宽和出峰速度变慢。这与快速分析，良好的分离度目标是违背的。另一方面，所谓的“快速气相色谱”，要求使用短的，小内径的、薄液膜的毛细管柱，需要 MS 扫描速度为50～$500s^{-1}$的采集速度，只有 TOF-MS 仪器才能实现。

GC-MS 分析的主要问题是接口。质谱仪是在高度减压下形成和移动离子，质量分析器需要在更低的压力（1025torr 或 1026torr）下工作。气相色谱仪的流出物是具有 1 个大气压（760torr，1torr＝133.322Pa）的载气流出物。MS 真空泵必须快速除去载气，以保持必要的低压。填充柱的流速为 10～40mL/min，和质谱联用分析是不可行的。毛细管气相色谱载气流速为 1mL/min 左右，使用扩散泵，或者更好一点的强有力的涡轮分子泵，能够实现 GC-MS 联用。另一个原因是气相色谱分离更适用使用毛细管柱而不是填充柱。许多现代的 GC-MS 仪器简单地把毛细管柱引入到离子源（直接耦合），使用一个泵对离子源的有限空间抽真空，当离子通过小孔进入质量分析仪区域时，使用另一个泵对该区抽真空以保持较低压力。这样的 MS 设计称为差动泵送系统。为了连接填充柱，喷射分离装置（图 3.14）首先除去流出物中轻的载气如氦气或氢气，而较少量浓缩的分析物以较小的流速进入离子源。质量越轻的载气在分离室真空条件下运行的速度越快，真空分布得越广，而不能通过锥形孔进入离子源。现在 GC-MS 中喷气分离器使用不多。

原则上，可以使用非破坏性检测器如 TCD 监测气相色谱流出物，使流出物通过一个合适的接口进入质谱，根据洗脱顺序，打印出色谱峰的质谱图。借助快速的数字数据转换器和现代台式电脑的速度，GC-MS 运行后能获得较好的连续序列质谱数据或选定的质量碎片信号。在计算机内存储的数据文件，包含一系列按保留时间顺序排列的 MS 扫描。获取和使用这种数据文件有以下三种主要模式。

(1) 全扫描

如果文件包含一系列全扫描光谱，根据每次扫描中所有质量的离子数进行加和得到总扫描数，这些扫描数对连续扫描数（或等同于 GC 保留时间）进行作图。这种图称为总离子色谱图（TIC）。这里质谱仪的作用就像一个离子检测器，可以选择任何一个 MS 离子源电离模式（EI、CI 等）。总离子色谱图与 FID 检测器获得的色谱图相似。不同的是，FID 数字化输出的色谱数据系统有可能用更多的点确定每个峰的宽度，这比 MS 扫描数多（除非它是快速扫描 TOF-MS）。为此，FID 色谱峰定量比 TIC 峰更精确。总离子色谱图数据文件的优点在于通过选择总离子色谱峰可以显示峰中任何一个化合物的质谱图。如果峰尖锐且窄，分析物的浓度在扫描时变化显著，会扭曲光谱质量碎片的比例，因此最好选择峰值的顶部。更好的方法是，通过对整个色谱峰的扫描得到平均光谱。比这更好的方法是在峰的附近选择没有明显洗脱物的背景质谱，并计算出等数目的扫描数。然后从色谱峰值的平均扫描数中减去这个值，消除柱流失的杂散碎片。这说明连续质谱扫描文件操作的灵活性。因此把毛细管气相色谱强大分离能力与质谱 MS 的表征和识别能力结合起来是可行的，因为在气相色谱中样品呈蒸气状态，进行质谱离子化和碎片识别处理也必须是气态的。EI-MS 谱图库是基于这种操作模式建立起来的。在许多情况下，它可以自动进行峰值检测，光谱选择，谱图库检索，并对分离的化合物匹配到谱图库鉴定。可能有人认为用 GC-MS 能自动分析就不需要人工分析了，实际上不全是这样的。但对于用于气态分离的 GC 需要与气态 MS 的接口问题，再与流畅的数字数据文件的 PC 处理联合使用就是一个极好的组合。

（2）质谱图（XIC）

可以尝试用气相色谱对大量复杂的混合物进行分析，从中确定和识别其中的某些特定类别。这可能需要特征性强的质量碎片。举个例子，作为燃料油的石油烃类复杂混合物中含有单、双和三甲基萘异构体。质谱图主要包括分子离子峰（M^+），质量分别为 142、156 或 170。我们可以对 GC-MS 数据系统编程，以便从全扫描的 GC-MS 文件提取并在三个独立的色谱图中仅仅标示的这些质量离子。这样的图从产生方式上看称为提取离子色谱图（XIC），或质量色谱图。因为它们显示峰的光谱包含选定离子质谱（ES）。如果一个石油样品色谱图中 $M^+=142$，会看到一对大峰靠得很近，很可能会怀疑它们是两个单甲基异构体，可以通过在每个峰的中心扫描完整的光谱证实我们的怀疑。因此，我们可以在不用标准物的情况下确定不同化合物的保留时间。但是，化合物的同分异构体的质谱是没有区别的，因此，我们需要标准物质在特定的气相色谱固定相的保留信息，确定异构体具体是什么物质。这个过程可能在许多其他不同类别烃的特征质量测定中讲过。在扫描范围内所有的质量数据和总离子图的每一点的数据都保存在全扫描质谱数据文件中。

（3）选择性离子监测（SIM）

对于全扫描模式，质谱仪在有限的扫描范围内对任意特定离子质量进行采集计数。例如，如果一秒中扫描质量从 50 至 550，同时要求单位质量分辨率，那么检测器在一次扫描过程每个质量碎片花费不到 2ms 的时间。如果只希望分析少数几类物质，而且也知道主要特征质量碎片，通过编写 MS 程序，仅获取选定的质量碎片的数量，增加每个质量停留时间，提高了信噪比，并提高灵敏度。

如果想要三个连续的甲基萘异构体的分布图，仅需检测 $M^+=142$、156 和 170 的质量碎片。在 1s 中这三种离子中每个离子监测花费的时间略小于 330ms，而不是 2ms，大大提高了在 50～550 全扫描采集的灵敏度。实际上，可以把循环周期缩短到 1s 到 0.2s，每个周期仍需 66ms，但现在取样确定峰形需要 5 次/秒，而不是每秒一次。如果我们知道每类异构体洗脱物在一个特定范围内的保留时间，可以在此期间设置 MS 监测最具有特征的大量离子，获得更高的灵敏度。这种模式被称为 SIM。SIM 在特征质量处通过收集多个点提高了灵敏度，并且通过确定 GC-MS 峰和综合使用多个点提高定量精确度。GC-MS 两个部件的灵敏度使 SIM 和全扫描的线性范围扩展到较低值。SIM 最终的灵敏度是特定分析物电离效率、检测的不同质量碎片，停留时间和循环时间的复杂函数。在最有利的情况下，GC-MS-SIM 可以达到 FID 的和 ECD 检测限以下。它比一般的气相色谱检测器的选择性更强。这种分析是对目标化合物特征的分析，因为它通过调谐或编程系统对特定化合物在特定保留时间内选择特定特征离子。

XIC 和 SIM 色谱之间的差别在于，前者的离子质量是从全扫描数据文件中提取，而在后一种情况下，MS 仅针对那些选定质量采集数据。SIM 能提高灵敏度和定量精度主要是由于使用了磁性和四极杆质谱仪（包括高分辨率 MS 使用）。离子阱质谱和 TOF-MS 仪器操作模式能产生接近全扫描模式的近似最佳灵敏度。因此，这些仪器一般不提供 SIM 采集，用 XIC 质谱图定量。GC-MS 中的 MS 是破坏性的质量型检测器。只要用质谱检测，分析者就要使用完美的内标物，即具有较高原子的质量（如 ^{2}H、^{13}C、^{15}N、^{18}O 等）的稳定同位素相同的化学物质。此过程中，同位素稀释质谱法（IDMS），可以校正样品制备回收率、衍生效率、MS 离子化效率的差异等，因此并不是任何一个与化合物不同的物质都可以作为内标物使用的。然而，IDMS 的 IS 材料是很难制备的，价格昂贵，而商业化产品仅限于几千种特别重要的目标分析物。

3.3.1 气相色谱-质谱法

对于 GC-MS 的操作有两类不同的认识：一类认为质谱仪是气相色谱仪的最有选择性的检测器，另一类认为气相色谱仪是质谱仪昂贵的进样装置。精通这两个领域的知识，并能融汇贯通的从业者不常见。Settlage 和 Jaeger 指出，不管是什么原因，GC-MS 技术远没达到其潜力。然而，GC-MS 已经发展成为分析化学中最富有成果的技术之一。GC-MS 能够通过光谱解释程序方式提供定量和定性数据。图 3.15 是一个框图，显示了 GC-MS-计算机系统的组件是如何相互关联的。

质谱仪接收分离的色谱流出物进入离子源，保持压力在 10^{-2} Pa，以确保中性分子高效产生分子离子。离子的产生可以通过电子轰击（电子电离）、化学电离（CI）、场电离或场解吸技术实现。在质量分析器中各种离子必须分开，并将它们的质荷比（m/z）和相对丰度记录为离子谱。各种类型的分离方法，包括已被使用的高扫描速率的四极杆和磁性/静电光谱已经开始使用，但是，在一般情况下，对于 GC-MS，会优先选择四极杆仪器，因其具有相对高的灵敏度，分辨率能达到大约 1000amu，操作压力高达 10^{-2} Pa，10^{-1} s 扫描速率。使用四极杆的唯一缺点是相对低的分辨能力。例如，在怀疑干扰时需要使用高的分辨率，必须使用高分辨率磁扇或双聚焦仪器。最近，开始使用傅利叶变换质谱仪和离子阱检测器。

质谱是线性光谱，展示的是分子离子峰和相对应的各个碎片信号。图 3.16 是 1,1,1-三氯乙烷的质谱图。分子离子表示母体化合物的分子量。围绕此峰的线是各种同位素的自然丰度。从质谱图上的碎片离子和它们的高度可以进一步获得结构信息。

GC-MS 产生的有用数据量非常大，不借助电脑，得到或分析这些数据是不实际的，也是不可能的。下面是一个典型的环境样品开管毛细管柱分离的气相色谱分析例子：假设通过 60min 程序升温和 2s 的质谱扫描速率得到色谱图，在色谱运行完成后，保存了 1800 个质谱图。在质谱图中每个峰由质量和丰度值组成，所以约为 100 个峰的典型质谱需要 200 个数值的存储。因此，整个分析需要超过 200×1800＝360000 数据的存储空间。还要有额外的存储空间用于存储保留时间这样的数据。因此，现代仪器拥有如同仪器校准和控制一样自动数据采集、存储和有效地处理数据系统。

在特定的实验条件，化合物的质谱图类似于指纹。把质谱峰与存储在基准谱图库中的其他光谱比较，确定一个未知峰的程序会大大简化，虽然专业的谱图库较少，仅仅包含几千种谱图，但这些都是可用的。原因很明显，质量控制措施是必要的，以确保只含有已经证实的、准确无干扰的质谱图。收藏最多的谱图库是 Wiley-NBS 质谱数据库，目前有不到 100000 种化合物。此数据库看起来可能很大，但已知的有机化合物的数目超过 5000000。与参考谱图比对，手工检索比较繁琐，所以通常用计算机进行比对。计算机比对通过计算可分为好、中和差的匹配。然而，没有一个系统可以仅根据搜索的数据对未知的物质能够做到准确无误识别。因此对质谱数据还需要进行人工的解释。对从不同的商业系统获得的数据质量进行评估，需要了解不同的检索系统。

3.3.2 接口

为了避免分子反应，质谱仪必须保持高真空。然而从气相色谱仪出来的载气直接进入离子源，离子源内压力将升高。为了确保气相色谱仪和质谱仪之间的样品转移最大化，同时除去大部分载气，目前出现了许多解决气相色谱仪和质谱仪接口问题的方法。接口设计不当，可能是抵消了色谱柱、质谱仪或两者的性能。一个理想的接口应该是这样的：峰增宽通过最小死体积最小化；在离子源处柱出口压力降低到大约 10^{-2} Pa；对那些对热或化学稳定化合物，不存在受接口的活性位点或接口被加热作用而也不产生歧视效应；并且能将整个样品有效地传输到离子化室。

理想情况下，为了使真空对柱效的影响降到最低，该接口应当使色谱柱与真空装置隔离。隔离对填充柱来说是必不可少的，早期的仪器，是一个全玻璃接口，称为喷射分离器，作为最有效的接口被广泛使用。由于窄孔和长的开管柱（>15m）要求低载气流速，隔离变得不再那么重要了。由于开管柱的普遍使用和真空泵设计的改进，开管柱与离子源“直接”连接是现在通常的做法。使用高容量扩散泵，离子源要根据开管柱低流量的特点进行设计（<5mL/min）。直接连接可以采取多种形式。最简单的方法是色谱柱直接连接到离子源，没有对色谱柱真空隔离。这种方法适用于较长的窄径柱，但即使这样，仍然存在这样的缺点：质谱仪没放空，不能拆除色谱柱；或者，在输送管线中一些分离被小口径的毛细管或者铂-铱或熔融二氧化硅影响。几个报道表明，铂-铱合金会促进某些物质的催化分解。第三个也是最成功的直接配置方法是开口分流。

3.3.3 记录和分析

气相色谱可用于定性和定量分析。定性分析时，保留时间是识别一种化合物的良好参数。保留时间是给定的化合物从色谱柱洗脱出来需要的时间，定义为从进样到柱后出现最大浓度所需要的时间。这个属性是样品和液相在给定温度条件下的特性。精确控制温度和流量能再现，且精度优于1%。

洗脱物信号通常被记录在长条纸记录器上，其中时间为横坐标，信号单位（毫伏）是纵坐标。目前，该信号更有可能以数字化形式收集，然后用电子计算机作图。

给定的样品的峰面积与其浓度成正比。根据这个原理就可以对给定的组分的浓度进行测定，物质的含量通过与已知的标准物质的含量对比计算。定量的准确性与样品浓度、制备方法、进样方式、检测方法和综合方案有关。用电子数字积分仪或计算机得到优于1%的精度是可能的。

3.3.4 分辨率

峰之间的分离情况是由指定体系的分辨度决定的。分离度方程如下：

$$R_{\mathrm{S}}=\frac{1}{4}\left(\alpha-\frac{1}{\alpha}\right)^{N^{1/2}}\left(\frac{k'}{1+k'}\right)$$

式中，R_{S} 为分离度；N 为理论塔板数；α 为分离因子；k'为容量因子，可以通过下面的公式求得：

$$\alpha=\frac{k'_2}{k'_1}$$

$$k'=\frac{保留时间-死时间}{死时间}$$

$$R=\frac{t_2-t_1}{(W_1+W_2)/2}$$

式中，t_2 和 t_1 分别为峰 2 和峰 1 的保留时间；W_2 和 W_1 分别为色谱峰 2 和色谱峰 1 的峰宽。指定体系的分离所需的塔板数可用分离度方程计算如下：

$$N_{req}=16R_S^2\left(\frac{\alpha}{\alpha-1}\right)\left(k'_2+\frac{1}{k'_2}\right)^2$$

式中，N_{req}为给定分离度时的塔板数；k'_2为第二个色谱峰的峰容量。

3.4 固定相的选择

固定相一般是根据这个领域的综合知识、文献检索和实际工作实验来选择的。合适的固定相的选择应符合以下标准：

• 对样品各组分具有良好的溶解能力，因为溶解性差的成分易被载气快速带走起不到分配作用。

• 与样品组分作用具有差异性，这与分配系数有关。

• 在高温下热稳定性好，如果在高温不稳定，会产生固体载体催化作用。

• 挥发性小；即固定相应是非挥发性的，在操作温度下蒸气压为 0.01～0.1mmHg。

• 在操作温度下不与样品组分发生反应。

对样本组分的化学性质有一定的了解是非常重要的。如果不能获得这些信息，要对样品可能的性质进行大致的猜测。或者，要分析样品中所有组成成分，以确保不错过任何一种化学成分。根据组分性质，尽量把样品划分为以下五类。

3.4.1 样品分类

根据极性，可以将样品分为以下几组。

(1) 第一组（极性）

该组化合物含有一个供体原子，如氮、氧或氟，和一个活性氢原子。包括：

• 醇

• 脂肪酸

• 酚

• 胺（伯胺和仲胺）

• 含有 α-H 原子的硝基化合物

• 含有 α-H 原子的腈类化合物

(2) 第二组（强极性）

能形成氢键的化合物。包括：

• 水

• 甘油
• 氨基醇
• 羟基酸
• 二元酸
• 多酚

(3) 第三组（中等极性）

包含一个供体原子，但不含有活性氢原子的化合物。

• 醛
• 醚
• 酯
• 酮
• 叔胺
• 不含有 α-H 原子的硝基化合物
• 不含有 α-H 原子的腈类化合物

(4) 第四组（弱极性）

含有活性氢原子，但不含有供体原子的化合物。

• 卤代烃
• 芳烃
• 烯烃

(5) 第五组（非极性）

包括不能形成氢键的化合物。

• 饱和烃类化合物
• 硫醇
• 硫化物
• 卤代烃，如四氯化碳

下一节将给出适合这些样品的固定相。

3.4.2 选择合适的固定相

固定相可以类似地按照极性分类，以便于根据给定的化合物基团的极性选择合适的固定相。

(1) 适用于极性化合物的固定相（1 组）

• β,β'-氧二丙腈
• 25%氰乙基-75%甲基聚硅氧烷（XE-60）

(2) 适用于强极性化合物的固定相（2 组）

• 聚乙二醇类化合物（不同分子量的）
• 聚乙二醇 20M 与 2-硝基对苯二甲酸反应物（Carbowax 20M-TPA）
• 改性的聚乙二醇（FFAP）
• 甘油二酯

(3) 适用于中等极性化合物的固定相（3 组）

• 四氯邻苯二甲酸二丁酯

• 磷酸三甲酚酯

• 50％苯基-50％甲基聚硅氧烷（OV-17）

（4）适用于弱极性化合物和非极性化合物的固定相（4 组和 5 组）

• 二甲基聚硅氧烷（SE-30）

• 角鲨烷

• 阿皮松

• 十六烷，鲸蜡烷

固定液推荐的最高使用温度见表 3.1。

3.5 气相色谱分析的范围

配置合适的气相色谱仪可以分离和识别分子量最小的氢气到分子量为 1000Da 的物质。质量上限范围只适用于高温稳定性好、低极性低沸点的化合物。这类化合物的原型是正构烷烃。科瓦茨的 RI 参考体系提出了在线性程序升温洗脱时，同系物具有“栅栏”轮廓特征。轻质燃料油的氢火焰离子化检测器色谱图说明了这些。较重的石油原油可能会拓宽了这一模式达到或超过 n-C_{60}（富勒烯）。限制因素如下：

• 有效挥发化合物的能力。

• 固定相在足够高的温度下不分解，而且在较长时间内能使分析物充分分配到流动相中。分析物能够在合理的时间内通过色谱柱，分析物快速地从色谱柱上解吸并产生高于噪声背景信号的、峰形良好的色谱峰。洗脱缓慢保留时间长的化合物是逐渐被洗脱出色谱柱进入检测器的，致使其色谱峰扩张，最终成为不太高和难以分辨的背景噪声和基线漂移。

• 色谱柱管在使用温度下必须是稳定的。熔融石英毛细管外的聚酰亚胺防刮涂层仅能承受 380℃。某些特殊的固定相可以超过这个温度限制，它们可以使用金属柱，在那个温度下对一些物质来说，金属柱的内壁必须用硅胶来处理以失活，但是其缺点是不能做成合适的线圈形状。

3.6 气相色谱法的局限性

气相色谱法可以用于分析监测达到污染环境的许多重要化合物，检测合成化学工业的工序和污染物，分析复杂石油烃的混合物，当然也能检测低挥发性化合物。在 21 世纪，应用色谱仪器多的领域主要有生物化学、医学和制药研究。生命科学的化学主要是水介质中的化学反应，大部分化合物具有相对极性，可以在极性介质中发挥作用。复杂的生命过程需要大量的复合分子：酶、蛋白质、受体、激素、双重螺旋结构的脱氧核糖核酸、核酸、核苷酸等。这些物质通常都是极性的，具有酸性和碱性性质，这些分子太大，如果不经过强烈的热分解或甚至于借助衍生化都无法挥发进入气相色谱。在接近生理温度值时液态的色谱法，能对复杂混合物进行分离、定性和定量。气相色谱法推动了色谱仪器的发展，然而这些发展正在应用于液态色谱分析中。在基因组学和蛋白质组学实验室，一般不使用气相色谱。

Chapter 4

High-Performance Liquid Chromatography

High-performance liquid chromatography (HPLC) has become an indispensable analytical tool. This chapter considers the theory and practice of HPLC, including partition, adsorption, ion-exchange, size-exclusion, affinity, and chiral chromatography. HPLC has applications not only in forensics but also in biochemistry, environmental science, food science, pharmaceutical chemistry, and toxicology.

High-performance liquid chromatography is the most versatile and widely used type of elution chromatography. The technique is used by scientists for separating and determining species in a variety of organic, inorganic, and biological materials.

In liquid chromatography, the mobile phase is a liquid solvent containing the sample as a mixture of solutes. The types of high-performance liquid chromatography are often classified by the separation mechanism or by the type of stationary phase. These include: partition or liquid-liquid chromatography; adsorption or liquid-solid chromatography; ion-exchange or ion chromatography; size-exclusion chromatography; affinity chromatography; and chiral chromatography.

Early liquid chromatography was performed in glass columns having inside diameters of perhaps 10 to 50mm. The columns were packed with 50 to 500cm lengths of solid particles coated with an adsorbed liquid that formed the stationary phase. To ensure reasonable flow rates through this type of stationary phase, the particle size of the solid was kept larger than 150 to 200μm. Even with these particles, flow rates were a few tenths of a milliliter per minute at best. Attempts to speed up this classic procedure by application of vacuum or pressure were not effective because increases in flow rates were accompanied by increases in plate heights and accompanying decreases in column efficiency.

Early in the development of the theory of liquid chromatography, it was recognized that large decreases in plate heights would be realized if the particle size of packings were reduced. This effect is shown by the data in Figure 4. 1.

indispensable [ˌɪndɪˈspensəbl] *adj*.不可缺少的，绝对必要的

forensics [fəˈrensɪks] *adj*.法医学

versatile [ˈvɜːsətaɪl] *adj*.多用途的，多功能的

affinity [əˈfɪnɪtɪ] *n*.亲和

packings [ˈpækɪŋz] *n*.填料

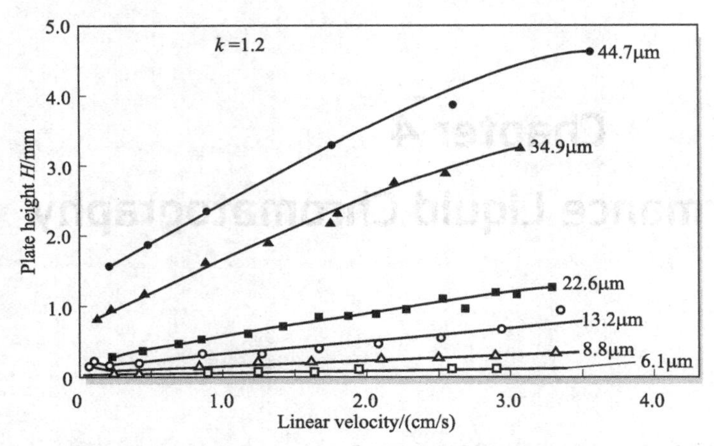

Figure 4.1 Effect of particles size of packing and flow rate on plate height in liquid chromatography

Not until the late 1960s, was the technology developed for producing and using packings with particle diameters as small as 3 to 10μm. This technology required instruments capable of much higher pumping pressures than the simple devices that preceded them. Simultaneously, detectors were developed for continuous monitoring of column effluents. The name high-performance liquid chromatography is often used to distinguish

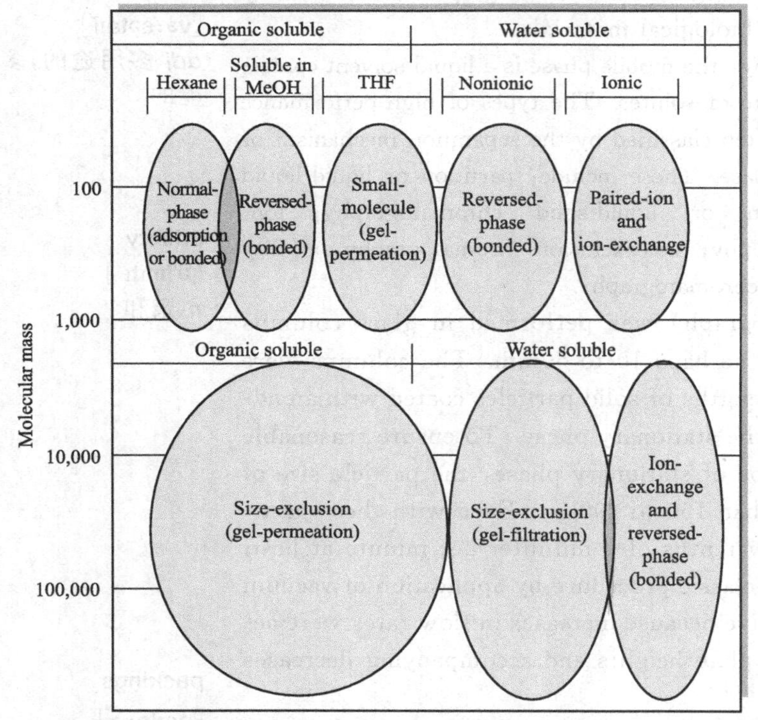

Figure 4.2 Applications of liquid chromatography. Methods can be chosen based on solubility and molecular mass. In many cases, for small molecules, reversed-phase methods are appropriate. Techniques toward the bottom of the diagram are best suited for high molecular mass ($M>2000$)

this technology from the simple column chromatographic procedures that preceded them. Simple column chromatography, however, still finds considerable use for preparative purposes.

Applications of the most widely used types of HPLC for various analyte species are shown in Figure 4. 2. Note that the various types of liquid chromatography tend to be complementary in their application. For example, for analytes having molecular masses greater than 10,000, one of the two size-exclusion methods is often used: gel permeation for non-polar species and gel filtration for polar or ionic compounds. For ionic species, ion-exchange chromatography is often the method of choice. In most cases for nonionic small molecules, reversed-phase methods are suitable.

High-performance liquid chromatography is a type of chromatography that combines a liquid mobile phase and a very finely divided stationary phase. In order to obtain satisfactory flow rates, the liquid must be pressurized to several hundred or more pounds per square inch. Schematic of a modular HPLC instrument are shown in Figure 4. 3.

preparative
[prɪ'pærətɪv]
adj.制备级的

permeation
[ˌpɜːmɪ'eɪʃn]
n.渗透

filtration
[fɪl'treʃən]
n.过滤

satisfactory
[ˌsætɪs'fæktərɪ]
adj.令人满意的，符合要求的

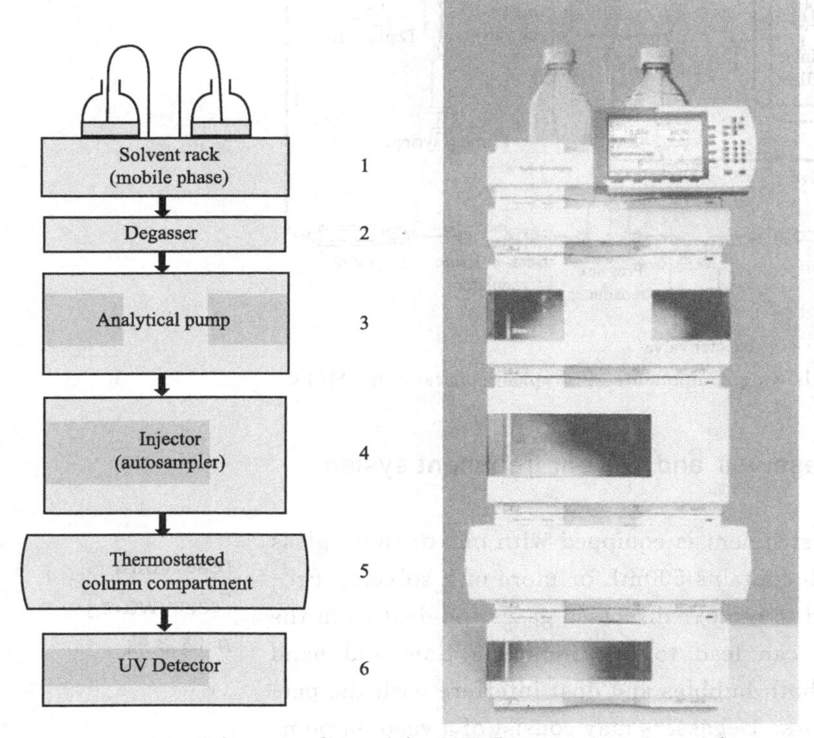

Figure 4.3 Schematic of a modular HPLC instrument. A modular system allows users to adapt the installation according to the applications to be carried out. The vertical assembly of the different modules affords an economy of space. Here the chromatograph, model HP 1200, comprises an auto-sampler that allows continuous operation and a thermostatically-controlled column to improve the reproducibility of the separations

thermostatically
[ˌθɜːmə'stætɪklɪ]
adv.恒温地，自动调节温度地

4.1 Instrumentation

Pumping pressures of several hundred atmospheres are required to achieve reasonable flow rates with packings in the 3- to 10-μm size range, which are common in modern liquid chromatography. Because of these high pressures, the equipment for high-performance liquid chromatography tends to be considerably more elaborate and expensive than that encountered in other types of chromatography. Figure 4. 4 is a diagram showing the important components of a typical HPLC instrument.

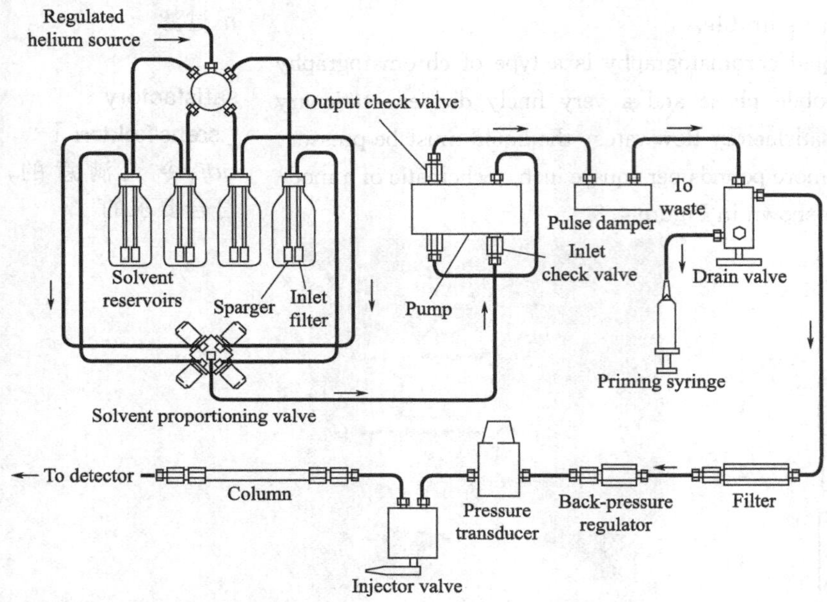

Figure 4.4 Block diagram showing components of a typical apparatus for HPLC

4.1.1 Mobile phase reservoir and solvent treatment system

A modern HPLC instrument is equipped with one or more glass reservoirs, each of which contains 500mL or more of a solvent. Provisions are often included to remove dissolved gases and dust from the liquids. Dissolved gases can lead to irreproducible flow and band spreading. In addition, both bubbles and dust interfere with the performance of most detectors. Degassers may consist of a vacuum pumping system, a distillation system, a device for heating and stirring, or, as shown in Figure 4. 4, a system for sparging in which the dissolved gases are swept out of solution by fine bubbles of an inert gas that is not soluble in the mobile phase.

An elution with a single solvent or solvent mixture of constant composition is termed an isocratic elution. In gradient elution, two

reservoirs
[ˈrezəvwɑːz]
*n.*储液瓶

distillation
[ˌdɪstɪˈleɪʃn]
*n.*蒸馏

(and sometimes more) solvent systems that differ significantly in polarity are used and varied in composition during the separation. The ratio of the two solvents is varied in a preprogrammed way, sometimes continuously and sometimes in a series of steps. As shown in Figure 4.5, gradient elution frequently improves separation efficiency, just as temperature programming helps in gas chromatography. Modern HPLC instruments are often equipped with proportioning valves that introduce liquids from two or more reservoirs at ratios that can be varied continuously (see Figure 4.4). An isocratic elution in HPLC is one in which the solvent composition remains constant. A gradient elution in HPLC is one in which the composition of the solvent is changed continuously or in a series of steps.

gradient
[ˈgrediənt]
adj 梯度的

composition
[ˌkɑmpəˈzɪʃn]
n.组成

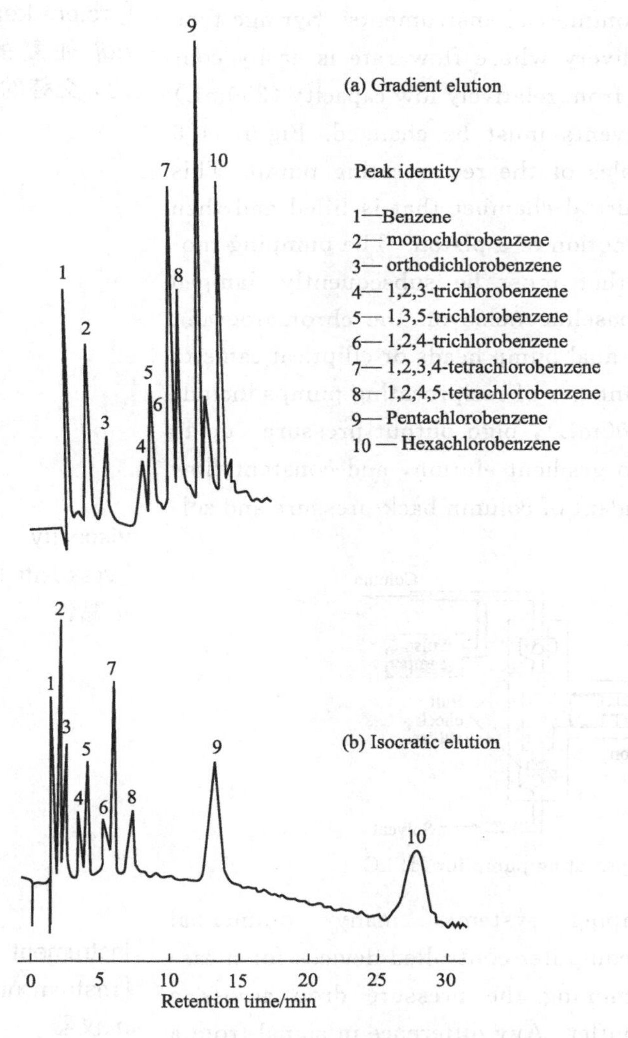

Figure 4.5 Improvement in separation effectiveness by using gradient elution

4.1.2 Pumping system

The requirements for liquid chromatographic pumps include ① the generation of pressures of up to 6000psi (lb/in^2), ② pulse-free output, ③ flow rates ranging from 0.1 to 10mL/min, ④ flow reproducibilities of 0.5% relative or better, and ⑤ resistance to corrosion by a variety of solvents. The high pressures generated by liquid chromatographic pumps are not an explosion hazard because liquids are not very compressible. Thus, rupture of a component results only in solvent leakage. Such leakage may constitute a fire or environmental hazard with some solvents, however.

reproducibility [rɪprədjuːsə'bɪlɪtɪ] *n*.可重现性

leakage['likɪdʒ] *n*.泄漏

Two major types of pumps are used in HPLC instruments: the screw-driven syringe type and the reciprocating pump. Reciprocating types are used in almost all commercial instruments. Syringe-type pumps produce a pulse-free delivery whose flow rate is easily controlled. They suffer, however, from relatively low capacity (250mL) and are inconvenient when solvents must be changed. Figure 4.6 illustrates the operating principles of the reciprocating pump. This device consists of a small cylindrical chamber that is filled and then emptied by the back-and-forth motion of a piston. The pumping motion produces a pulsed flow that must be subsequently damped because the pulses appear as baseline noise on the chromatogram. Modern HPLC instruments use dual pump heads or elliptical cams to minimize such pulsations. Advantages of reciprocating pumps include small internal volume (35 to 400mL), high output pressure (up to 10000psi), ready adaptability to gradient elution, and constant flow rates, which are largely independent of column back-pressure and solvent viscosity.

reciprocating [rɪ'sɪprəˌkeɪtɪŋ] *adj*.往复的，来回的，交替的

viscosity [vɪ'skɑsɪtɪ] *n*.黏性

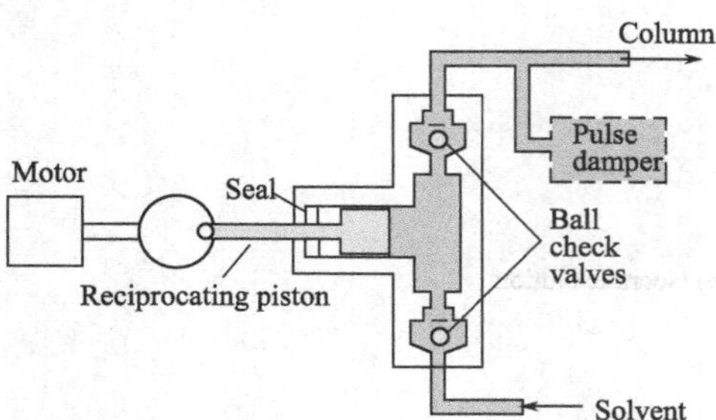

Figure 4.6 A reciprocating pump for HPLC

As part of their pumping systems, many commercial instruments are equipped with computer-controlled devices for measuring the flow rate by determining the pressure drop across a restrictor located at the pump outlet. Any difference in signal from a preset value is then used to increase or decrease the pump motor. Most instruments also have a means for varying the composition of

instrument ['ɪnstrəmənt] *n*.仪器

the solvent either continuously or in a stepwise speed of the fashion. For example, the shown in Figure 4.4 contains a proportioning valve that permits mixing of up to four solvents in a preprogrammed and continuously variable way.

4.1.3 Sample injection system

The most widely used method of sample introduction in liquid chromatography is based on a sampling loop, such as that shown in Figure 4.7. These devices are often an integral part of liquid chromatography equipment and have interchangeable loops capable of providing a choice of sample sizes ranging from 1 to 100mL or more. The reproducibility of injections with a typical sampling loop is a few tenths of a percent relative. Many HPLC instruments incorporate an autosampler with an automatic injector. These injectors can introduce continuously variable volumes from containers on the autosampler.

interchangeable [ˌɪntəˈtʃeɪndʒəbl] *adj*.可交换的，可交替的，可互换的

autosampler [ˌɔːtəˈsæmplə] *n*.自动取样器

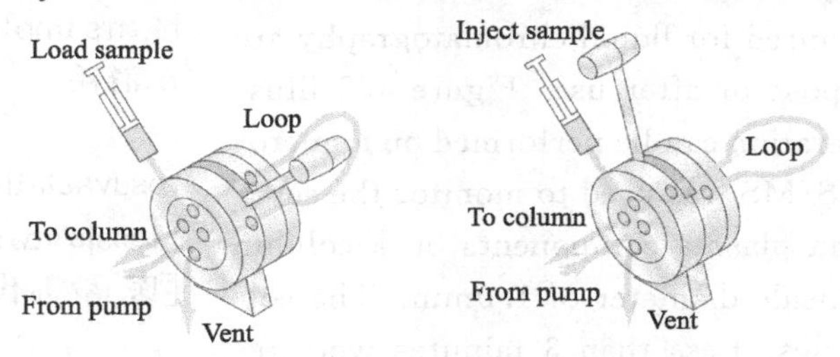

Figure 4.7 A sampling loop for liquid chromatography

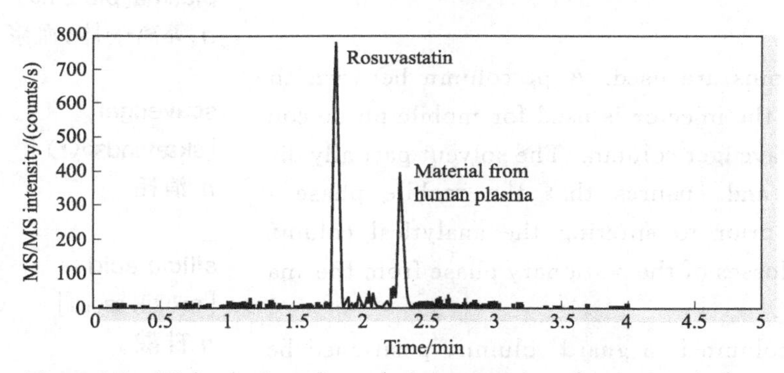

Figure 4.8 High-speed gradient elution separation of rosuvastatin from human plasma-related components. Column: 5cm × 1.0mm i. d. Luna C_{18} 3μm. Monitored by MS/MS at m/z 5488.2 and 264.2

A guard column between the injector and the column removes particulates and other solvent impurities.

4.1.4 Columns for HPLC

Liquid chromatographic columns are usually constructed from stainless steel tubing, although glass and polymer tubing, such as polyetheretherketone (PEEK), are sometimes used. In addition, stainless steel columns lined with glass or PEEK are also available.

stainless [ˈsteɪnlɪs] *adj*.不锈的

Hundreds of packed columns differing in size and packing can be pur chased from HPLC suppliers. The cost of standard-size nonspecialty columns ranges from $200 to more than $500. Specialized columns, such as chiral columns, can cost more than $1000.

(1) Analytical Columns

Most columns range in length from 5 to 25cm and have inside diameters of 3 to 5mm. Straight columns are invariably used. The most common particle size of packings is 3 or 5μm. Commonly used columns are 10 or 15cm long, 4.6mm in inside diameter, and packed with 5μm particles. Columns of this type provide 40000 to 70000 plates/m. In the 1980s, microcolumns became available with diameters of 1 to 4.6mm and lengths of 3 to 7.5cm. These columns, which are packed with 3μm or 5μm particles, contain as many as 100000 plates/m and have the advantage of speed and minimal solvent consumption. This latter property is of considerable importance because the high-purity solvents required for liquid chromatography are expensive to purchase and to dispose of after use. Figure 4.8 illustrates the speed with which a separation can be performed on a microbore column. In this example, MS/MS was used to monitor the separation of rosuvastatin from human plasma components on a column that was 5cm in length with an inside diameter of 1.0mm. The column was packed with 3μm particles. Less than 3 minutes were required for the separation.

(2) Precolumns

Two types of precolumns are used. A precolumn between the mobile phase reservoir and the injector is used for mobile-phase conditioning and is termed a scavenger column. The solvent partially dissolves the silica packing and ensures that the mobile phase is saturated with silicic acid prior to entering the analytical column. This saturation minimizes losses of the stationary phase from the analytical column.

A second type of precolumn is a guard column, positioned between the injector and the analytical column. A guard column is a short column packed with a similar stationary phase as the analytical column. The purpose of the guard column is to prevent impurities, such as highly retained compounds and particulate matter, from reaching and contaminating the analytical column. The guard column is replaced regularly and serves to increase the lifetime of the analytical column.

(3) Column Temperature Control

For some applications, close control of column temperature is not necessary, and columns are operated at room temperature. Often, however, better, more reproducible chromatograms are ob-

consumption
[kən'sʌmpʃn]
n.消耗

rosuvastatin
[rəʊsjʊ'vɑːsteɪtɪn]
[医]罗苏伐他汀

plasma[plæzmə]
n.等离子体，血浆

scavenger
['skævɪndʒə(r)]
n.消耗

silicic acid
[sɪ'lɪsɪk'æsɪd]
n.硅酸

guard[gɑːd]
n.保护

tained by maintaining constant column temperature. Most modern commercial instruments are equipped with heaters that control column temperatures to a few tenths of a degree from near room temperature to 150℃. Columns can also be fitted with water jackets fed from a constant-temperature bath to give precise temperature control. Many chromatographers consider temperature control to be essential for reproducible separations.

(4) Column Packings and Hardware

Two types of packings are used in HPLC, pellicular and porous particle. The original pellicular particles were spherical, nonporous, glass or polymer beads with typical diameters of 30 to 40μm. A thin, porous layer of silica, alumina, a polystyrene-divinyl benzene synthetic resin, or an ion-exchange resin was deposited on the surface of these beads. Small porous microparticles have completely replaced these large pellicular particles. In recent years, small (<5μm) pellicular packings have been reintroduced for separation of proteins and large biomolecules.

pellicular [ˈpelɪkjʊlə] *adj*.薄壳型的

polystyrene-divinyl benzene 聚苯乙烯-二乙烯基苯

resin[ˈrezɪn] *n*.合成树脂

deposit [dɪˈpɒzɪt] *v*.吸附，沉淀

The typical porous particle packing for liquid chromatography consists of porous microparticles having diameters ranging from 3 to 10μm; for a given size particle, a very narrow particle size distribution is desirable. The particles are composed of silica, alumina, the synthetic resin polystyrene-divinylbenzene, or an ion-exchange resin. Silica is by far the most common packing in liquid chromatography. Silica particles are often coated with thin organic films, which are chemically or physically bonded to the surface. Column packings for specific chromatographic modes are discussed in later sections of this chapter.

The three major contributions to band broadening in LC are eddy diffusion, molecular diffusion and resistance to mass transfer, as described in the van Deemter equation. The effect of each of these processes on column performance can be related to a large number of variables including mobile phase flow-rate (or linear velocity), column length, average particle size, particle size distribution, column configuration, porosity of packing, solvent viscosity, shape of packing, integrity of column bed, mass and volume of injection and nature of the chemical interaction. In this section, general aspects of packings, configuration, hardware and their effect on chromatographic performance are discussed.

Modern HPLC packings can be broadly classified according to the type, size, shape, nature and material of the particle. A wide variety of different particle types is used in HPLC columns. Pellicular particles have a solid, inner core and a thin outer surface layer of stationary phase. Alternatively, the outer layer of coated stationary phase can be porous, creating a superficially porous particle. These pellicular materials consist of solid, spherical glass beads of relatively large diameter (e. g. 30μm), with a thin (approximately 1μm) layer of porous silica on the surface, as typified by the Zipax range of columns from Du Pont. Porous particles are materials of relatively large diameter (e. g. 30μm), which are fully porous and can be either irregular or spherical in shape. Microparticulate materials are of smaller diameter (e. g. 3-10μm), fully porous, and again can be either irregular or spherical in shape. Figure 4. 9 illustrates some of the particle types that are used in modern HPLC columns.

superficially [ˌsjuːpəˈfɪʃəlɪ] *adv*.表面地

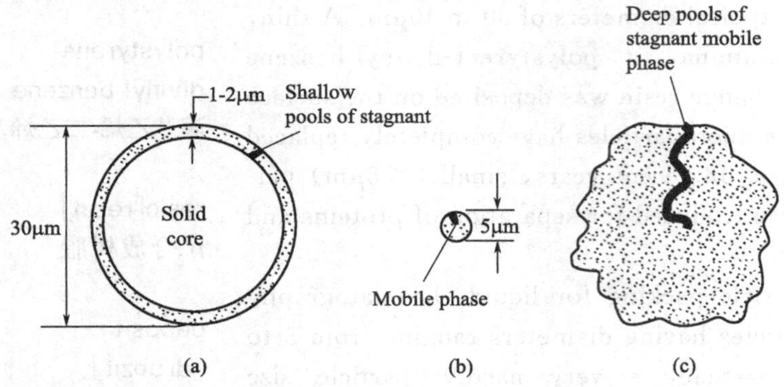

Figure 4.9 Particle types used in HPLC columns: (a) Superficially porous particle; (b) Very small totally porous particle; (c) Totally porous particle

Pellicular materials give significantly higher efficiencies (or lower HETP values) than porous particles of the same size as a result of reducing the *C* term (resistance to mass transfer) in the van Deemter equation. Microparticulate (i.e.smaller diameter) materials give higher efficiencies than larger porous materials as the resistance to mobile phase mass transfer decreases with particle diameter. Smaller particles also reduce the contribution to the A term (eddy diffusion), again resulting in increased column efficiency. Pellicular and microparticulate materials can provide similar chromatographic efficiencies; however, pellicular materials are restricted to small sample loadings because of their low active surface area, but are more easily packed into columns than are microparticulate materials.

restrict [rɪˈstrɪkt] *vt*.限制，束缚

The chromatographic properties of spherical and irregular microparticulate materials are essentially similar in terms of efficiency, but the structure of a bed of irregular particles is less stable than a dense, well-packed bed of spherical particles. This makes preparation of columns packed with irregular particles more difficult and also limits their stability when operated continuously at high pressures. In general, highest column efficiencies are obtained if the stationary phase thickness is minimized, columns are packed with small particles and the packing procedure results in a dense, tightly compacted and uniform column bed. The majority of HPLC columns today are packed with spherical, microparticulate (3-10μm diameter) materials, which results in stable, high-efficiency columns which can be used for relatively large sample loadings.

In addition to the physical nature of the particle (i. e. microporous or pellicular) and its size and shape, the particle material also significantly affects the performance of the HPLC packing. Column packing alternatives include the use of rigid solids (most commonly silica), resins (usually polystyrene divinylbenzene) and soft gels.

Rigid solids based on silica are the most common HPLC packings used today, particularly for adsorption and reversed-phase modes of chromatography. Silica packings can withstand the high pressures generated when 10-30cm columns packed with 3-10μm particles are used. Silica is abundant and available in a variety of shapes, sizes and degrees of porosity. Most importantly, silica can be readily functionalized and the chemistry of its bonding reactions is relatively well understood. Silica also has a number of disadvantages, largely as a result of its pH instability and its high surface activity. Other rigid particles which have been used as HPLC column packings include alumina, zircon and carbon, although these materials have not yet to find widespread acceptance.

Resin-based packings are increasingly widely used in HPLC columns. These packings are predominantly used in gel permeation chromatography (GPC) and ion-exchange chromatography, but resin-based, reversed-phase columns are also available commercially. Resin-based columns have the advantage that they can be used over a wide pH range, although most resin types are limited to moderate operating pressures (1000-2000psi) and mobile phases are often restricted to those with low concentrations of organic modifiers. Soft gels, such as agarose and Sephadex, are used almost exclusively for the separation of aqueous proteins; however, they cannot tolerate very high backpressure.

(5) Column packing methods

The two most common methods for packing HPLC columns are the dry-fill procedure and the wet-fill (or slurry-packing) procedure. The dry-fill procedure is recommended for the packing of rigid solids and resins with particle diameters >20μm, such as pellicular materials. This procedure involves first degreasing then drying the interior of the tubing which will form the column blank. A porous screen (typically 2μm) is then placed in the outlet fitting of the column and a small amount of the packing material added into the verti-

uniform [ˈjuːnɪfɔːm] n.均匀

loading [ˈləʊdɪŋ] v.进样

functionalize [ˈfʌŋkʃənəlaɪz] vt.使功能化

bonding [ˈbɒndɪŋ] n.键合

agarose [ˈɑːgərəʊs] n.琼脂糖

cally held column via a funnel.

The column is tapped to settle the packing and more material is added and so on until the column is full. The packing is then levelled off and the inlet fitting with screen is screwed onto the top of the column. While dry-fill columns can be prepared to give relatively reproducible characteristics for large porous particles and pellicular materials, packing smaller size particles in this manner leads to poor column efficiency. Small particles tend to agglomerate because of their high surface energy to mass ratio, resulting in a poorly compacted column bed. This results in widely varying flow velocities along the column, creating significant band broadening and hence poor efficiency. The dry-fill method is used routinely for preparing columns for preparative chromatography and also for solid phase extraction cartridges, which are used for sample clean-up.

agglomerate
[əˈglɒməreɪt]
v.凝结

The wet-fill or slurry packing method of column preparation uses a suitable liquid to suspend the particles, which are then pumped under high pressure into the column blank. The suspending solvent must be chosen to maintain a uniform particle distribution without agglomeration and must also wet the packing thoroughly. High surface energy materials, such as unfunctionalized silica, require polar solvents, while lower surface energy packings, such as C_{18} functionalized silica, may be packed in less polar solvents. The slurrying solvent density should also be considered, particularly when packing particles$>10\mu m$ in diameter. For these larger particles, the solvent density should be similar to that of the particles themselves in order to reduce settling, particularly if the particle size distribution is large. In order to pack a column, the packing is slurried at modest concentrations with the solvent and placed in a specially designed reservoir, often called a column bomb, which fits on the inlet end of the column blank. A porous screen (1-2μm) is placed at the outlet end of the column and the solvent is pumped into the bomb at high velocity with a constant-pressure HPLC pump. This forces the slurried packing into the column and produces a compact bed. The packing is complete when a constant flow-rate from the column is finally obtained, after which the bomb is removed, the packing levelled off and the inlet fitting with screen screwed onto the top of the column. The column may be packed in either the upward or downward direction, with the majority of columns being packed in the downward direction.

slurry packing
湿法填充

unfunctionalized
未官能化

bomb
[bɒm]
n.高压储罐

Both rigid solid and resin-based columns can be prepared in this fashion, although resins must first be allowed to swell in the solvent before packing. Owing to their decreased rigidity, lower packing pressures are normally used for resin-based columns. A standard 25cm×4.6mm i.d. HPLC column contains approximately 2-3g of ma-

rigidity
[rɪˈdʒɪdətɪ]
n.刚度

terial if microparticulate silica particles are used. While column blanks and packing apparatus are available from column suppliers, such as Alltech, the vast majority of HPLC columns are purchased prepacked by the supplier. The two main reasons for this are that it takes both time and considerable skill to produce reproducible, efficient columns, and most column manufacturers do not sell their analytical scale packings in bulk, preferring to sell prepacked columns.

(6) Column configuration and hardware

hardware [ˈhɑːdweə(r)] n.硬件

The design of the HPLC column has been the subject of much research over many years and reviews on the subject are still published frequently. The column must be constructed of materials that withstand the pressures used in HPLC and are also chemically resistant to the mobile phase. As is the case with HPLC pumps, the majority of columns are constructed from 316 stainless steel. Stainless steel columns may be constructed using either a rigid single wall or a double wall. Columns are also available in inert materials, such as glass, teflon and PEEK. Such columns are used in cases where chemically aggressive mobile phases, such as HCl, are used or when the sample may adsorb onto the stainless steel surface, as is the case with some proteins. Stainless steel materials offer great rigidity and can be operated at high pressures, hence they are most widely used for silica and bonded phase packings. Polymeric column materials are commonly used for ion-exchange packings, while glass columns are used mostly for protein separations. Both polymeric and glass columns are subject to stringent pressure limitations and care should be taken not to exceed the manufacturer's recommended operating pressures when using such columns.

HPLC columns are available in a wide variety of configurations, shapes and sizes. Most columns are packed in straight sections of tubing fitted with zero dead volume end couplings. Care must always be taken to ensure that compression screws and ferrules being used are compatible with the column end fittings. Generally, compression screws from one manufacturer cannot be used with another's column without permanently damaging the column. Cartridge columns are available as an alternative to the standard steel column, as typified by the Brownlee range of columns. With such columns, the body of the column may be replaced and the old end fittings reused. While the initial purchase price of a cartridge column is usually higher than an equivalent steel column, the replacement cartridges are less expensive, hence they become cost-effective if the user frequently need to replace columns.

cartridge [ˈkɑːtrɪdʒ] n.筒

replacement [rɪˈpleɪsmənt] n.置换,代替

Yet another alternative to the conventional steel column is the radial compression cartridge. In this format, the stationary phase is packed into a teflon cartridge, which is compressed by applying an external pressure radially to the cartridge before use. The pressure is applied by compressing either a liquid or a gas surrounding the cartridge. Reforming the packing bed and removing any voids and channels each time the column is used leads to greater stability, improved durability and efficiency. This type of cartridge reduces the 'wall effects' which result from the fact that there is less resistance to solvent flow near the rigid walls of a conventional column because of the lower density of the packed bed. This wall effect is a significant source of band broadening and the efficiency of radial compression columns has been shown to be up to 25% higher than equivalent steel columns.

In addition to choosing the packing material and column hardware, it is also necessary to select the appropriate column dimensions. Column manufacturers provide their packing materials in columns of various lengths and diameters. Generally, the longer the column, the higher its efficiency and resolution; the shorter the column, the faster the separation. Also, the larger the column diameter, the greater is its loading capacity, while the narrower the column, the greater is its mass sensitivity. Most analytical columns range from 50 to 300mm in length and 2.0 to 5.0mm in diameter. The use of shorter columns provides rapid separations but limits resolving power; hence, this approach should only be used for simple samples. Some resin-based materials, such as polymethacrylate, are often packed only into short columns in order to keep the pressure drop across the column as low as possible. Longer columns provide greater resolving power, although at the expense of increased chromatographic run times. Narrow bore (or microbore) columns with diameters of 1-2mm or less are used for applications where high sensitivity is required, where the amount of sample is limited or where solvent (and disposal costs) are significant. Flow rates of 0.05 to 1.0mL/min are used with narrow-bore columns. However, to obtain the best results from such columns, it is critical that sources of bandspreading within the system be minimized. Figure 4.10 compares the sensitivity obtained for a mixture of synthetic fragments of ACTH using 3.9mm and 2.0mm i. d. C_{18} reversed-phase columns.

dimension [daɪ'menʃn] *n*.尺寸

bandspreading [bændsp'redɪŋ] *n*.频带展宽

Conversely, if larger sample loadings are required than conventional columns allow, larger internal diameter columns should be used. Sample load is proportional to the column length and the square of the column diameter. Large-bore columns are most frequently used in preparative chromatography. In summary, while HPLC columns can be prepared with efficiencies over 1 million theoretical plates, the columns dimensions (7000 × 0.02cm i. d.) and

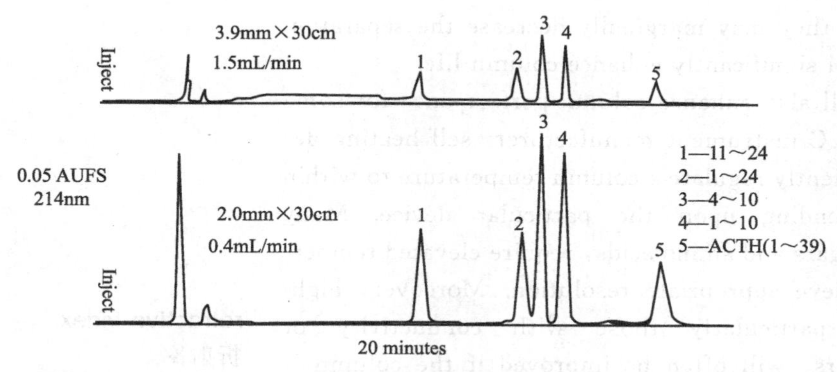

Figure 4.10 Comparison of sensitivity obtained using 3.9mm and 2.0mm i.d. columns. A Waters ILBondapak C_{18} column was used with a linear gradient of 10%-60% acetonitrile/water over 20min Two-hundred and fifty picomoles of ACTH peptides were injected

excessive run times (>7days) make them unsuitable for routine use. In practical terms, the most commonly used columns are 250-300mm in length with a diameter of 3. 9-4. 6mm, although 150mm columns are becoming increasingly popular. Columns in this size range offer the best overall compromise in terms of efficiency, sample loading and chromatographic run times, without the need for excessive attention to instrument performance in regard to bandspreading.

compromise [ˈkɒmprəmaɪz] n 协调

(7) Measuring column performance

Column performance can be evaluated using a number of parameters, although traditionally the number of theoretical plates (N) is used as a measure of column efficiency. Other parameters of importance include peak asymmetry, capacity factor, selectivity, resolution of a critical peak pair, pressure drop across the column, etc.

Ideally, all of the above parameters should be monitored continuously and the data recorded. The concept of system suitability is based on careful monitoring and documentation of the column and overall system performance in order to set up control specifications and limits for given assays. This quality approach is becoming increasingly important, particularly in pharmaceutical laboratories. The latest versions of chromatography data software now calculate and automatically document a wide range of system parameters, such as those listed above. In addition to indicating if the system meets the assay specifications, continual monitoring of the column and system performance provides a benchmark to consult when troubleshooting a chromatographic problem. It also allows the user to gain longer term data, such as column lifetime, which aids in maintaining stock levels and even deciding if a column has lasted for a reasonable number of injections.

benchmark [ˈbentʃmɑ:k] n.基准

troubleshooting [ˈtrʌblʃu:tɪŋ] n.修理故障

Guard columns are intended to be used only for short periods and should be replaced after a limited number of injections, usually in the order of 50-200. While they may marginally decrease the separation efficiency, their use will significantly enhance column life.

Column heaters will also enhance column performance for some applications. Most HPLC instrument manufacturers sell heating devices which will conveniently regulate a column temperature to within 0. 1 to 1. 0℃, depending upon the particular device. Many separations, such as sugars and amino acids, require elevated temperatures in order to achieve appropriate resolution. Moreover, high-sensitivity analyses, particularly those with conductivity or refractive index detectors, will often be improved if the column is thermally stabilized. In conclusion, a well-treated column can last for many thousands of injections, but the use of one inappropriate mobile phase or untreated sample may drastically reduce a column's performance. Careful attention to the points listed above will enhance the column life. In addition, a great deal of help is readily available regarding column information and troubleshooting. Many manufacturers of HPLC instrumentation now provide free phone support lines to answer questions regarding column and system performance criteria. Virtually all columns are supplied with a manual which should be followed carefully. Instrument manuals are also a useful source of troubleshooting information.

refractive index
折射率

4.1.5 HPLC detector

The ideal detector for HPLC should have all the characteristics of the ideal GC detector except that it need not have as great a temperature range. In addition, an HPLC detector must have low internal volume (dead volume) to minimize extra-column band broadening. The detector should be small and compatible with liquid flow. Unfortunately, no highly sensitive, universal detector system is available for high-performance liquid chromatography. Thus, the detector used will depend on the nature of the sample. Table 4. 1 lists some of the common detectors and their properties.

The most widely used detectors for liquid chromatography are based on absorption of ultraviolet or visible radiation (see Figure 4. 11). Both photometers and spectrophotometers, specifically designed for use with chromatographic columns, are available from commercial sources. Photometers often make use of the 254nm and 280nm lines from a mercury source because many organic functional groups absorb in the region. Deuterium sources or tungsten-filament sources with interference filters also provide a simple means of detecting absorbing species. Some modern instruments are equipped with filter wheels that contain several interference filters, which can be rapidly switched into place. Spectrophotometric detectors are consid-

ultraviolet
[ˌʌltrə'vaɪəlɪt]
adj.紫外的,紫外线的

deuterium
[dju'tɪərɪəm]
n.氘,重氢

erably more versatile than photometers and are also widely used in high-performance instruments. Modern instruments use diode-array detectors that can display an entire spectrum as an analyze exits the column.

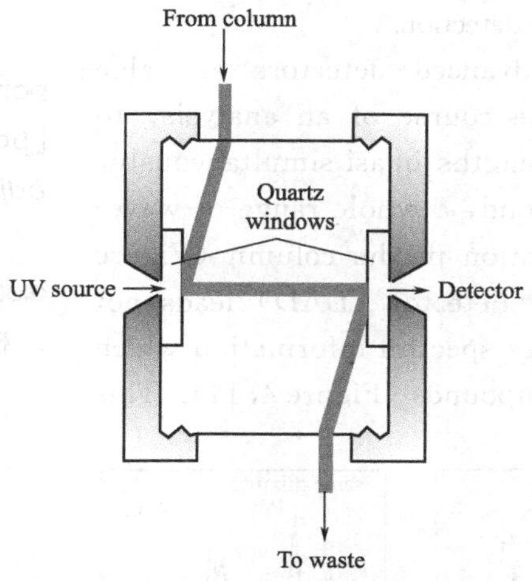

Figure 4.11 A UV-visible absorption detector for HPLC

(1) Spectrophotometric detectors

Detection is based upon the Lambert-Beer Law ($A=\varepsilon_{\lambda}lc$): The absorbance A of the mobile phase is measured at the outlet of the column, at one or several wavelengths in the UV or visible spectrum. The intensity of the absorption depends upon the molar absorption coefficient of the species detected. It is essential that the mobile phase be transparent or possess only a very little absorption (Figure 4.12).

transparent [træns'pærənt] *adj*.清澈的，透明的

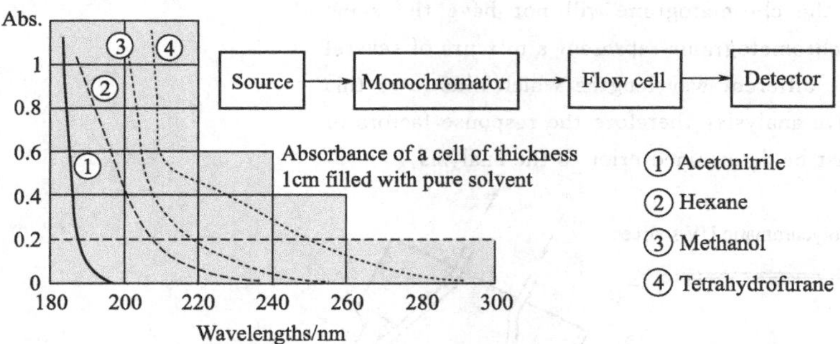

Figure 4.12 Photometric detection at a single wavelength. Principle of a photometric detector along with the absorption spectra of several solvents used in liquid chromatography. Here the transparence limit of a solvent corresponds to an absorbance of 0.2 for 1cm of optical path in the cell

The area of a peak, without taking into account this specific parameter, renders the direct calculation of concentration unfeasible by a simple check of the chromatogram. Spectrophotometric detectors are examples of selective detection. For compounds that do not possess a significant absorption spectrum it is possible to perform derivatization of the analytes prior to detection.

unfeasible [ʌn'fi:zəbl] *adj*.不可行的

derivatization [də'rɪvəˌtɪzeɪʃən] *n*.衍生化

Monochromatic detection: The basic model comprises a deuterium or mercury vapour light source, a monochromator for isolating a narrow bandwidth (10nm) or a characteristic spectral line (e. g. 254nm if the source is a mercury lamp), a flow cell with a volume of a few μL (optical path of 0.1 to 1cm) and a means of optical detection.

Polychromatic detection: More advanced detectors are able either to change wavelength during the course of an analysis, to record the absorbance at several wavelengths quasi-simultaneously, or even to capture in a fraction of a second, a whole range of wavelengths without interrupting the circulation in the column (Figure 4. 13 to Figure 4. 15). The diode array detector (DAD) leads not only to a chromatogram but also provides spectral information which can be used to identify the separated compounds (Figure 4. 14). This is called specific detection.

monochromatic [ˌmɒnəkrəˈmætɪk] *adj*.单色的

flow cell 流通池

polychromatic [pɒlɪkrəʊˈmætɪk] *adj*.多色的

diode array detector 二极管阵列检测器

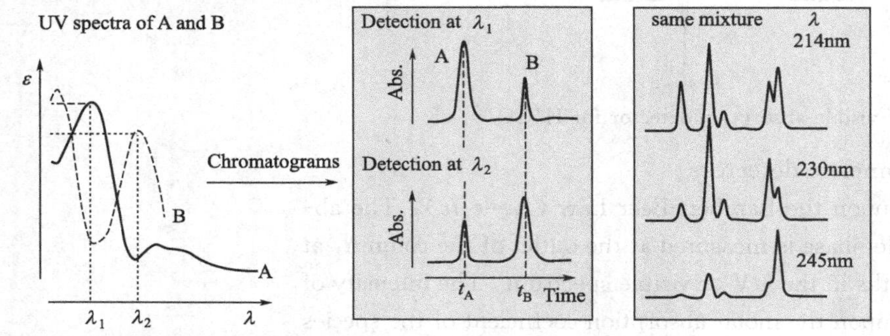

Figure 4.13 Chromatograms of a sample containing two compounds A and B, for which the UV spectra are different. According to the choice of detection wavelength, the chromatogram will not have the same aspect. On the right, the chromatograms represent a mixture of several pesticides recorded at three different wavelengths which illustrates this phenomenon. In quantitative analysis, therefore the response factors of each of the compounds must be determined prior to the analysis

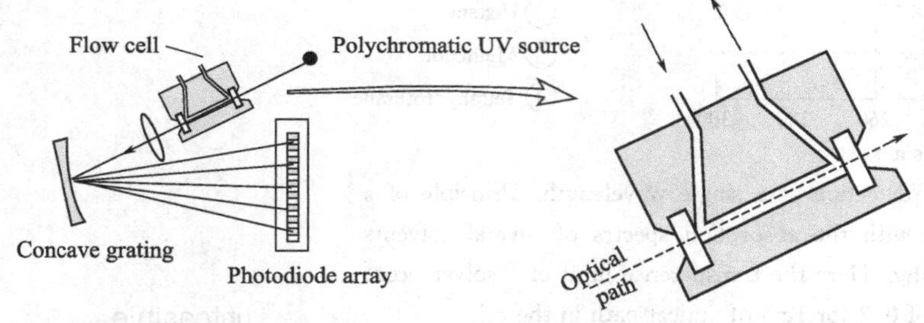

Figure 4.14 Optical schematic of the diode array detector. The flow cell is irradiated with a polychromatic UV/Vis light source. The light transmitted by the sample is dispersed by a concave grating towards a detector which comprises a diode array. The number of diodes can attain several hundred, each one monitoring the mean absorption of a very narrow interval of wavelengths (e. g. 1nm)

The successive spectra of the compounds eluted with the mobile

phase are recorded continuously and stored in the memory of the instrument, to be treated later using appropriate software. Often spectacular chromatograms can be obtained (Figure 4.15). The ability to record thousands of spectra during a single analysis increases the potential of these detectors systems. A topographic representation of the separation can be conducted, $A=f(\lambda t)$ (iso-absorption diagrams).

The rapid development of biotechnologies, as in biochemistry, requires the analysis of amino acids (proteins hydrolysates); photometric detectors can be used with the condition that prior to passage in the measuring cell a post-column reaction with ninhydrin is carried out.

hydrolysate [haɪ'drɒlɪseɪt] n.水解产物

ninhydrin [nɪn'haɪdrɪn] n.茚三酮

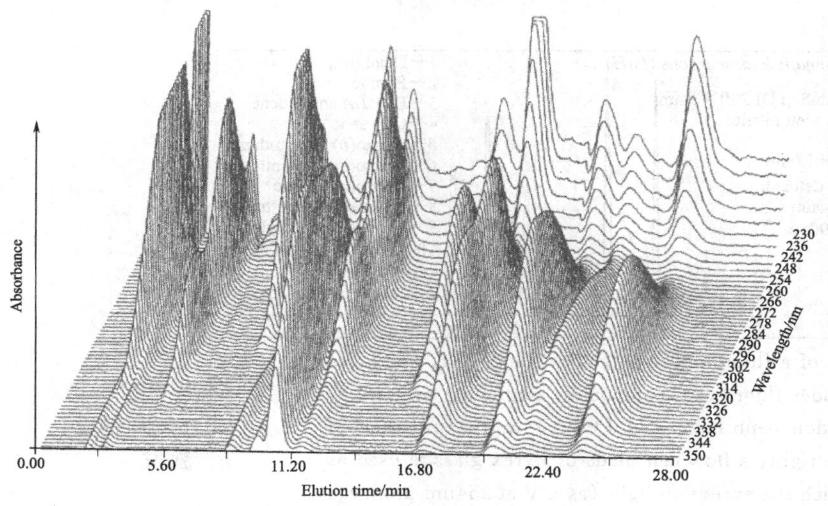

Figure 4.15 Three-dimensional representation, $I=f(\lambda t)$, of a chromatographic separation obtained by a rapid recording method

(2) Fluorescence detector

About 10 percent of organic compounds are fluorescent, in that they have the ability to re-emit part of the light absorbed from the excitation source. The intensity of this fluorescence is proportional to the concentration of the analyte, as long as this concentration is kept low. Application to LC chromatography gave rise to fluorescence detectors, very sensitive and for this reason often used for trace analysis. Unfortunately, the response is only linear over a relatively limited concentration range of two orders of magnitude.

fluorescence [flu:ə'resəns] n.荧光,荧光性

The domain of this selective detector (Figure 4.16), can be extended by the application of a derivatization procedure pre- or post-column to make the substances of interest detectable. There are a number of reagents that have been developed specifically to synthesized fluorescent derivatives. For this methodology an automatic reagent dispenser is placed either prior to the column or between the column and the detector to effect one, or several reactions, upon the sample analytes in order to render them fluorescent before or after injection.

derivative [dɪ'rɪvətɪv] n.衍生物

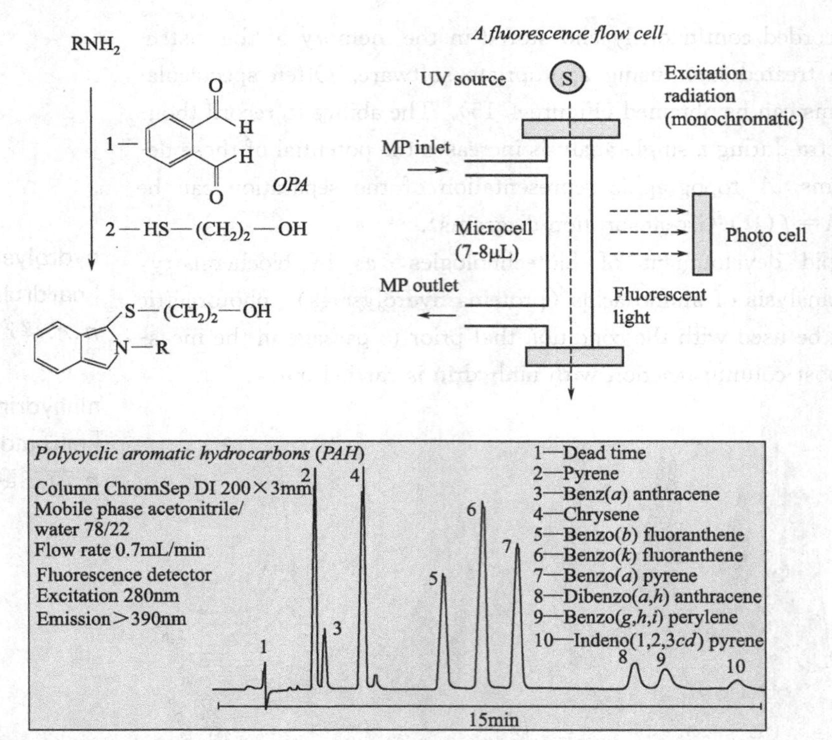

Figure 4.16 Flow-cell of a fluorometric detector. Above left, examples of reagents used to render fluorescent compounds containing primary amines through the action *o*-phthalic acid (OPA) in the presence of monothioglycol. Above right, a flow cell made of Pyrex glass is used as the sensor through which the excitation light (as UV at 254nm produced by the mercury lamp) passes axially. A photocell is situated at the side of the cell to receive radially emitted light (perpendicular to the direction of the primary light). Below, chromatogram of a mixture of several polycyclic aromatic hydrocarbons (PAH). The intensity of the fluorescence varies from one compound to another because there are variations in fluorescent quantum yields. To remedy this, the excitation wavelength should be adjusted for each compound. This is done with programmable detectors, able to select the optimal technical conditions for each species

excitation [ˌeksaɪ'teɪʃən] *n*. 激发

By following this principle, traces of carbamates (pesticides) in the environment can be measured by saponification with sodium hydroxide and then reaction with *o*-phthalic aldehyde to transform the methylamine into a fluorescing derivative (Figure 4.16).

saponification [səˌpɒnɪfɪ'keɪʃən] *n*.皂化

o-phthalic aldehyde 邻苯二甲醛

(3) Refractive index detector (RI detector)

This type of detector relies on the Fresnel principle of light transmission through a transparent medium of refractive index *n*. It is designed to measure continuously the difference in the refractive index between the mobile phase ahead of and following the column. So, a differential refractometer is used. Schematically, a beam of light travels through a cell that has two compartments: one is filled with the pure mobile phase while the other is filled with the mobile

refractive [rɪ'fræktiv] *adj* 折射的

differential refractometer 示差折光仪

phase eluting the column (Figure 4.17). In practice, the optical dispersions of the media are likely to differ, and consequently the refractive index will only match at one particular wavelength. As a result, the fully transmitted light will be largely monochromatic. The change in refractive index between the two liquids, which appears when a compound is eluting the column, is visualized as an angular displacement of the refracted beam. In practice, the signal corresponds to a continuous measurement of the retroaction that must be provided to the optical element in order to compensate the deviation of the refracted beam.

dispersion
[dɪˈspɜːʃn]
n.分散，色散

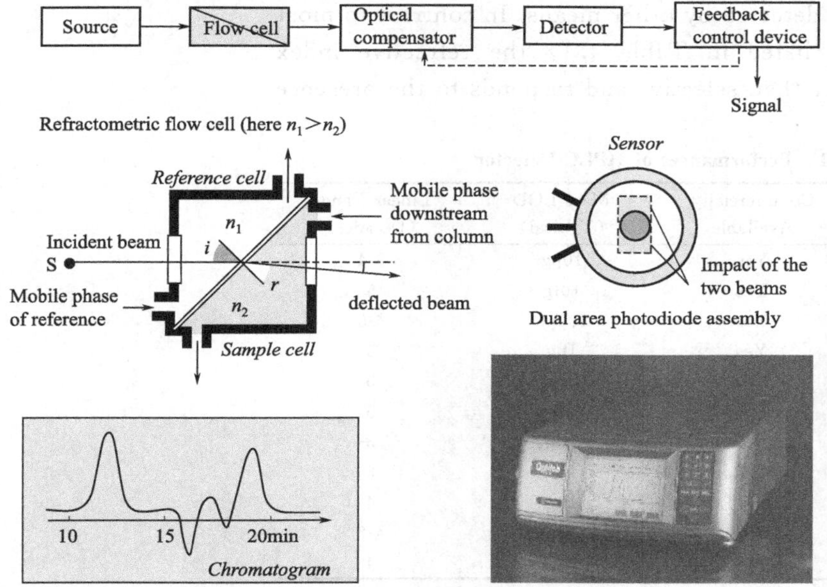

Figure 4.17 Schematics of a differential refractive index detector. The measurement of refractive index in liquid chromatography detection systems can be done in a number of ways. On this schematic, as sample elutes through one side, the changing angle of refraction moves the beam. This results in a change in the photon current falling on the dual stage photodiode which unbalances it. The responses of the two areas are maintained equal by the optical assembly (not displayed). A chromatogram of a mixture of sugars obtained with this type of detector. A commercial instrument, the model Optilab rEX

photodiode
[ˈfəʊtəʊˌdɪəʊd]
n. 光电二极管

The refractive index detector is one of the least sensitive LC detectors. It is very affected by changes in ambient temperature, in pressure and in flow-rate. The temperature of the detector must be regulated with precision (to 0.001℃) and the column thermostatically controlled. This detector leads to both positive and negative peaks, which requires that the baseline is fixed at the mid-height of the graph (Figure 4.17). Also, this detector can only be used in the isocratic mode because in gradient elution, the composition of the mobile phase evolves with time, as does its refractive index. The compensation, easily obtained in the case of a mobile phase of constant composition, is no longer attainable when the composition of the eluent at the outlet of the column differs from that at the inlet. Consequently, it is often arranged in series with other detectors, in the isocratic mode, to give a supplementary chromatogram.

Despite these disadvantages, this detector, considered to be almost universal, is extremely useful for detecting compounds that are non-ionic, do not adsorb in the UV, and do not fluoresce. It finds use in the recognition of those substances (fatty acids, alcohols, sugars, etc.) that are not easily detected by other means. In contrast to most of the other detectors listed in Table 4.1, the refractive index detector is general rather than selective and responds to the presence of all solutes

fluoresce [fluə'res] vi.发荧光

Table 4.1 Performances of HPLC Detector

HPLC Detector	Commercially Available	Mass LOD (Typical)	Linear Range (Decades)
Absorbance	Yes	10pg	3-4
Fluorescence	Yes	10fg	5
Electrochemical	Yes	100pg	4-5
Refractive index	Yes	1ng	3
Conductivity	Yes	100pg-1ng	5
Mass spectrometry	Yes	<1pg	5
FTIR	Yes	1μg	3
Light scattering	Yes	1μg	5
Optical activity	No	1ng	4
Element selective	No	1ng	4-5
Photoionization	No	<1pg	4

4.1.6 LC/MS and LC/MS/MS

The combination of liquid chromatography and mass spectrometry would seem to be an ideal merger of separation and detection. Just as in gas chromatography, a mass spectrometer could identify species as they elute from the chromatographic column. There are major problems though in the coupling of these two techniques. A gas-phase sample is needed for mass spectrometry, while the output of the LC column is a solute dissolved in a solvent. As a first step, the solvent must be vaporized. When vaporized, however, the LC solvent produces a gas volume that is 10 to 1000 times greater than the carrier gas in GC. Hence, most of the solvent must also be removed. There have been several devices developed to solve the problems of solvent removal and LC column interfacing. Today, the most popular approaches are to use a low flow-rate atmospheric pressure ionization technique. The block diagram of a typical LC/MS system is shown in Figure 4.18. The HPLC system is typically a nano scale capillary LC system with flow rates in the μL/min range. Alternatively, some interfaces allow flow rates as high as 1 to 2mL/min, which is typical of conventional HPLC conditions. The most common ionization sources are electrospray ionization and atmospheric pressure chemical ionization. The combination of HPLC and mass spectrometry gives high selectivity since unresolved peaks can be isolated by monitoring only a selected mass. The LC/MS technique can provide fingerprinting of a particular elute instead of relying on retention time as in conventional HPLC. The combination also can give molecular mass and structural information as well as accurate quantitative analysis.

elute
[ɪˈljuːt]
n.洗提液

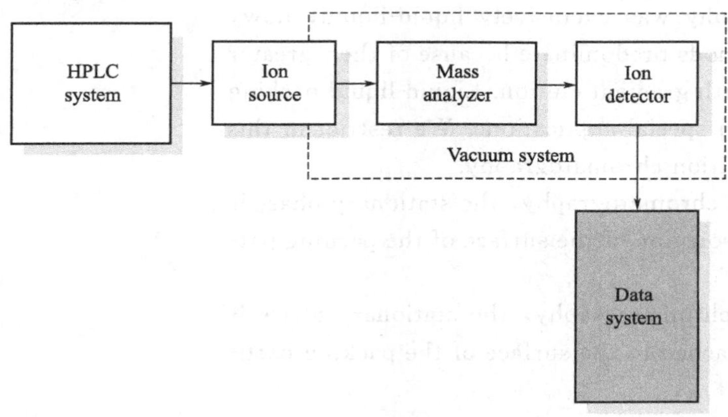

Figure 4.18 Block diagram of an LC/MS system. The effluent from the LC column is introduced to an atmospheric pressure ionization source, such as an electrospray or a chemical ionization source. The ions produced are sorted by the mass analyzer and detected by the ion detector

For some complex mixtures, the combination of LC and MS does not provide enough resolution. In recent years, it has become feasible to couple two or more mass analyzers together in a technique known as tandem mass spectrometry. When combined with LC, the tandem mass spectrometry system is called an LC/MS/MS instrument. Tandem mass spectrometers are usually triple quadruple systems or quadruple ion trap spectrometers.

tandem
[ˈtændəm]
n.串联

To attain higher resolution than that can be achieved with a quadruple, the final mass analyzer in a tandem MS system can be a time-of-flight mass spectrometer. Sector mass spectrometers can also be combined to give tandem systems. Ion cyclotron resonance and ion trap mass spectrometers can be operated in such a way as to provide not only two stages of mass analysis but n stages. Such MS systems provide the analysis steps sequentially within a single mass analyzer. These spectrometers have been combined with LC systems in LC/MS instruments.

cyclotron
[ˈsaɪkləʊtrɒn]
n.回旋加速器

4.2 Partition chromatography

The most widely used type of HPLC is partition chromatography in which the stationary phase is a second liquid that is immiscible with the liquid mobile phase. Partition chromatography can be subdivided into liquid-liquid and liquid- bonded-phase chromatography. The difference between the two lies in the way that the stationary phase is held on the support particles of the packing. The liquid is held in place by physical adsorption in liquid-liquid chromatography, while it is attached by chemical bonding in bonded-phase chromatography. Early partition chromatography was exclusively liquid-liquid; now, however, bonded-phase methods predominate because of their greater stability and compatibility with gradient elution. Liquid-liquid packing are today relegated to certain special applications. We restrict in this section to bonded-phase partition chromatography.

partition
[pɑːˈtɪʃn]
n. 分配

In liquid-liquid partition chromatography, the stationary phase is a solvent held in place by adsorption of the surface of the packing particles.

In liquid-bonded-phase chromatography, the stationary phase is an organic species that is attached to the surface of the packing particles by chemical bonds.

4.2.1 Bonded-phase packings

$$\equiv Si{-}OH + Cl{-}Si(CH_3)_2{-}R \longrightarrow \equiv Si{-}O{-}Si(CH_3)_2{-}R$$

Most bonded-phase packings are prepared by reaction of an organochlorosilane with the OH groups formed on the surface of silica particles by hydrolysis in hot dilute hydrochloric acid. The product is an organosiloxane. The reaction for one such SiOH site on the surface of a particle can be written as where R is often a straight chain octyl- or octadecyl-group. Other organic functional groups that have been bonded to silica surfaces include aliphatic amines, ethers, and nitriles as well as aromatic hydrocarbons. Thus, many different polarities for the bonded stationary phase are available.

organochlorosilane
有机氯硅烷

octyl
[ˈɒktl]
n.辛基

octadecyl
十八烷基

4.2.2 Normal and reversed-phase packings

Two types of partition chromatography are distinguishable based on the relative polarities of the mobile and stationary phases. Early work in liquid chromatography was based on highly polar stationary phases such as triethylene glycol or water; a relatively nonpolar solvent such as hexane or *i*-propyl ether then served as the mobile phase. For historic reasons, this type of chromatography is now

glycol
[ˈglaɪkəʊl]
n.乙二醇，甘醇

called normal-phase chromatography. In reversed-phase chromatography, the stationary phase is non-polar, often a hydrocarbon, and the mobile phase is a relatively polar solvent (such as water, methanol, acetonitrile, or tetrahydrofuran).

In normal-phase chromatography, the least polar component is eluted first; increasing the polarity of the mobile phase then decreases the elution time. In contrast, with reversed-phase chromatography, the most polar component elutes first, and increasing the mobile phase polarity increases the elution time.

It has been estimated that more than three-quarters of all HPLC separations are currently performed with reversed-phase, bonded, octyl- or octyldecyl siloxane packings. With such preparations, the long-chain hydrocarbon groups are aligned parallel to one another and perpendicular to the surface of the particle, giving a brushlike, nonpolar hydrocarbon surface. The mobile phase used with these packings is often an aqueous solution containing various concentrations of such solvents as methanol, acetonitrile, or tetrahydrofuran.

perpendicular [pɜːpənˈdɪkjələ(r)] *adj*.垂直的

brushlike *n*.刷子

Ion-pair chromatography is a subset of reversed-phase chromatography in which easily ionizable species are separated on reversed-phase columns. In this type of chromatography, an organic salt containing a large organic counterion, such as a quarternary ammonium ion or alkyl sulfonate, is added to the mobile phase as an ion-pairing reagent. Two mechanisms for separation are postulated. In the first, the counterion forms an uncharged ion pair with a solute ion of opposite charge in the mobile phase. This ion pair then partitions into the nonpolar stationary phase giving differential retention of solutes based on the affinity of the ion pair for the two phases. Alternatively, the counterion is retained strongly by the normally neutral stationary phase and imparts a charge to this phase. Separation of organic solute ions of the opposite charge then occurs by formation of reversible ion-pair complexes with the more strongly retained solutes forming the strongest complexes with the stationary phase. Some unique separations of both ionic and nonionic compounds in the same sample can be accomplished by this form of partition chromatography. Figure 4. 19 illustrates the separation of ionic and nonionic compounds using alkyl sulfonates of various chain lengths as ion-pairing agents. Note that a mixture of C_5- and C_7-alkyl sulfonates gives the best separation results.

counterion [ˈkaʊntəaɪən] *n*.带相反电荷的离子，抗衡离子，补偿离子

complex [ˈkɒmpleks] *n*. 复合物，配合物

sulfonate [ˈsʌlfəneɪt] *n*.硅酸盐

In normal-phase partition chromatography, the stationary phase is polar and the mobile phase nonpolar. In reversed-phase partition chromatography, the polarity of these phases is reversed.

In normal-phase chromatography, the least polar analyze is

eluted first. In reversed-phase chromatography, the least polar analyte is eluted last.

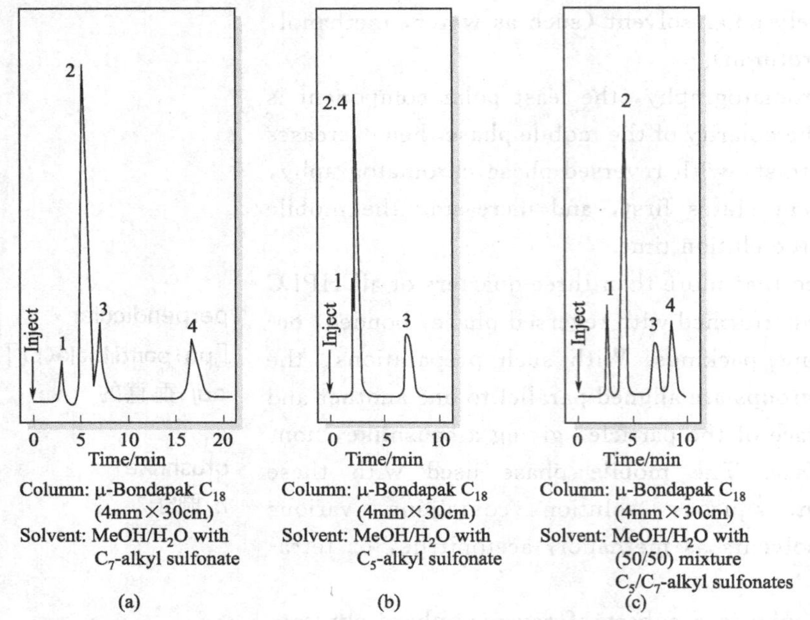

Figure 4.19 Chromatograms illustrating separations of mixtures of ionic and nonionic compounds by ion-pair chromatography. Compounds: 1—niacinamide, 2—pyridoxine, 3—riboflavin, and 4—thiamine. At pH 3. 5, niacinamide is strongly ionized, while riboflavin is nonionic. Pyridoxine and thiamine are weakly ionized. Column: *μ*-Bondapak, C_{18}, 4mm×30cm. Mobile phase: (a) MeOH/H_2O with C_7-alkyl sulfonate, (b) MeOH/H_2O with C_5-alkyl sulfonate, and (c) MeOH/H_2O with 1 : 1 mixture of C_5- and C_7-alkyl sulfonate

4.2.3 Choice of mobile and stationary phases

Successful partition chromatography requires a proper balance of intermolecular forces among the three participants in the separation process: the analyte, the mobile phase, and the stationary phase. These intermolecular forces are described qualitatively in terms of the relative polarity possessed by each of the three components. In general, the polarities of common organic functional groups in increasing order are hydrocarbons, ethers, ketones, aldehydes, amides, alcohols. Water is more polar than compounds containing any of the preceding functional groups.

Often, in choosing a column and mobile phase, the polarity of the stationary phase is matched roughly with that of the analyzes; a mobile phase of considerably different polarity is then used for elution. This procedure is generally more successful than one in which the polarities of the analyze and the mobile phase are matched but are different from that of the stationary phase. In this latter case, the sta-

ether
[ˈi:θə(r)]
n.醚类

ketone
[ˈki:təʊn]
n.酮类

aldehyde
[ˈældɪhaɪd]
n.醛类

tionary phase cannot compete successfully for the sample components; retention times then become too short for practical application. At the other extreme is the situation where the polarities of the analyze and stationary phase are too much alike; then, retention times become inordinately long.

tetrahydrofuran
[tetrəhaɪdrə'fjʊərən]
n. 四氢呋喃

The order of polarities of common mobile phase solvents are water, acetonitrile, methanol, ethanol, tetrahydrofuran, propanol, cyclohexane and hexane.

cyclohexane
[saɪkləʊ'heksem]
n. 环已烷

Acetonitrile (CH_3CN) is a widely used organic solvent. Its use as an LC mobile phase stems from its being more polar than methanol but less polar than water.

4.2.4 Applications

Figure 4.20 illustrates typical applications of bonded-phase partition chromatography for separating soft drink additives and organophosphate insecticides. Table 4.2 further illustrates the variety of samples to which the technique is applicable.

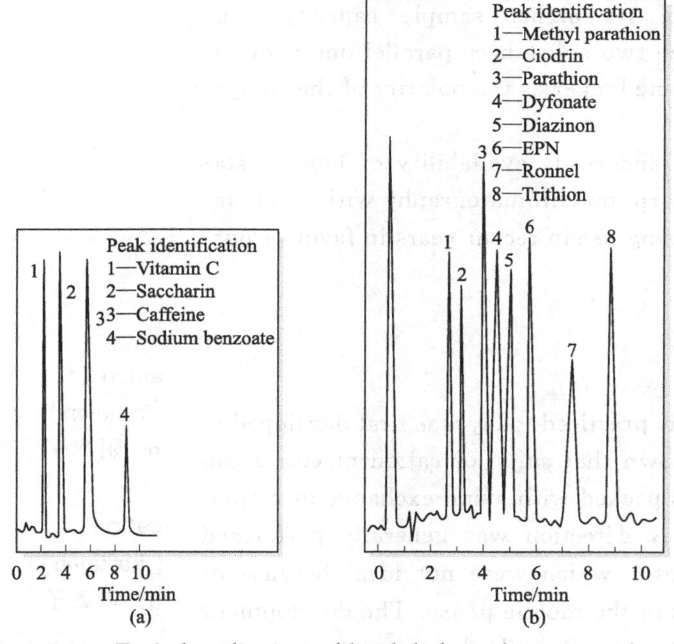

Figure 4.20 Typical applications of bonded-phase chromatography. (a) Soft drink additives. Column: 4.6mm×250mm packed with polar (nitrile) bonded- phase packing. Isocratic elution with 6% HOAc/94% H_2O. Flow rate: 1.0mL/min. (Courtesy of BTR Separations, a DuPont ConAgra affiliate.) (b) Organophosphate insecticides. Column: 4.5mm ×250mm packed with 5μm C_8 bonded-phase particles. Gradient elution: 67% CH_3OH/4% H_2O to 80% CH_3OH/20% H_2O. Flow rate: 2mL/min. Both used 254nm UV detectors

organophosphate
[ɔ:gənəʊ'fɒsfeɪt]
n.有机磷酸酯

Adsorption, or liquid-solid, chromatography is the classic form of liquid chromatography first introduced by Tswett at the beginning of the twentieth century. Because of the strong overlap between normal-phase partition chromatography and adsorption chromatography, many of the principles and techniques used for the former apply to adsorption chromatography. In fact, in many normal-phase separations, adsorption/displacement processes govern retention.

Table 4.2 Typical Applications of High-Performance Partition Chromatography

Field	Typical Mixtures Separated
Pharmaceuticals	Antibiotics, sedatives, steroids, analgesics
Biochemical Food	Amino acids, proteins, carbohydrates, lipids
products	Artificial sweeteners, antioxidants, aflatoxins, additives
Industrial chemicals	Condensed aromatics, surfactants, propellants, dyes
Pollutants	Pesticides, herbicides, phenols, polychlorinated biphenyls (PCBs)
Forensicscience	Drugs, poisons, blood alcohol, narcotics
Clinical chemistry	Bile acids, drug metabolites, urine extracts, estrogens

biochemical [ˌbaɪəʊˈkemɪkl] *adj*. 生物化学的

Finely divided silica and alumina are the only stationary phases that find use for adsorption chromatography. Silica is preferred for most applications because of its higher sample capacity. The adsorption characteristics of the two substances parallel one another. For both, retention times become longer as the polarity of the analyze increases.

alumina [əˈluːmɪnə] *n*. 氧化铝

Because of the versatility and ready availability of bonded stationary phases, traditional adsorption chromatography with solid stationary phases has seen decreasing use in recent years in favor of normal-phase chromatography.

4.3 Ion chromatography

Ion chromatography as it is practiced today was first developed in the mid-1970s when it was shown that anion or cation mixtures can be resolved on HPLC columns packed with anion-exchange or cation-exchange resins. At that time, detection was generally performed with conductivity measurements, which were not ideal because of high electrolyte concentrations in the mobile phase. The development of low-exchange-capacity columns allowed the use of low-ionic-strength mobile phases that could be further deionized (ionization suppressed) to allow high sensitivity conductivity detection. Currently, several other detector types are available for ion chromatography, including spectrophotometric and electrochemical.

anion [ˈænaɪən] *n*. 阴离子

cation [ˈkætaɪən] *n*. 阳离子

eletrochemical [ɪˌlektrəʊˈkemɪkəl] *adj*. 电化学的

Two types of ion chromatography are currently in use: suppressor-based and single-column. They differ in the method used to prevent the conductivity of the eluting electrolyte from interfering with the measurement of analyze conductivities.

The conductivity detector is well suited for ion chromatography.

4.3.1 Ion chromatography based on suppressors

Conductivity detectors have many of the properties of the ideal detector. They can be highly sensitive, they are universal for charged species, and as a general rule, they respond in a predictable way to concentration changes. Furthermore, such detectors are simple to operate, inexpensive to construct and maintain, easy to miniaturize, and usually give prolonged, trouble-free service. The only limitation to the use of conductivity detectors, which delayed their general application to ion chromatography until the mid-1970s, was due to the high electrolyte concentrations required to elute most analyze ions in a reasonable time. As a result, the conductivity from the mobile-phase components tends to swamp that from the analyze ions, greatly reducing the detector sensitivity.

predictable [prɪ'dɪktəbl] *adj*. 可预见的

In 1975, the problem created by the high conductance of eluents was solved by the introduction of an eluent suppressor column immediately following the ion-exchange column. The suppressor column is packed with a second ion-exchange resin that effectively converts the ions of the eluting solvent to a molecular species of limited ionization without affecting the conductivity due to analyze ions. For example, when cations are being separated and determined, hydrochloric acid is chosen as the eluting reagent, and the suppressor column is an anion-exchange resin in the hydroxide form. The product of the reaction in the suppressor is water, that is

suppressor [sə'presə(r)] *n*. 抑制物

$$H^+(aq) + Cl^-(aq) + resin^+OH^-(s) \longrightarrow resin^+Cl^-(s) + H_2O$$

The analyze cations are not retained by this second column.

For anion separations, the suppressor packing is the acid form of a cation-exchange resin, and sodium bicarbonate or carbonate is the eluting agent. The reaction in the suppressor is

$$Na^+(aq) + HCO_3^-(aq) + resin^-H^+(s) \longrightarrow resin^-Na^+(s) + H_2CO_3(aq)$$

In suppressor-based ion chromatography, the ion-exchange column is followed by a suppressor column, or a suppressor membrane, that converts an ionic eluent into a nonionic species that does not interfere with the conductometric detection of analyte ions.

conductometric [kən'dʌktɒmetrɪk] *adj*. 热导计的，电导计的

The largely undissociated carbonic acid does not contribute significantly to the conductivity

undissociated ['ʌndɪ'səʊʃɪeɪtɪd] *adj*.（分子）未解离的

An inconvenience associated with the original suppressor columns was the need to regenerate them periodically (typically every 8 to 10h) in order to convert the packing back to the original acid or base form. In the 1980s, however, micromembrane suppressors that operate continuously became available. For example, where sodium carbonate or bicarbonate is to be removed, the eluent is passed over a series of ultrathin cation-exchange membranes that separate it from a stream of acidic regenerating solution that flows continuously in the opposite direction. The sodium ions from the eluent exchange with hydrogen ions on the inner surface of the exchanger membrane and then migrate to the other surface for exchange with hydrogen ions from the regenerating reagent. Hydrogen ions from the regeneration solution migrate in the reverse direction, thus preserving electrical neutrality. Micromembrane separators are capable of removing essentially all the sodium ions from a 0.1mol/L NaOH solution with an eluent flow rate of 2mL/min (Figure 4.21).

ultrathin
[ˈʌltrəθɪn]
adj.超薄的,极细的

Concentrations/ppm	
F^-	3
Formate	8
BrO_3^-	10
Cl^-	4
NO_2^-	10
HPO_4^{2-}	30
Br^-	30
NO_3^-	30
SO_4^{2-}	25

Concentrations/ppm	
Ca^{2+}	3
Mg^{2+}	3
Sr^{2+}	10
Ba^{2+}	25

Figure 4.21 Two applications of ion chromatography based on a suppressor column and conductometric detection. In each, the ions were present in the parts-per-million range; the sample size was 50mL in one case and 100mL in the other. The method is particularly important for anion analysis because there is no other rapid and convenient method for handling mixtures of this type

4.3.2 Single-column ion chromatography

Commercial ion chromatography instrumentation that requires no suppressor column is also available. This approach depends on the small differences in conductivity between sample ions and the prevailing eluent ions. To amplify these differences, low-capacity exchangers are used that permit elution with solutions with low electrolyte concentrations. Furthermore, eluents of low conductivity are chosen. Single-column ion chromatography offers the advantage of not requiring special equipment for suppression. However, it is a somewhat less sensitive method for determining anions than suppressor-column methods.

In single-column ion-exchange chromatography, analyte ions are separated on a low-capacity ion exchanger by means of a low-ionic strength eluent that does not interfere with the conductometric detection of analyte ions.

In size-exclusion chromatography, fractionation is based on molecular size.

4.4 Size-exclusion chromatography

Size-exclusion, or gel chromatography, is a powerful technique that is particularly applicable to high-molecular-mass species. Packings for size-exclusion chromatography consist of small (10mm) silica or polymer particles containing a network of uniform pores into which solute and solvent molecules can diffuse.

While in the pores, molecules are effectively trapped and removed from the flow of the mobile phase. The average residence time of analyte molecules depends on their effective size. Molecules that are significantly larger than the average pore size of the packing are excluded and thus suffer no retention, that is, they travel through the column at the rate of the mobile phase. Molecules that are appreciably smaller than the pores can penetrate throughout the pore maze and are thus entrapped for the greatest time; they are last to elute. Between these two extremes are intermediate-size molecules whose average penetration into the pores of the packing depends on their diameters. The fractionation that occurs within this group is directly related to molecular size and, to some extent, molecular shape. Note that size-exclusion separations differ from the other chromatographic procedures in the respect that there are no chemical or physical interactions between analytes and the stationary phase. Indeed, such interactions are avoided because they lead to lower column efficiencies. Note also that, unlike other forms of chromatography, there is an

maze
[meɪz]
n. 混乱

entrap
[ɪn'træp]
vt. 诱使,诱捕

fractionation
[ˌfrækʃən'eɪʃən]
n. 分馏法

upper limit to retention time because no analyte species is retained longer than those small molecules that totally permeate the stationary phase.

size
[saɪz]
n. 尺寸

4.4.1 Column packing

Two types of packing for size-exclusion chromatography are encountered: polymer beads and silica-based particles, both of which have diameters of 5 to 10mm. Silica particles are more rigid, which leads to easier packing and permits higher pressures to be used. They are also more stable, allowing a great range of solvents to be used and exhibiting more rapid equilibration with new solvents.

exclusion
[ɪk'sklu:ʒn]
n. 排除

Gel permeation is a type of size-exclusion chromatography in which the packing is hydrophobic. It is used to separate nonpolar species. Numerous size-exclusion packings are on the market. Some are hydrophilic for use with aqueous mobile phases; others are hydrophobic and are used with nonpolar organic solvents. Chromatography based on the hydrophilic packings is sometimes call gel filtration, while that based on hydrophobic packings is termed gel permeation. With both types of packings, many pore diameters are available. Generally, a given packing will accommodate a 2- to 2.5-decade range of molecular mass. The average molecular mass suitable for a given packing may be as small as a few hundred or as large as several million.

hydrophilic
[hərdrə'fɪlɪk]
adj. 亲水的

hydrophobic
[hərdrə'fəubɪk]
adj. 疏水的

4.4.2 Applications

Figure 4.22 illustrates typical applications of size-exclusion chromatography. Both chromatograms were obtained with hydrophobic packings in which the eluent was tetrahydrofuran. In Figure 4.22(a), the separation of fatty acids with molecular mass M from 116 to 344 is shown. In Figure 4.22(b), the sample was a commercial epoxy resin in which each monomer unit had a molecular mass of 280 (5 number of monomer units).

Another important application of size-exclusion chromatography is the rapid determination of the molecular mass or the molecular mass distribution of large polymers or natural products. The key to such determinations is an accurate molecular mass calibration. Calibrations can be accomplished by means of standards of known molecular mass (peak position method) or by the universal calibration method. The latter method relies on the principle that the product of the intrinsic molecular viscosity h and molecular mass M is proportional to hydrodynamic volume (effective volume including solvation sheath). Ideally, molecules are separated in size-exclusion chromatography according to hydrodynamic volume. Hence, a universal cali-

accurate
['ækjərət]
adj. 精确的，准确的

bration curve can be obtained by plotting lg[hM] versus the retention volume, V_r. Alternatively, absolute calibration can be achieved by using a molar mass-sensitive detector such as a low-angle, light-scattering detector.

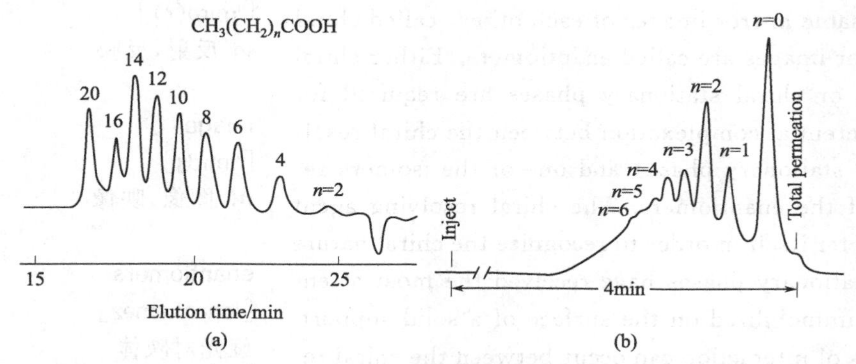

Figure 4.22 Applications of size-exclusion chromatography. (a) Separation of fatty acids. Column: polystyrene based, 7.5mm×600mm. Mobile phase: tetrahydrofuran. (b) An analysis of a commercial epoxy resin. (n=number of monomeric units in the polymer). Column: porous silica 6.2mm×250mm. Mobile phase: tetrahydrofuran

4.5 Affinity chromatography

In affinity chromatography, a reagent called an affinity ligand is covalently bonded to a solid support. Typical affinity ligands are antibodies, enzyme inhibitors, or other molecules that reversibly and selectively bind to analyte molecules in the sample. When the sample passes through the column, only the molecules that selectively bind to the affinity ligand are retained. Molecules that do not bind pass through the column with the mobile phase. After the undesired molecules are removed, the retained analytes can be eluted by changing the mobile phase conditions.

The stationary phase for affinity chromatography is a solid, such as agarose, or a porous glass bead to which the affinity ligand is immobilized. The mobile phase in affinity chromatography has two distinct roles to play. First, it must support the strong binding of the analyte molecules to the ligand. Second, once the undesired species are removed, the mobile phase must weaken or eliminate the analyte-ligand interaction so that the analyte can be eluted. Often, changes in pH or ionic strength are used to change the elution conditions during the two stages of the process.

ligand
[ˈlɪgənd]
n.配合基,向心配合(价)体,配体

Affinity chromatography has the major advantage of extraordinary specificity. The primary use is in the rapid isolation of biomolecules during preparative work.

4.6 Chiral chromatography

Tremendous advances have been made in separating compounds that are nonsuperimposable mirror images of each other, called chiral compounds. Such mirror images are called enantiomers. Either chiral mobile-phase additives or chiral stationary phases are required for these separations. Preferential complexation between the chiral resolving agent (additive or stationary phase) and one of the isomers results in a separation of the enantiomers. The chiral resolving agent must have chiral character itself in order to recognize the chiral nature of the solute. Chiral stationary phases have received the most attention. A chiral agent is immobilized on the surface of a solid support. Several different modes of interaction can occur between the chiral resolving agent and the solute. In one type, the interactions are due to attractive forces such as those between π bonds, hydrogen bonds, or dipoles. In another type, the solute can fit into chiral cavities in the stationary phase to form inclusion complexes. No matter what the mode, the ability to separate these very closely related compounds is of extreme importance in many fields. Figure 4.23 shows the separation of racemic mixtures of an ester on a chiral stationary phase. Note the excellent resolution of the R and S enantiomers.

A chiral resolving agent is a chiral mobile-phase additive or a chiral stationary phase that preferentially complexes one of the enantiomers.

nonsuperimposable
不重合

mirror
[ˈmɪrə(r)]
vt.反射，反映

image
[ˈɪmɪdʒ]
n. 图像，映像

enantiomers
[ˈenəntɪəməz]
旋光对映体

inclusion
[ɪnˈkluːʒn]
n.包含，内含物

racemic
[rəˈsiːmɪk]
adj.外消旋的

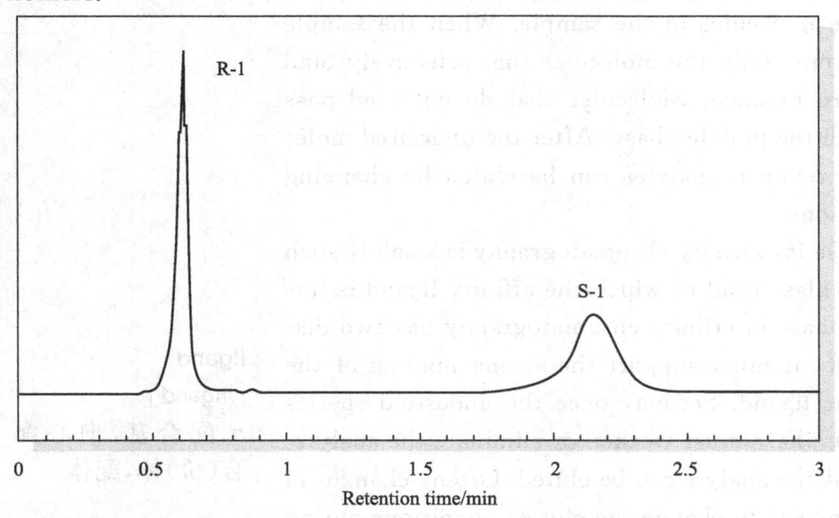

Figure 4.23 Chromatogram of a racemic mixture of *N*-(1-Naphthyl) leucine ester 1 on a dinitrobenzene- leucine chiral stationary phase. The *R* and *S* enantiomers are seen to be well separated. Column: 4.6mm× 50mm. Mobile phase: 20% 2-propanol in hexane. Flow rate: 1.2mL/min; UV detector at 254nm

Table 4.3 provides a comparison between high-performance liquid chromatography and gas-liquid chromatography. When either is applicable, GC offers the advantage of speed and simplicity of equipment. On the other hand, HPLC is applicable to nonvolatile substances (including inorganic ions) and thermally unstable materials, but GC is not. Often the two methods are complementary.

nonvolatile
[nɒn'vɒlətaɪl]
adj. 非挥发性的，不挥发的

Table 4.3 Comparison of High Performance Liquid Chromatography and Gas-Liquid Chromatography

Characteristics of Both Methods	Efficient, highly selective, widely applicable
	Only small sample required
	May be nondestructive of sample
	Readily adapted to quantitative analysis
Advantages of HPLC	Can accommodate nonvolatile and thermally unstable compounds
	Generally applicable to inorganic ions
Advantages of GC	Simple and inexpensive equipment
	Rapid
	Unparalled resolution (with capillary columns)
	Easy to interface with mass spectrometry

nondestrutive
[nɒndɪstr'ʌktɪv]
adj. 非破坏性的

Problems

4.1 List the types of substances to which each of the following chromatographic methods is most applicable:

(a) gas-liquid

(b) ion

(c) gel permeation

(d) chiral

4.2 Define

(a) isocratic elution

(b) normal-phase packing

(c) bonded-phase packing

(d) ion-pair chromatography

(e) gel filtration

4.3 Indicate the order in which the following compounds would be eluted from an HPLC column containing a reversed-phase packing: benzene, diethyl ether, *n*-hexane.

4.4 Indicate the order of elution of the following compounds from a normal-phase packed HPLC column: ethyl acetate, acetic acid, dimethylamine.

4.5 Describe the fundamental difference between adsorption and partition chromatography.

4.6 Describe the difference between gel-filtration and gel permeation chromatography.

4.7 Describe the differences between single-column and suppressor-column ion chromatography.

4.8 Although temperature does not have nearly the effect on HPLC separations that it has on GC separations, it nonetheless can play an important role. Discuss how and why temperature might or might not influence the following separations:

(a) a reversed-phase chromatographic separation of asteroid mixture.

(b) an adsorption chromatographic separation of a mixture of closely related isomers.

第 4 章
高效液相色谱法

高效液相色谱法（HPLC）已成为不可缺少的分析工具。本章内容涉及 HPLC 的理论和实践，包括分配色谱、吸附色谱、离子交换色谱、体积排阻色谱、亲和色谱和手性色谱法。HPLC 不仅应用在法医学科，而且应用于生物化学、环境科学、食品科学、药物化学和毒理学。

高效液相色谱法是最灵活和使用最广泛的洗脱色谱类型。科学家将该技术应用于不同的有机、无机和生物材料的分离、定性和定量。在液相色谱中，流动相是液体溶剂，是包含样品的混合溶液。高效液相色谱法根据分离机制或固定相类型分类，包括：分配或液-液色谱，吸附或液-固色谱，离子交换或离子色谱，体积排阻色谱，亲和色谱，手性色谱。

早期的液相色谱，在内径为 10～50mm 的玻璃柱中进行，柱子填充 50～500cm 长涂敷有吸附液体的固体颗粒组成的固定相。为确保流动相能以合适的流速通过固定相，固体颗粒的尺寸应控制在 150～200μm。即使用这类大小的颗粒，流速最好为十分之几毫升/分钟。因为流速增加，伴随着塔板高度增加和柱效降低，所以，对液相色谱，通过运用真空或加压来加快流速没有意义。

液相色谱理论发展的早期人们已经认识到，填料粒径减小，塔板高度会降低。这一作用可由图 4.1 看出。

直到 20 世纪 60 年代末，由于科技的发展，可以生产和使用直径为 3～10μm 的颗粒填料，但需要设备能够承受更高的泵压。同时，连续监测柱流出物的检测器也问世了。经常用高效液相色谱（HPLC）将这种技术区别于以前简单的色谱技术。

但是，简单的色谱柱仍然能够用于以制备为目的的领域。图 4.2 为高效液相色谱法在各种类型分析物中的应用。值得注意的是，各种类型的液相色谱在应用上往往互补。例如，对于分子量大于 10000 的样品，经常使用以下两种体积排阻方法中的一种：对于非极性物质使用凝胶渗透色谱，对于极性或离子化合物使用凝胶过滤色谱。对于离子型物质，离子交换色谱法是经常选用的方法。在大多数情况下，对于非离子型小分子，反相色谱法较为合适。

高效液相色谱法是一种将液体流动相和精细分类的固定相结合的色谱技术。为了获得理想的流速，液体必须加压到几百磅每平方英寸或更高。模块式高效液相色谱仪的原理参见图 4.3。

4.1 仪器

为了在 3～10μm 尺寸范围内的填料中达到理想的流速，几百个大气压的泵压是必不可少的，这在现代液相色谱中很常见。由于高压，应用于高效液相色谱的设备比其他类型色谱法所使用的设备更加精密且昂贵。图 4.4 为 HPLC 仪器上的重要组成部分。

4.1.1 流动相储液瓶和溶剂处理系统

现代 HPLC 仪器配备一个或多个玻璃储瓶，每个都盛有 500mL 或更多的溶剂。溶剂需预处理以除去溶解的气体和尘埃。溶解的气体可导致流速不稳定及色谱峰扩展。另外，气泡和灰尘干扰很多检测器的性能。脱气设备包括真空泵系统、蒸馏系统、一种用于加热和搅拌的装置，或者如图 4.4 所示，一种喷射系统，该系统使用惰性气体小气泡将可溶性的气体赶出流动相，这些惰性气体不溶于流动相。

单一溶剂的洗脱或具有恒定组成溶剂混合物的洗脱称为等度洗脱。在梯度洗脱下，分离过程中经常使用两种（或多种）极性差别很大的溶剂体系，并经常改变其组成。在预编程方法中，两种溶剂的比是不同的，有时连续，有时是一系列的步骤。如图 4.5 所示，梯度洗脱可以提高分离效率，正如温度程序在气相色谱中的作用。现代 HPLC 仪器通常配备比例阀，可以从两个或多个储瓶中连续地引入比例不断变化的液体（见图 4.4）。HPLC 等度洗脱中溶剂组分保持不变。HPLC 梯度洗脱中溶剂的组成连续地改变或按照一系列步骤改变。

4.1.2 泵系统

用于液相色谱的泵需要达到如下要求：①产生的压力高达 6000lb/in^2（psi）；②无脉冲输出；③流动相的流速为 0.1～10mL/min；④流动相的重复性相对偏差为 0.5%或更小；⑤耐多种溶液腐蚀。因为液体是不可压缩的，液相色谱泵产生的高压不会发生爆炸危险。组件破裂仅会导致溶剂泄漏，但溶剂泄漏可能造成火灾或环境危害。

HPLC 仪器主要使用两种类型的泵：螺杆驱动注射泵和往复泵。商品化设备中几乎都使用往复式类型。注射泵工作产生无脉冲输送，且流速容易控制。但是，由于注射型泵容量（250mL）相对较低，因此须改变溶剂时会非常不方便。往复泵的工作原理如图 4.6 所示。往复泵装置由一个小的圆柱形腔室组成，这个圆柱形腔室通过一个活塞的往复运动使其“先充满再清空”。泵的抽吸运动会产生脉冲流，随后必须将其抑制，否则脉冲会使色谱图上出现基线噪声。现代 HPLC 仪器采用双泵头或椭圆形凸轮尽量减少这种波动。往复泵的优点包括内部容积小（35～400mL），输出压力高（高达 10000lb/in^2），梯度洗脱适应性强，流速恒定，这在很大程度上与色谱柱背压和溶剂黏度无关。

计算机控制装置作为泵系统的一部分在许多商业仪器中都有配备，通过测量放置在泵出口限流器的压力降来确定流量。通过测量值与预设值的信号差异来增大或减小泵电动机的速度。大多数仪器还能够连续地改变溶剂组成或逐步改变溶剂组成。例如，在图 4.4 所示的仪器包括一个比例阀，它能够通过预编程和连续可变的方式将四种溶剂混合。

4.1.3 进样系统

如图 4.7 所示，在液相色谱中，进样环是引入样品最常用的方法。通常，这些装置是液相色谱设备的一个组成部分，并有可互换的进样环，能提供范围为 1～100mL 或更高体积的样品。典型进样环的进样重现性相对偏差低于百分之零点几。许多 HPLC 仪器将自动进样装置与自动进样器结合。进样器能够从进样瓶引入不同体积的样品。

处于进样器和柱子之间的保护柱用来去除小颗粒和其他溶剂中的杂质。

4.1.4 HPLC 柱子

液相色谱柱通常采用不锈钢管，有时还使用玻璃和聚合物管材，如聚醚醚酮（PEEK）。此外，也使用内衬玻璃或 PEEK 的不锈钢柱。数百种不同尺寸和不同填料的填充柱都可以从 HPLC 供应商处购得。标准尺寸但非专用的色谱柱的费用为 200 美元到 500 美元及以上。专用色谱柱，如手性柱，费用超过 1000 美元。

（1）分析柱

大多数柱子的长度为 5～25cm，内径为 3～5mm，并且使用的柱子都是直柱。最常用的填料粒径为 3μm 或 5μm。常用柱子的长度为 10cm 或 15cm，内径为 4.6mm，填充颗粒的大小为 5μm。这种类型的柱子提供 40000～70000 塔板数/米的柱效。

到 20 世纪 80 年代，开始使用内径为 1～4.6mm 和长为 3～7.5cm 的微柱。微柱中填充有 3μm 或 5μm 的颗粒，理论塔板数达 100000 板/米，并具有速度快、溶剂消耗小的优点。溶剂消耗小这个性质是相当重要的，因为液相色谱用高纯度溶剂非常昂贵并且使用后需要处理。图 4.8 显示了微柱分离的速度。在这个例子中，MS/MS 用于监测从人血浆成分中分离罗苏伐他汀，这个柱子的长度为 5cm，内径为 1.0mm。该柱填充有 3μm 的颗粒。分离完成仅需不到 3min 的时间。

（2）预柱

预柱的类型有两种。一种存在于流动相储瓶和喷射器之间，用于控制流动相，被称为冲洗柱。溶剂能溶解部分二氧化硅填料，并确保流动相在进入分析柱之前为硅酸所饱和。这种饱和减少了分析柱内固定相的损失。

第二种类型的预柱是保护柱，位于进样器和分析柱之间。保护柱是一种短柱，里面装填了相似于分析柱的固定相。保护柱的目的是为了防止杂质到达而污染分析柱，这些杂质包括长时间保留的化合物和颗粒物质。为了增加分析柱的寿命，要定期更换保护柱。

（3）柱温控制

对于一些应用，没有必要严格控制柱温，操作大多在室温下进行。然而，通常维持恒定的柱温可以延长色谱柱的寿命。大多数现代商业仪器配备有温度控制装置来控制柱温在室温至 150℃之间。也可以给柱子装上水套，水套可以由恒温水浴来提供水，能够准确地控制温度。许多色谱分析学家认为温度的控制是分离重现性必不可少的。

（4）色谱柱填料和硬件

HPLC 使用的填料有两种，薄壳和多孔颗粒。原始薄壳颗粒是内径为 30～40μm 的球状无孔玻璃或聚合物珠粒。薄且多孔的二氧化硅、氧化铝、聚苯乙烯-二乙烯基苯合成树脂或离子交换树脂附着在这些颗粒的表面上。小而多孔的微粒已经完全取代了这些大而薄壳的颗粒。近几年，粒径为 5μm 的薄壳填料已再次应用到蛋白质和生物大分子的分离上。

典型的液相色谱多孔颗粒填料是直径范围为 3～10μm 的多孔微粒；对于给定尺寸的颗粒，尺寸范围波动越小的颗粒越理想。这些颗粒包括二氧化硅、氧化铝、合成树脂的聚苯乙烯-二乙烯基苯或离子交换树脂。迄今为止，二氧化硅是液相色谱中最常用的填料，有机薄膜通过化学或物理键合的方式涂覆到二氧化硅表面。特殊模式的色谱柱填料会在本章后面的章节中讨论。

在范第姆特方程中描述的液相色谱中导致谱带展宽的三种主要原因是涡流扩散、分子扩散和传质阻力。每个过程对柱效的影响与很多变量有关，包括流动相流速（或线速度）、柱长、平均粒径、粒径分布、柱结构、填料孔隙率、溶剂黏度、填料形状、柱床的完整性、进样的质量和体积、化学相互作用的性质等。在本节中，将讨论柱填料、结构、硬件等情况以及它们对色谱性能的影响。

现代高效液相色谱的填料可以根据颗粒的类型、大小、形状、性质和材料进行广泛的分类。色谱柱中使用了各种不同的颗粒类型。薄壳粒子有一个坚实的内核和外表面的一薄层固定相。或者，涂覆固定相的外层可以是多孔的，形成表面多孔颗粒。这些薄壳材料由较大直径的球形玻璃珠（例如 30μm）和薄的多孔二氧化硅表面层（大约 1μm）组成，如典型的杜邦 Zipax 系列柱。多孔颗粒这样的材料有比较大的直径（例如 30μm），它们是全多孔的结构，形状可以是不规则的，也可以是球形的。微粒材料有较小的直径（如 3～10μm），全多孔，形状不规则或球形。图 4.9 展示了现代高效液相色谱中使用的一些颗粒类型。

相同尺寸的薄壳材料相比多孔颗粒具有显著提高的效率（或更低的 HETP 值），因此可以减小范第姆特方程的 C 项（传质阻力）。由于流动相传质阻力随粒径的减小而减小，微粒（即小直径）材料相比大的多孔材料拥有更高的效率。较小的颗粒也减少了 A 项（涡流扩散）的贡献，再次导致柱效率的提高。薄壳和微粒材料可以提供类似的色谱效率；然而，薄壳材料因为低的活性表面积仅适用于小的上样量，但与微粒材料相比，更容易填入色谱柱。球形和不规则微粒材料的色谱性能与效率方面基本相同，但不规则颗粒床层的结构不如致密的良好填充的球形颗粒床层稳定，这使得不规则颗粒填充柱的制备更加困难，而且也限制了在高压下连续操作时它们的稳定性。一般来说，如果最小化固定相的厚度，能得到最高的柱效率。使用小颗粒填充色谱柱，填充过程会形成致密、结实和均匀的柱床。目前大多数的液相色谱柱都填充了球形微粒材料（直径 3～10μm），以获得稳定高效的色谱柱，并可用于比较大的上样量。

此外，颗粒（即微孔或薄壳）的物理性质、大小和形状、颗粒材料也显著影响高效液相色谱填料的性能。备选的柱填料包括刚性固体（最常见的是二氧化硅）、树脂（通常是聚苯乙烯-二乙烯基苯）和软凝胶。

目前，最常用的 HPLC 填料是基于二氧化硅的刚性固体，特别是吸附和反相色谱模式。硅胶填料可以承受使用填充 3～10μm 颗粒的 10～30cm 长色谱柱时带来的高压力。各种形状、大小和孔隙度的二氧化硅均来源丰富。最重要的是，二氧化硅可以很容易地功能化，其键合反应的机理比较清楚。二氧化硅也有许多缺点，主要包括 pH 的不稳定性和高的表面活性。虽然包括氧化铝、氧化锆和碳等材料尚未被广泛接受，但是这些刚性颗粒已被用作 HPLC 柱填料。

树脂基填料在高效液相色谱柱中得到越来越广泛的应用。这些填料主要用于凝胶渗透色谱（GPC）和离子交换色谱，但树脂基反相柱也已商品化。树脂基柱具有优势，它们可以在很宽的 pH 范围内使用，虽然大多数的树脂类型只适合中等的操作压力（1000～2000psi）和流动相只能含低浓度的有机改性剂。软凝胶，如琼脂糖和葡聚糖，几乎完全用于分离含水蛋白质，当然，它们不能耐受非常高的背压。

(5) 柱填充方法

用于填充 HPLC 柱的两种最常用的方法是干法填充和湿法填充（或匀浆填充）过程。干法填充过程主要用于粒径大于 20μm 的刚性固体和树脂填料，如薄壳材料。这个过程包括脱脂，将柱管的内部干燥处理，形成空白柱。然后将多孔筛板（通常为 2μm）放置在柱的出口配件上，然后通过漏斗将少量填料放入垂直固定的柱体内。

敲击色谱柱使填料下沉，再加上更多的填料，直到色谱柱填满为止。然后平整填料，将带有筛板的进口端配件拧紧到柱的顶部。对于大尺寸多孔颗粒和薄壳材料，干法填充色谱柱可获得重复性的特点，但以这种方式填充尺寸较小的微粒时会导致柱效变差。因为高的表面能量与质量的比值，小颗粒倾向于结块，导致很差的压缩柱床，这导致沿柱方向的流速变化很大，产生明显的谱带展宽，因而效率低。干法填充通常用于制备用于制备色谱的色谱柱，也可制备用于样品净化的固相萃取柱。

使用湿法或匀浆法制备色谱柱的过程，是通过一种合适的液体来悬浮颗粒填料，然后在高压下将其泵入空柱中。必须选择合适的匀浆液，以保持颗粒的均匀分布避免聚集，而且能彻底地湿润填料。高表面能量材料，如未官能化硅胶，需要使用极性溶剂，而低表面能量填料，如 C_{18} 官能化的硅胶，可以在较低极性的溶剂中填充。匀浆液的密度也应考虑，特别是当填料粒子直径大于 10μm 时。对于这些较大的微粒，溶剂密度应该与粒子本身相似，以减少沉降，特别是当颗粒大小分布较大时。为了填充色谱柱，填料配成合适浓度的匀浆，并放置在一个特别设计的通常称为匀浆罐的容器中，匀浆罐与空柱的入口匹配。多孔筛板（1～2μm）放置在色谱柱的出口端，使用高效液相色谱恒压泵将溶剂高速注入匀浆罐，迫使填料匀浆进入色谱柱并产生一个紧凑的柱床。当柱中的流速恒定时，填充就完成了，在此之后，移走匀浆罐，平整填料，然后将带筛板的入口配件安装在色谱柱的顶端。色谱柱可以沿向上或向下的方向填充，大多数的色谱柱使用向下的方向填充。

刚性固体柱和树脂基柱都可以这种方式制备，但是树脂在填充之前必须先在溶剂中溶胀。由于树脂刚性降低，树脂基柱通常使用较低的填充压力。如果使用二氧化硅微粒，一个标准的 25cm 长、4.6mm 内径的高效液相色谱柱要用 2～3g 的填料。虽然空柱管和填充设备已经可以从像 Alltech 这样的色谱柱供应商处获得，但是绝大多数的液相色谱柱都是从供应商处购买预装填好的色谱柱。这样做主要有两个原因：一是制备重现性好、柱效高的色谱柱费时且需要具备相当的技巧；二是大多数色谱柱制造厂家不愿意大量销售散装分析级填料，而更倾向于卖预装柱。

（6）色谱柱构造和硬件

高效液相色谱柱的设计多年来一直是众多研究的目标，目前仍在不断发表关于这一目标的综述文章。色谱柱的构成材料必须能耐受液相色谱的高压，并且对流动相具有耐腐蚀性。与高效液相色谱泵一样，大多数色谱柱是由 316 不锈钢制成的。不锈钢柱可以采用刚性单壁或双壁结构。柱色谱也可使用惰性材料，如玻璃、聚四氟乙烯和聚醚醚酮（PEEK）。这种柱可用于化学腐蚀性流动相，如 HCl，或当样品（如某些蛋白质）可能吸附到不锈钢表面时。不锈钢材料具有很强的刚性，可以在高压下操作，因此在二氧化硅和键合相填料中应用最为广泛。聚合物柱材料通常用于离子交换填料，而玻璃柱主要用于蛋白质分离。聚合物和玻璃柱都会有严格的压力限制，在使用这种柱时，应注意不要超过生产商推荐的操作压力。

高效液相色谱柱可有各种各样的构造、形状和尺寸。大多数柱子都是用无死体积接头的直管填充的。必须始终注意确保加压螺钉和垫圈与柱端配件兼容。一般来说，为确保不会永久损坏柱子，一个制造商的加压螺钉不能与另一个制造商的色谱柱一起使用。卡套柱可作为标准钢柱的替代，例如 Brownlee 系列柱，这类色谱柱柱体可以被替换，旧的端口配件可以重复使用。虽然卡套柱的初始购买价格通常高于同等的钢柱，但更换卡套的成本较低，因此，如果用户经常需要更换色谱柱，使用卡套柱成本效益较高。

另一种替代传统钢柱的是径向压缩柱。在这种类型柱中，固定相装入聚四氟乙烯套筒，在使用前向套筒施加径向压力来压缩，通过压缩套筒周围的液体或气体来加压。对填料床层进行改造，每次使用时除去空隙和通道，获得更高的稳定性，提高耐久性和效率。这种类型的套筒减少了“壁效应”，这是因为在填充床密度较低的情况下，传统柱的刚性壁附近对溶剂流动的阻力较小，导致产生“壁效应”。这种壁效应是谱带展宽的一个重要来源，径向压缩柱的效率比等效钢柱高出25%。

除了选择填料和色谱柱硬件外，还必须选择适当的柱尺寸。生产商为它们的填料提供不同长度和直径的柱子。一般来说，柱越长，其效率和分辨率越高；柱越短，分离速度越快。另外，柱直径越大，其承载能力越高，而柱子越窄，其质量灵敏度越高。大多数分析柱的长度为50～300mm，直径为2～5mm。短柱的使用提供了快速分离，但限制了分辨力，因此，这种方法只应用于简单的样品。一些树脂基材料，例如聚甲基丙烯酸酯，为了保持柱压降尽可能低，常常仅装填为短柱。较长的色谱柱提供了更强的分辨能力，但以增加色谱运行时间为代价。当样品量是有限的或溶剂（和处置费用）量大时，应用直径1～2mm或更小的窄孔（或微孔）柱，以满足高灵敏度的要求。使用窄径柱时流速为0.05～1mL/min。当然，为了在这样的柱子上获得最好的结果，最小化系统内谱带展宽的来源是关键。图4.10针对ACTH合成片段混合物比较了使用3.9mm和2mm内径色谱柱时获得的灵敏度。

相反，如果需要比常规柱更大的上样量，则应使用内径较大的色谱柱。上样量与柱长和柱直径的平方成正比。大口径色谱柱是制备色谱中最常用的。总之，高效液相色谱柱柱效超过100万理论塔板数时，柱尺寸（7000cm长，0.02cm内径）和过长的运行时间（大于7d）使它们不适合日常使用。在实际应用中，虽然150mm长色谱柱越来越受欢迎，但最常用的色谱柱长度为250～300mm、直径为3.9～4.6mm。这个尺寸范围内的色谱柱提供了在效率、上样量和色谱运行时间等方面的最好折中，无需过度关注关于谱带展宽的仪器性能。

(7) 测量柱的性能

柱的性能可以用一些参数来评价，虽然传统的理论塔板数（N）是用来衡量柱效率的。其他重要参数包括峰不对称性、容量因子、选择性、关键峰的分离度、柱压降等。

理想情况下，应连续监测上述所有参数并记录数据。为给定的测试设置控制规范和限制，系统适配性的概念基于对色谱柱和整个系统性能的仔细监视和记录。这种质量方法变得越来越重要，特别是在制药实验室。色谱数据软件的最新版本现在能计算和自动记录各种系统参数，如上面列出的那些参数。除了指示系统是否符合检测规范外，对色谱柱和系统性能的连续监测提供了在色谱问题排除时参考的基准。它还允许用户获得更长期的数据，如色谱柱寿命，这有助于保持库存水平，甚至决定一支色谱柱是否持续了一个合理的进样次数。

保护柱仅适用于短时间应用，通常在50～200次的有限次数的进样后应更换。虽然它们可能会轻微地降低分离效率，但它们的使用将显著提高柱寿命。

在某些应用中色谱柱加热设备也将提高色谱柱性能。大多数高效液相色谱仪器制造商出售加热设备，根据具体设备，可以方便地在0.1～1℃范围内调节柱温。许多分离，如糖和氨基酸，需要升高的温度以达到适当的分辨率。此外，如果柱温稳定，特别是那些应用电导检测器或示差折光检测器时，高灵敏度的分析往往会获得改善。总之，经过良好处理的色谱柱可以持续几千次进样，但使用一个不适当的流动相或未经处理的样品可能会大大降低色谱柱的性能，仔细注意上面列出的要点将提高柱寿命。此外，对于色谱柱信息和故障排除，可以获得大量的帮助。许多高效液相色谱仪器制造商现在提供免费的电话支持来回答关于色谱柱和系统性能标准的问题。几乎所有的色谱柱都有一个（维护）手册，应该仔细遵循。仪器手册也是故障诊断信息的有用来源。

4.1.5 HPLC 检测器

除了不需要具有同样宽的温度范围，理想的高效液相色谱检测器应该具有理想气相色谱检测器的所有特征。此外，高效液相色谱检测器必须具有较低的内部容积（死体积），以最大限度地减小柱外的谱带展宽。检测器要小，并与液体流兼容。用于高效液相色谱的普通检测系统的不足之处在于灵敏度不高。因此，所使用的检测器取决于样品的性质。表 4.1 列出了一些常见检测器及其属性。

最广泛使用的液相色谱检测器是以紫外线或可见光吸收为基础的（见图 4.11）。专为色谱柱设计的光度计和分光光度计可从市场上购买。光度计通常使用波长为 254nm 和 280nm 的汞射线，因为很多有机官能团的吸收区域在该波段范围内。氘源或有干涉滤光片的钨丝灯源还可以提供一种简单的检测物质的方法。一些现代仪器设备包含多个干涉滤光片的滤光轮，滤光片可以迅速地切换位置。分光光度检测器要比光度计更灵活，也被广泛应用于高性能仪器。当分析物离开柱子时，现代仪器使用的二极管阵列检测器，它可以显示流出柱子的分析物的完整光谱。

（1）分光光度检测器

检测所依据的原理是朗伯-比耳定律（$A=\varepsilon_\lambda lc$）：在色谱柱的出口端使用一个或多个紫外或可见波长检测流动相的吸光度 A。吸收强度由被检测物质的摩尔吸光系数所决定。流动相必须没有吸收或者吸收很小（图 4.12）。

峰面积没有考虑摩尔吸光系数，所以通过简单地检查色谱图峰面积直接计算浓度是不行的。分光光度检测器是选择性检测的一种。对于不具有显著吸收光谱的化合物，可以在检测前进行分析物的衍生化。

单色检测：基本模块包括氘或汞蒸气光源、用于分离窄带宽（10nm）或特征谱线（例如，光源为汞灯时的 254nm）的单色仪、体积为几个微升的流通池（光路为 0.1～1cm）和光学检测装置。

多色检测：更高级的检测器能够在分析过程中改变波长，可以在几个波长上同时记录吸光度或者甚至以几分之一秒的速度记录整个波长范围的吸光度，而不用中断柱中的循环（见图 4.13～图 4.15）。二极管阵列检测器（DAD）不仅生成色谱图，还提供可用于识别分离化合物的光谱信息（见图 4.14），被称为特定检测。

连续记录用流动相洗脱的化合物的连续光谱并用仪器存储起来，之后使用合适的软件进行处理，通常可以获得很好的色谱图（见图 4.15）。在单次分析期间，记录数千个光谱的能力增加了这些检测器系统的应用潜力。据此可导出表示分离的等高线图（等吸收图），$A=f(\lambda t)$。

生物技术的快速发展（如生物化学）使得分析氨基酸（蛋白质水解产物）成为必需。可以使用光度检测器，氨基酸在通过测量池之前，与茚三酮进行柱后反应来检测。

（2）荧光检测器

约 10％的有机化合物是发荧光的，因为它们具有重新发射一部分从激发源吸收的光的能力。只要化合物的浓度较低，该荧光的强度与分析物的浓度成正比。应用于液相色谱法的荧光检测器非常灵敏，因此经常用于痕量分析。缺点是只在两个数量级的有限的浓度范围内的响应才是线性的。

这种选择性检测器的应用范围（见图 4.16）可以通过在柱前或柱后应用衍生化程序来扩展，使得感兴趣目标物被检测到。现在有许多专门用于合成荧光衍生物的试剂。对于该方法，将自动试剂分配器放置在柱之前或柱与检测器之间，以对样品中的被分析物进行一次或多次反应（取决于样品），以便在进样之前或之后使其发出荧光。

通过这一原理，可以使用皂化来测量环境中痕量的氨基甲酸酯（农药），然后与邻苯二甲醛反应以将甲胺转化为荧光衍生物（见图 4.16）。

(3) 示差折光检测器（RI 检测器）

这种类型的检测器利用通过折射率为 n 的透明介质的光透射的菲涅耳原理。其使用示差折光计连续测量柱前和柱后的流动相之间折射率的差异。图中光束穿过具有两个隔室的检测池：一个填充有纯流动相，而另一个填充有柱上洗脱的流动相（见图 4.17）。实际上介质的光学分散可能是不同的，因此折射率将仅在一个特定波长处匹配，所以完全透射的光将在很大程度上是单色的。当化合物从柱上洗脱时出现的两种液体之间的折射率变化被可视化为折射光束的角位移。在实际操作中，对反作用连续测量的相应信号必须提供给光学元件，以补偿折射光束的偏差。

示差折光检测器是最不灵敏的液相色谱检测器之一，它受环境温度、压力和流速变化的影响非常大。检测器的温度必须精确（至 0.001℃）和对色谱柱恒温控制。该检测器既有正峰也有负峰，这要求基线固定在图的中间高度（图 4.17）。此外，该检测器只能用于等度洗脱模式，因为在梯度洗脱中，流动相的组成与其折射率一样随时间而变化。当柱出口处洗脱液的组成与入口处洗脱液的组成不同时，不能获得在恒定组成的流动相情况下容易获得的补偿。因此，通常以等度模式与其他检测器串联使用，以获得增补的色谱图。

尽管存在这些缺点，这种检测器被认为几乎是通用型检测，特别适用于检测非离子化合物、没有紫外吸收和不发射荧光的物质。它用于识别那些不容易被其他方法检测到的物质（脂肪酸、醇、糖等）。与表 4.1 中所列的大多数其他检测器相反，通常示差折光检测器不会选择性地而是会响应所有存在的溶质。

4.1.6 LC/MS 和 LC/MS/MS

液相色谱和质谱的结合被视为分离和检测的理想结合。正如在气相色谱中，当物质从色谱柱洗脱出来时，质谱仪可以将其鉴别。这两种技术的结合也存在一些主要问题。质谱需要使用气相样品，而 LC 柱流出物是溶解在溶剂中的溶质。因此，第一步，溶剂必须被汽化。当溶剂汽化之后，由 LC 溶剂产生的气体体积比 GC 上载气多 10～1000 倍。因此，必须去除大部分溶剂。现已开发了解决去除 LC 柱接口溶剂问题的若干设备。现在最常用的方法是使用低流速大气压电离技术。典型的 LC/MS。系统框图如图 4.18 所示。HPLC 系统通常是纳升级的毛细管液相色谱系统，流速在 nL/min 范围内。另外，普通高效液相条件下，一些接口允许流速高达 1～2mL/min。最常见的离子源是电喷射离子源和大气压化学离子源。高效液相色谱法和质谱法的组合选择性高，仅仅通过检测一种选定的质量使得未分离的峰得到分离。LC/MS 技术可提供特定洗脱液的指纹图谱，和传统高效液相色谱法不一样的是，它与保留时间无关。LC/MS 技术还可以提供分子量和结构信息，还能够准确地进行定量分析。

对于一些复杂的混合物，LC/MS 组合不能提供足够的分离度。近年来，将两个或多个质量分析仪组合在一起已经成为可能，这种技术称为串联质谱技术。当与 LC 组合，所述的串联质谱系统称为 LC/MS/MS 仪。串联质谱仪通常也称为三重四极杆系统或四极杆离子阱质谱仪。

为了获得比四极杆更高的分离度，串联在质谱系统中的终极质量分析仪可以是飞行时间质谱仪。扇形质谱仪也可以合并成串联系统。以这样的方式操作离子回旋共振和离子阱质谱仪，不仅可以提供两个阶段的质量分析，而且还可以提供更多阶段的质量分析。这些质谱系统提供的分析步骤按顺序在单一的质量分析器中进行。这些光谱仪已与液相色谱系统合并到 LC/MS 仪器中。

4.2 分配色谱

HPLC 中使用最广泛的是分配色谱。其中，固定相是与液体流动相不相溶的第二液相。分配色谱可细分成液-液分配色谱和键合相色谱。液-液分配色谱和键合相色谱的差异主要是固定相附着在填充颗粒上的方式不同。液-液分配色谱中，液体通过物理吸附在适当的位置，而在键合相色谱中是以化学键连接的。早期的分配色谱只有液-液分配色谱；但是由于键合相色谱方法具有高稳定性和与梯度洗脱的兼容性，现在键合相色谱法更占优势。目前液-液分配色谱填料主要在某些特殊应用上。本节将学习键合相分配色谱。

液-液分配色谱法中的固定相是溶剂通过吸附作用附着在填料颗粒表面。

键合相色谱中的固定相是有机物通过化学键键合在填料颗粒表面。

4.2.1 键合相填料

大多数键合相填料是通过有机氯硅烷和二氧化硅颗粒上的羟基反应来制备的，二氧化硅颗粒上的羟基是在热的稀盐酸中水解得到的。该产品是一种有机硅氧烷。该反应中颗粒表面上 Si—OH 的位置可以写成 R，R 常是直链辛基或十八烷基。其他已键合到二氧化硅表面上的有机官能团包括脂肪族胺、醚和腈以及芳烃。因此，可以得到多种极性不同的键合型固定相。

4.2.2 正相和反相填料

两种分配色谱法的区别是流动相和固定相的相对极性不同。早期液相色谱主要是用高极性固定相，如甘油或水；非极性溶剂如已烷或异丙醚作为流动相。由于历史原因，现在称这种类型的色谱法为正相色谱。在反相色谱中，通常固定相是一种非极性烃类物质；流动相是相对极性溶剂（例如水、甲醇、乙腈或四氢呋喃）。

在正相色谱中，极性最小的组分首先被洗脱出来；流动相极性越高，需要的洗脱时间越短。与之相反，反相色谱中，极性最大的组分首先被洗脱出来，且流动相极性越高，洗脱时间越长。

据估计，目前超过四分之三的高效液相分离都是使用反相键合辛基或十八烷基硅氧烷填料进行的。填料中相互平行的长链烃基垂直于粒子表面形成呈刷状的非极性烃类表面。通常，这些填料中，流动相是不同浓度的甲醇、乙腈或四氢呋喃的水溶液。

离子对色谱是反相色谱的一个分支，离子对色谱可以较容易地在反相色谱柱上分离电离物质。这种类型的分配色谱，含有大量的有机平衡离子，如季铵离子或烷基磺酸盐的有机盐，加入到流动相中作为离子对试剂。现在人们提出了两种分离机制。首先，平衡离子与流动相中带有相反电荷的溶质离子形成不带电荷的离子对。基于离子对和两相的亲和力的差异，离子对在非极性的固定相中进行分配表现出不同的溶质保留。另一种说法是，平衡离子强烈地被中性固定相保留，从而使固定相带上电荷。通过反离子对配合物与强保留溶质的结合，和固定相形成最强的络合物，然后将相反电荷的有机溶质粒子分离。一些特殊的分离，如同一样品中既含有离子化合物也有非离子化合物，此类样品的分离可以通过分配色谱法来完成。图 4.19 显示出了用不同链长的烷基磺酸盐作为离子对试剂，对离子和非离子化合物进行分离。注意，C_5 和 C_7 烷基磺酸盐的混合物能够达到最好的分离效果。

正相分配色谱中，固定相是极性的，流动相是非极性的。反相分配色谱法与之相反，固定相是非极性的，流动相是极性的。

在正相色谱，极性最小的分析物首先被洗脱。在反相色谱，极性最小的分析物最后被洗脱。

4.2.3 流动相和固定相的选择

成功的分配色谱要求在分离过程中流动相和固定相中的分析物保持适当分子间力的平衡。这些分子间力可用三种组分所具有的相对极性来定性描述。一般情况下，一些普通有机官能团，如烃、醚、酮、醛、酰胺、醇，它们的极性按照顺序逐渐增强。水比含有任何上述官能团的化合物的极性都要强。

通常情况下，在选择色谱柱和流动相时，固定相的极性要大致与分析物匹配；其次，用于洗脱的流动相应与固定相显示出不同的极性。分析物与流动相的极性一致，但与固定相的极性不同，通常前一种情况比这一种情况更成功。后一种情况下，固定相不能较好地分离样品组分，并且在实际应用上保留时间太短。另一个极端是分析物和固定相的极性过于相近，使得保留时间变得过长。

普通流动相溶剂的极性顺序：

水＞乙腈＞甲醇＞乙醇＞四氢呋喃＞丙醇＞环已烷＞正已烷

乙腈（CH_3CN）是一种使用广泛的有机溶剂，经常作为比甲醇极性大、比水极性小的液相色谱流动相使用。

4.2.4 应用

图 4.20 表明键合相分配色谱法在分离软饮料中添加剂和有机磷杀虫剂的典型应用。表 4.2 进一步表明不同的样品所适用的技术。

吸附色谱法和液-固色谱法是 Tswett 在 20 世纪初首次提出的液相色谱的经典模式。由于正相分配色谱法和吸附色谱的强烈重合，许多用于液-固色谱的原理和技术也可应用到吸附色谱。事实上，在正相分离中，吸附和置换的过程控制保留时间。

吸附色谱法唯一的固定相是精细的二氧化硅和氧化铝。实际应用中，大多数优先考虑精细的二氧化硅，这是因为二氧化硅样品容积大。这两种物质的吸附特性彼此相似。随着分析物极性的增加，这两种物质保留时间变长。

因为键合型固定相较强的通用性和易得性，近年来，使用固体作为固定相的传统吸附色谱的使用正在减少，这一趋势有利于正相色谱法的发展。

4.3 离子色谱法

目前应用于实践的离子色谱法在 20 世纪 70 年代中期首次被开发出来，离子色谱法表明，阴离子和阳离子混合物可以通过向 HPLC 柱填充阴离子交换树脂和阳离子交换树脂来实现分离。这时，通常通过测量电导率来进行检测，然而由于流动相中电解质浓度高，所以很不理想。低交换容量色谱柱的发展使低离子强度的流动相得以使用，低离子强度流动相可以进一步去离子（抑制离子化），以达到高灵敏度电导检测。目前，离子色谱法可用多种类型的检测器，包括光谱检测器和电化学方法检测器。

当前正在使用的离子色谱有两种类型：抑制型和单柱。它们的区别在于：防止洗脱电解质的导电性干扰分析物的电导率的测量方法不同。

电导检测器非常适合于离子色谱。

4.3.1 基于抑制柱的离子色谱

电导检测器具备很多理想检测器的性能。电导检测器不仅灵敏度高，而且普遍适用带电物质。通常情况下，它还可以根据预测的浓度变化作出反应。此外，这种检测器操作简单，制造和维护费用低廉，易于小型化，且使用寿命长。电导检测器的唯一缺点是要在合理的时间内洗脱大部分的分析离子需要使用高浓度电解质，这使其直到 70 年代中期才应用于离子色谱。其结果是，流动相组分的电导率远大于分析物离子的电导率，这极大地降低了检测器的灵敏度。

1975 年，通过引入洗脱剂抑制柱，能够很好地解决高电导洗脱液产生的问题。抑制柱填充有第二离子交换树脂，这种交换树脂能够有效地把洗脱剂离子转换为有限电离的分子种类，而不影响分析物离子的导电性。例如，当分离和确定阳离子时，盐酸作洗脱剂，抑制柱是氢氧化物形式的阴离子交换树脂。在抑制柱反应的产物是水，反应方程如下：

$$H^+(aq)+Cl^-(aq)+resin^+OH^-(s) \longrightarrow resin^+Cl^-(s)+H_2O$$

第二个柱子不保留分析物阳离子。

对于阴离子分离，抑制柱填料是阳离子交换树脂的酸形式，用碳酸氢钠或碳酸作洗脱剂。抑制柱中的反应如下：

$$Na^+(aq)+HCO_3^-(aq)+resin^-H^+(s) \longrightarrow resin^-Na^+(s)+H_2CO_3(aq)$$

基于抑制柱的离子色谱，离子交换柱之后是抑制柱或抑制膜，可以将离子洗脱液转换成非离子样品，非离子样品不干扰分析物离子的电导检测。大量未解离的碳酸并不导致高电导率。

为了将填充柱转换为原来的酸碱形式，虽然不方便，但需要将原来的抑制柱相连，定期再生它们（通常每 8～10h 再生一次）。然而，在 20 世纪 80 年代，可以使用能够连续操作的微膜。例如，假如除去碳酸钠或碳酸氢钠，洗脱剂经一系列的超薄阳离子交换膜，交换膜能够将它从相反方向不断流出的酸性再生液中分离出来。洗脱液的钠离子在交换膜的内表面与氢离子交换，然后转移到其他表面与再生试剂的氢离子交换。再生溶液中的氢离子向反方向迁移，从而保持电中性。在洗脱液流速为 2mL/min 的条件下，微膜分离器基本能够从 0.1mol/L 的氢氧化钠溶液中除去所有的钠离子（见图 4.21）。

4.3.2 单柱离子色谱

商业离子色谱仪，无需抑制柱也可使用。这种方法依赖于样品离子和主要洗脱液的电导之间的细微差异。为了扩大这些差异，使用低容量交换器，允许洗脱液含有低浓度电解质。此外，选择低电导率的洗脱液。

单柱离子色谱具有不需要为抑制准备特殊设备的优点。然而，对于确定的阴离子，这种方法没有抑制柱方法灵敏。

在单柱离子交换色谱中，凭借不干扰分析物离子电导检测的低离子强度分析物离子在低容量的离子交换器中得到分离。

在体积排阻色谱中，基于分子大小分离。

4.4 体积排阻色谱

体积排阻色谱，或凝胶色谱，是一种强大的技术，特别适用于高分子量样品的分离分析。体积排阻色谱的填料包括小尺寸（10mm）的二氧化硅颗粒和聚合物颗粒，这些颗粒含有均匀的网状结构，允许溶质与溶剂分子向内扩散。

而在孔隙内，分子被有效地困住，又能被流动相带走。分析物分子的平均停留时间取决于它们的有效尺寸。比填料平均孔径明显大的分子被排斥，因此没有得到保留，即，它们按照流动相的速率通过柱子。比填料平均孔径明显小的分子，可以穿过孔，并且因此得到最大的保留时间；它们是最后洗脱出来的。在这两个极端尺寸之间的是中间尺寸分子，其渗透进入填料孔隙的能力取决于它们的直径。发生在组分内的分离与分子大小直接相关，并在一定程度上和分子的形状有关。需要注意的是，体积排阻分离不同于其他色谱方法，在于分析物和固定相之间没有化学或物理的相互作用。实际上，应避免这种相互作用，因为它们会降低柱效。还要注意，体积排阻分离不同于其他色谱方法，它有一个上限保留时间，因为没有比能够完全穿透固定相的小分子的保留时间更长的分析物。

4.4.1 柱填料

会遇到两种类型的体积排阻色谱填料：聚合物珠和二氧化硅颗粒，这两者的直径为5～10mm。二氧化硅颗粒更坚硬，使得填料简单，并允许在较高的压力下使用。它们也更稳定，从而允许使用更大范围的溶剂，并与新的溶剂显示出更快速的平衡。

凝胶过滤是体积排阻色谱的一种类型，其中填料是亲水性的。它被用来分离极性物质。

凝胶渗透是体积排阻色谱法的一种类型，其中所述填料具有疏水性。它被用来分离非极性样品。

市场上有众多体积排阻色谱的填料。有些是亲水性的，可适用于含水流动相；有些是疏水性的，适用于非极性有机溶剂。基于所述亲水填料的色谱有时称之为凝胶过滤，同时，基于疏水填料的称为凝胶渗透。对于这两种类型填料，许多孔径是可使用的。一般情况下，一个给定的填料可以容纳二十到二十五倍的分子质量范围。适合于给定填料的平均分子质量可小至几百或大至几百万。

4.4.2 应用

图 4.22 所示为体积排阻色谱的典型应用。两个色谱图由疏水填料获得，其中洗脱液为四氢呋喃。图 4.22(a) 为分子量（M）116～344 的脂肪酸分离。图 4.22(b) 为商业化环氧树脂的分离，其中每个单体的分子量为 280（聚合度为 5）。

体积排阻色谱的另一个重要应用是快速测定分子量或高聚物或天然产物的分子量分布。测定的关键是精确的分子量校准。校准可以通过公认的分子质量（峰位置的方法）来完成，或通过普通的校准方法来实现。后一种方法依赖的原则是，产物的内在分子黏度 h 与分子量 M 是和流体动力学体积（有效容积包括溶剂鞘）成正比的。理想的情况下，在体积排阻色谱中，分子按流体动力学体积被分离。因此，通过绘制 lg ［hM］ 对保留体积 V_r 的曲线可以得到一个通用的校准曲线。或者，绝对校准可以通过使用摩尔质量感应探测器实现，例如一个低角度光散射检测器。

4.5 亲和色谱

在亲和色谱中，作为亲和配体的试剂共价结合到固体支撑物上。典型的亲和配体是抗体、酶抑制剂，或是在样品中能够可逆地和选择性地结合到分析物上的其他分子。当样品通过柱子时，只有能够选择性结合到亲和配体上的分子才能被保留。不能够结合的分子随着流动相流出柱子。不理想的分子被除去后，通过改变流动相条件，保留的分析物能够进行洗脱。

亲和色谱的固定相是一种固体，如琼脂糖或多孔玻璃珠，通过亲和配体被固定化。亲和色谱的流动相有两个不同的作用。首先，它必须确保分析物分子与配体强烈结合。其次，一旦不想要的样品被除去，必须减弱流动相或消除分析物与配体之间的相互作用，使得该分析物可以被洗脱下来。通常情况下，在该过程的两个阶段，通过改变 pH 或离子强度来改变洗脱条件。

亲和色谱的主要优势是具有非凡的特异性。主要用途是在准备工作中快速分离生物分子。

4.6 手性色谱

不可重叠的镜像化合物（又称为对映体）称为手性化合物，目前在拆分手性化合物方面取得了巨大进步。这些拆分要么需要手性流动相添加剂，要么需要手性固定相。手性拆分剂（添加剂或固定相）会和其中一个异构体优先结合，使得一种对映体得到分离。为了识别溶质的手性特征，手性拆分剂本身必须具有手性特征。对手性固定相的关注最多。手性剂被固定在固体支撑物的表面上。手性拆分剂和溶质产生几种不同模式的相互作用。有一种类型是由于吸引力产生相互作用，这种吸引力如 π 键、氢键、偶极作用。另一种类型是溶质可以进入固定相的手性空腔以形成包合物。不管是什么模式，在许多领域中，分离这些密切相关化合物的能力是极其重要的。图 4.23 为酯的外消旋混合物在手性固定相上的分离。注意 R 和 S 对映体优异的分离度。手性拆分剂是一种手性流动相添加剂或手性固定相，能够优先络合其中之一的对映体。

表 4.3 为高效液相色谱法和气-液相色谱法之间的比较。GC 有速度优势，其设备简易。另一方面，高效液相色谱法适用于不挥发物质（包括无机离子）和受热不稳定的材料，但 GC 并不适用于这样的物质。这两种方法常常是互补的。

Chapter 5

Ion Chromatography

Ion chromatography (IC) is a separation technique which shares numerous common features with HPLC, yet possesses sufficient novel aspects such as its principle of separation or modes of detection, to make it the object of a separate study. IC is adapted to the separation of ions and polar compounds. The mobile phase is composed of an aqueous ionic medium and the stationary phase is an ion-exchange resin. Besides the detection methods based on absorbance or fluorescence, IC also uses electrochemical methods based on the ionic nature of the species to be separated. Its greatest utility is for analysis of anions for which there are no other rapid analytical methods. Yet current applications of IC are far broader than the analysis of simple ions by which the technique first gained renown. The operating domain, comparable with that of capillary electrophoresis, concerns the separation of many kinds of inorganic or organic species such as amino acids, carbohydrates, proteins and peptides in complex matrices.

This chapter will also review the main methods of quantitative analysis from chromatographic data.

fluorescence [fluːəˈresəns] *n*. 荧光，荧光性

renown [rɪˈnaʊn] *n*. 名望，声誉

electrophoresis [ɪˌlektrəʊfəˈriːsɪs] *n*. 电泳

amino acid [əˌmiːnəʊˈæsɪd] *n*. 氨基酸

5.1 Basics of ion chromatography

This chromatographic technique concerned with the separation of ions and polar compounds. Stationary phases contain ionic sites that create dipolar interactions with the analytes present in the sample. If a compound has a high charge density, it will be retained a longer time by the stationary phase. This exchange process is much slower when compared with those found in other types of chromatography. This mechanism may be anion associated, for molecular compounds, with those already dealt with by HPLC when equipped with RP-columns.

For HPLC, some columns contain ion exchange packings but they are used in significantly different ways. They are not considered

protein [ˈprəʊtiːn] *n*.蛋白(质)

peptide [ˈpeptaɪd] *n*. 肽，缩氨酸

dipolar [ˈdaɪpəʊlə] *adj*. 偶极的

RP-column 反相色谱柱

to be IC columns. They require concentrated buffers that cannot be suppressed and so are not compatible with conductivity detection. Applications with these columns use more traditional HPLC detection methods (such as UV or fluorescence).

Ion chromatography instruments have the same modules as those found in HPLC (Figure 5.1). They can exist as individual components or as in an integrated model. The pieces into contact with the mobile phase should be made of inert materials capable of withstanding the corrosiveness of acid or alkaline entities, which serve as eluents. The detection of ionic species present in the sample is difficult because these analytes are in low concentrations in a mobile phase that contains high quantities of ions.

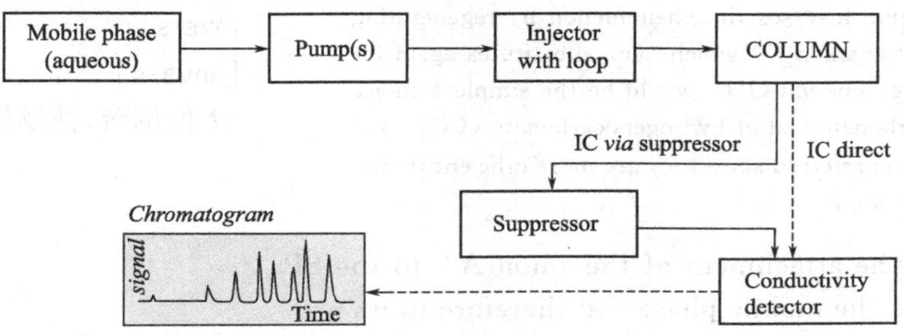

Figure 5.1 Schematic of an ion chromatograph instrument. The classic modular building design of liquid chromatography is seen here again, yet with the difference that the separation is generally performed isocratically. The configuration shows a suppressor' device installed after the column and series with a conductivity detector. The suppressor serves to eliminate the ions arising from the eluent to improve sensitivity

The separation of compounds within the sample is founded upon the occurrence of ion exchange, for which two classic examples are given below; If cationic species (type M^+) are to be separated, a cationic column with a stationary phase capable of exchanging cations will be employed. Such a phase is constituted, for example, of a polymer containing sulfonate ($-SO_3^-$) groups. Consequently, the stationary phase is the equivalent of a polyanion. Alternately, if anionic species (type A^-) are to be separated, an anionic column is selected capable of exchanging anions. This is achieved, for example, by employing a polymer containing quaternary ammonium groups.

To understand the mechanism of a separation, take for example an anionic column containing quaternary ammonium groups, in equilibrium with a mobile phase composed of a solution of hydrogenated carbonate anions (e.g. sodium counter ions). All of the cationic sites of the stationary phase find themselves paired with anions of the mobile phase (Figure 5.2).

When an anion A^- within the sample is taken up by the mobile phase, a series of reversible equilibria are produced which are directed

compatible [kəm'pætəbl] *adj*. 兼容的，相容的

conductivity [kɒndʌk'tɪvətɪ] *n*. 电导率

individual [ɪndɪ'vɪdʒuəl] *adj*.独特的，个别的

inert['ɪnə:t] *adj*.惰性的

corrosiveness [kə'rəusɪvnəs] *n*. 腐蚀性

isocratically ['aɪsəʊkrætɪkəl] *adv*. 等度洗脱

found[faʊnd] *vt*.创办，成立，建立

cationic [kæt'aɪənɪk] 阳离子的

consequently [kɒnsɪkwəntlɪ] *adv*.所以，因此

equivalent [ɪ'kwɪvələnt] *adj*.[化学]当量的；*n*.对等物，[化学]当量

alternately [ɔ:l'tɜ:nətlɪ] *adv*. 轮流地，交替地

anionic [ænaɪɒnɪk] *adj*. 阴离子的

quaternary [kwə'tənerɪ] *adj*.四元的，季的

mechanism ['mekənɪzəm] *n*. [生]机制，机能

equilibrium [i:kwɪ'lɪbrɪəm] *n*. 平衡

reversible [rɪ'və:rsəbl] *adj*. 可逆的

by an exchange equation giving the ion's distribution between the mobile phase (MP) and the stationary phase (SP).

distribution
[dɪstrɪbjuːʃn]
n.分配，分布

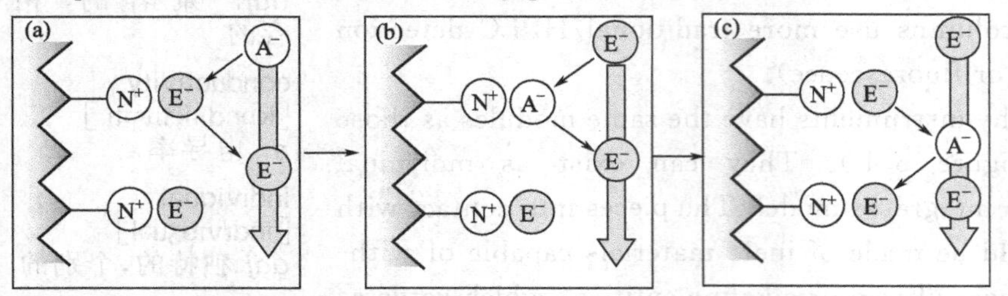

Figure 5.2 Scheme showing progression of an anion A^- by successive exchanges with the counter anion E^- in contact with an ammonium stationary phase. (a) Initially the counterion E^- fixed to the stationary phase is exchanged with the anionic species A^- present in the mobile phase. (b) Next, the elution inverses the phenomenon by regenerating the stationary phase with the anion E^- which, (c) substitutes again for A^- on the stationary phase. The ion OH^- would be the simplest choice for E^- but mixtures of carbonate and of hydrogenocarbonate (CO_3^{2-} and HCO_3^- at 0.003mol/L) are preferred since they are more efficient to displace the anions to be separated

inverse
[ˈɪnvɜːs]
vt.使倒转，使颠倒

Arrow 1 corresponds the attachment of the anion A^- to the SP and arrow 2 to its return to the mobile phase and therefore to its progression down the column.

$$A^-_{MP} + [HCO_3]^-_{SP} \underset{2}{\overset{1}{\rightleftharpoons}} [HCO_3]^-_{MP} + A^-_{SP}$$

$$\frac{[A^-_{SP}][(HCO_3)^-_{MP}]}{[A^-_{MP}][(HCO_3)^-_{SP}]} = K_{equ} \tag{5.1}$$

K_{equ} represents the selectivity between the two anions with respect to the cation of the stationary phase. As different anions have different K_{equ} they are therefore retained on the column during differrent times. The time at which a given ion elutes from the column can be controlled by adjusting the pH. Most of instruments use two mobile phase reservoirs containing buffers of different pH, and a programmable pump that can change the pH of the mobile phase during the separation.

selectivity
[səˌlekˈtɪvətɪ]
n.选择性

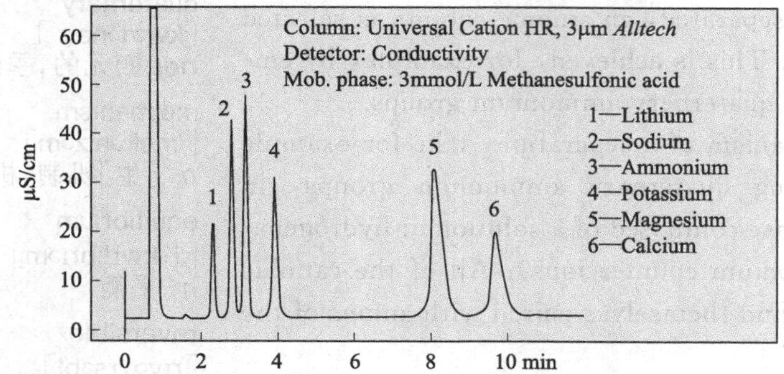

Figure 5.3 Separation fo several cations, mono and divalents with a cationic column

Using a cation exchange resin, a similar situation can be described (the stationary phase SP corresponds, for example, to Polym-SO_3H, strongly acidic):

$$M^+_{MP} + H^+_{SP} \underset{2}{\overset{1}{\rightleftharpoons}} M^+_{SP} + M^+_{MP}$$

This exchange phenomenon, which allows polar species to be retained on the resin, is known as solid phase extraction (Figure 5.3). If the sample contains two ions X and Y and if $K_Y > K_X$, Y will be retained more than X on the column.

5.2 Stationary phases

Ion chromatography can be subdivided into cation exchange chromatography, in which positively charged ions bind to a negatively charged stationary phase and anion exchange chromatography, in which the negatively charged ions bind to a positively charged stationary phase. The column packings consist of a reactive layer bonded to inert polymeric particles. Stationary phases must satisfy implicitly a number of requirements as narrow granulometric distribution (monodisperse), large specific surface area, mechanical resistance, stability under acid and basic pHs and rapid ion transfer.

5.2.1 Polymer-based materials

The best known stationary phases are issued from copolymers of styrene and divinylbenzene, in order to obtain packings hard enough to resist pressure in the column. They are made of spherical particles with diameters of 5 to 15μm (Figure 5.4) that are modified on the surface in order to introduce functional groups with acidic or basic properties.

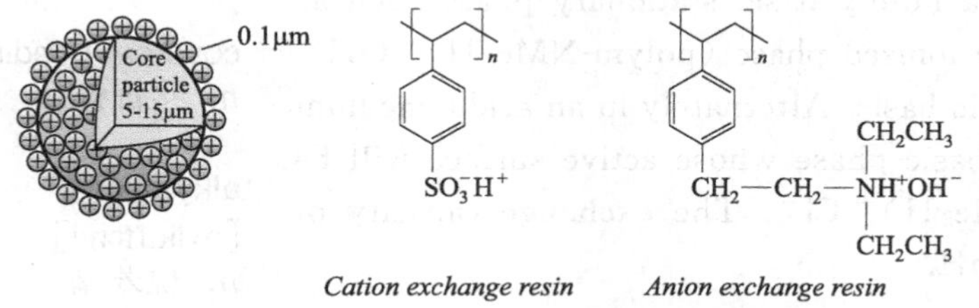

Figure 5.4 Stationary phases in IC. Cross-section of a spherical particle of polystyrene used as a cation exchanger. The polystyrene matrix is transformed either to a cation (ex. DOWEX@R 4) or to an anion (ex. DOWEX@R MSA-1) exchange resin. For anion separation, the resin is usually a quaternary ammonium group

For cation separation, the cation-exchange resin is usually a sulfonic or carboxylic acid. Thus, concentrated sulfuric acid is used to attack the accessible aromatic rings of the copolymer surface to link

solid phase extraction 固相萃取

implicitly [ɪmˈplɪsɪtlɪ] *adv*.无疑问地,无保留地

granulometric [grænjʊləˈmetrɪk] *adj*.(沙粒等的)颗粒测定的,粒度测定的

styrene [ˈstaɪriːn] *n*.苯乙烯

divinylbenzene [daɪvaɪnɪlˈbenziːn] *n*.二乙烯基苯

diameter [daɪˈæmɪtə] *n*.直径

sulfonic [səlˈfɒnɪk] *adj*.磺酸基的

carboxylic [kɑːbɒkˈsɪlɪk] *adj*.羧基的

accessible [əkˈsesəbl] *adj*.易接近的,可理解的,易相处的,易感的

SO_3H functional groups. A strongly acidic phase is obtained for cation exchange-on which the anion is fixed to the macromolecule while the cation can be reversibly exchanged with other cationic species present in the mobile phase. These materials are stable over a wide range of pHs and have an exchange capacity of a few mmol/g.

Another approach for obtaining these stationary phases is based on the copolymerization of a mixture of two acrylic monomers. One is anionic (or cationic), according to the nature of the phase desired, and the other is polyhydroxylated (Figure 5.5), in order to ensure the hydrophilic character of the stationary phase. There is, however, an inconvenience with these resins as their rate of swelling depends upon the composition of the mobile phase. They are normally reserved for medium pressure chromatography and some biochemical applications.

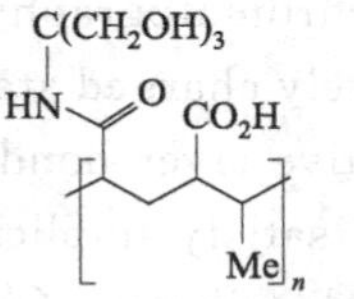

$C(CH_2OH)_3$
HN O CO_2H
Me$]_n$

Figure 5.5 Copolymerization of two monoethylenic (an acid and a trihydroxyamide). Example of the structure obtained (CM-TRISACRYLM of IBF-France). Arising from a weak acid the resultant phase will be unusable at acid pH, as it will no longer be in its ionized form.

Starting from the same copolymer, an anion exchange resin can be synthesized, first by chloromethylation, which binds $-CH_2Cl$ (Merryfield's resin), followed by reaction with a secondary or tertiary amine depending on the basicity required for the stationary phase.

On contact with water a mildly basic stationary phase such as Polym-NMe_2 yields a weakly ionized phase (polym-$NMe_2H)^+ OH^-$ especially when the medium is basic. Alternately in an acidic medium it will appear as a strongly basic phase whose active surface will be strongly ionized (polym-$NMe_2H)^+ Cl^-$. The exchange capacity of these resins varies with the pH.

5.2.2 Silica-based materials

Porous silica particles can serve to support, through covalent bonding, alkylphenyl chains carrying sulfonated groups or quaternary ammonium groups. This fixation step is similar to that used to obtain bonded silica phases developed in HPLC.

Some of these phases associate the properties of ion chromatography with those of HPLC. Separations depend simultaneously on both ionic coefficients and partition coefficients. Silica packings

macromolecule [mækrə'mɒləkjuːl] *n*. 巨大分子，高分子

copolymerization [kəʊˌpɒlɪməraɪ'zeɪʃn] 共聚合作用

acrylic [ə'krɪlɪk] *adj*.丙烯酸的，丙烯酸衍生物的

inconvenience [ˌɪnkən'viːnɪəns] *n*. 不方便，麻烦

swell [swel] *n* & *v*. 膨胀

monoethylenic *adj*. 单烯的

trihydroxyamide *n*. 三羟基酰胺

tertiary ['tɜːʃərɪ] *adj*. 第三的，叔的

basicity [be'sɪsətɪ] *n*. 碱性，碱度

covalent bonding *n*. 共价键合

alkylphenyl [ɔːlkɪl'fenɪl] *n*. 烷苯基

simultaneously [ˌsɪməl'teɪnɪəslɪ] *adv*.同时地

ionic coefficient 离子系数

partition coefficient 分配系数

usually display greater efficiency than their polymeric equivalents.

5.2.3 Resin films

A polymer called "latex", prepared from a monomer that contains organic groups, is deposited as an array of tiny beads (0.1-0.2μm in diameter) on a waterproof support to form a continuous film-like layer about 1-2μm thickness.

The support is made of micro-spheres of silica or glass or polystyrene of about 25μm diameter (Figure 5.6). This gives rapid equilibriums between stationary and mobile phases. Latex polymer results from the reaction of two unsaturated monomers such as 1,3-butadiene with maleic acid or 2-hydroxyethyl methacrylate.

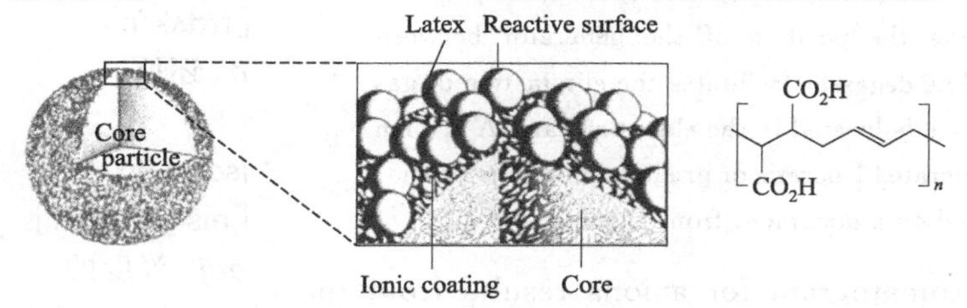

Figure 5.6 Film resins. Example of a resin made from a hard core onto which has been deposited a copolymer, derived from the reaction of maleic acid on 1,3-butadiene (Reproduced courtesy of the Dionex Company).

5.3 Mobile phases

IC mobile phases are usually 100 percent aqueous with organic or inorganic buffers to control selectivity and when necessary a small content of methanol or acetone used to dissolve certain samples having a low degree of ionization. Depending upon the type of stationary phase, the counter ions present in the mobile phase derived from acids (perchloric, benzoic, phthalic, methane sulfonic), or bases (the most popular for anion analyses are variants of sodium hydroxide and sodium carbonate/bicarbonate).

The pH is adjusted according to the separation to be achieved. The eluents can be prepared in advance remembering that basic solutions have a tendency to absorb atmospheric carbon dioxide, with for consequence a modification in the retention times.

To avoid these inconveniences, an eluent generator can be used (either acidic or basic) which is inserted, as a supplementary module, between the pump and the injector of the ion chromatograph (Figure 5.7). If the flow rate of water and the electrolytic current are known then the concentration gradients can be effected, a procedure seldom used in ion chromatography.

latex[ˈleɪteks]
n. 乳胶

waterproof
[ˈwɔːtəpruːf]
adj. 不透水的，防水的

unsaturate
[ʌnˈsætʃərɪt]
n. 不饱和化合物

1,3-butadiene
[bjuːtəˈdaɪiːn]
n. 1,3-丁二烯

hydroxyethyl
[haɪˈdrɒksjeθɪl]
n. 羟乙基

methacrylate
[meˈθækrəleɪt]
n. 甲基丙烯酸盐

maleic acid
[ˈmeɪlɪkˈæsɪd]
n. 顺丁烯二酸，马来酸

counter ion
[ˈkauntəaɪən]
n. 抗衡离子

perchloric
[pəˈklɔːrɪk]
adj. 高氯酸的

benzoic
[benˈzəʊɪk]
adj. 苯甲酸的

phthalic[ˈθælɪk]
adj. 邻苯二甲酸的

sulfonic[sʌlˈfɒnɪk]
adj. 磺酸基的

in advance
在前头，预先

tendency[ˈtendənsi]
n. 倾向，趋势

modification
[ˌmɒdɪfɪˈkeɪʃn]
n. 修改

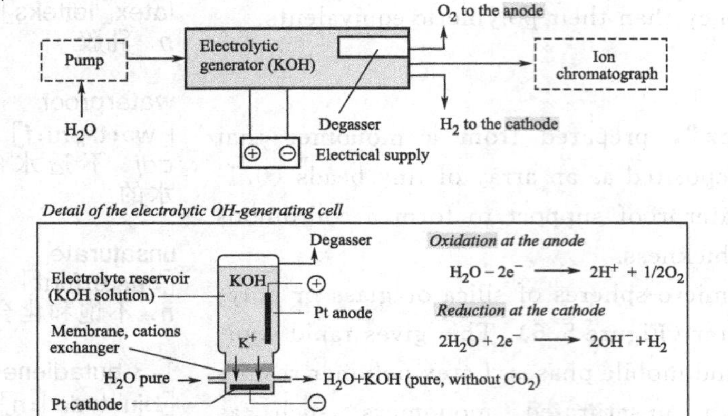

Figure 5.7 Ion chromatograph containing a high — purity OH^- generator. Schematic showing the position of the generator between pump and chromatograph. The degasser facilitates the elimination of gas which forms around the electrode located in the eluent stream. A K^+ ion is formed for every OH^- generated Isocratic or gradient elution is provided on demand (diagram based on a document from Dionex)

anode [ˈænəʊd] *n*. 阳极

cathode[ˈkæθəʊd] *n*. 阴极

oxidation [ˌɒksɪˈdeɪʃn] *n*. 氧化

reduction [rɪˈdʌkʃn] *n*. 还原

isocratic [ˌaɪsəʊˈkrætɪk] *adj*. 等度的

gradient [ˈgreɪdɪənt] *adj*. 梯度的

displace [dɪsˈpleɪs] *vt*.取代

The first peak in a chromatogram for anions results from the ionic strength of the injected sample being different than that of the eluent. The anions in the sample displace the anions (e. g carbonate/ bicarbonate or hydroxide) that are adsorbed onto the column packing. These displaced anions move forwards with the mobile phase and when passed through the detector appear as a positive peak (Figure 5. 8). If a suppressor (cf. section 5. 5) is installed at the column outlet and if carbonates make the mobile phase, a negative peak, called the "water dip" is often present. This peak is the result of carbon dioxide which is formed in the suppressed mobile phase (in the form of carbonic acid).

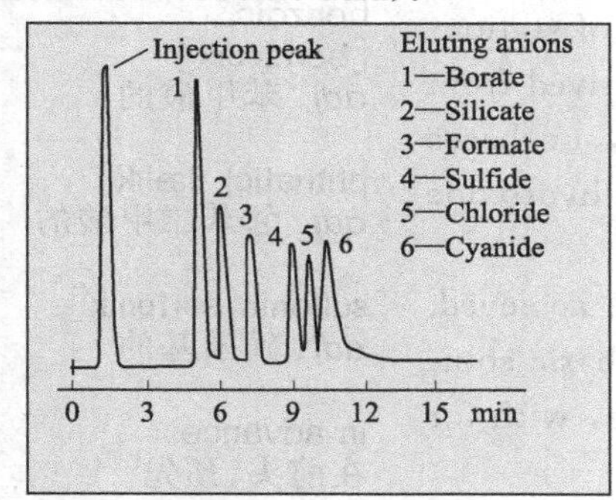

Column: Anion/R
Detector: Conductivity
Mob. phase: 5 mmol/L *p*-OH-benzoic acid, pH 8.5

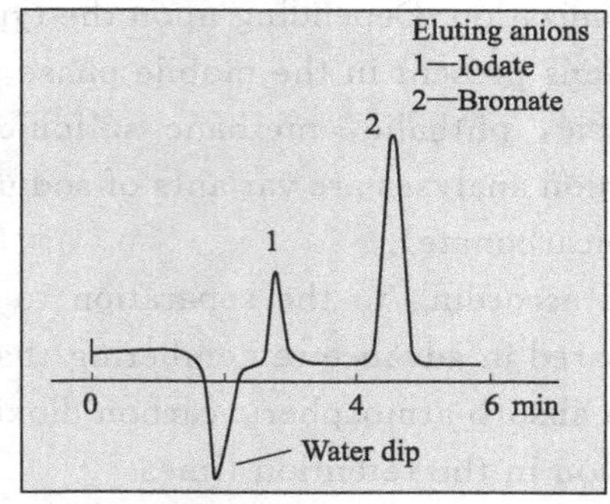

Column: Allsep Anion
Detector: Suppressed Conductivity
Mob. phase: 0.85 mmol/L $NaHCO_3$

Figure 5.8 Chromatograms displaying injection peak. The injection peak is the unretained peak that allows access to retention factors. This is normally the first peak on the chromatogram. It can interfere with other early-eluting anions such as fluoride (Chromatograms from Alltech)

If the ionic strength of the sample is greater than that of the eluent, there will be a positive peak. These peaks indicate the hold-up time of the chromatogram underway.

5.4 Conductivity detectors

Besides the spectrophotometric detectors based on absorbance or fluorescence of UV/visible radiation, and used when the mobile phase does not absorb appreciably, another mode of detection exists based upon electrolyte conductivity. Thus, at the outlet of the column, the conductance (the inverse of the resistance) of the mobile phase is measured between two microelectrodes. The measuring cell should be of a very small volume (approx. 2μL). The difficulty is to recognize in the total signal the part due to ions or ionic substances present in the sample. In order to do direct measurements, the ionic charge of the mobile phase has to be as low as possible and the measuring cell requires strict temperature control to within 0.01℃ because of the high dependence of conductance on temperature (~5%/℃).

conductance [kənˈdʌktəns] n. 电导

dependence [dɪˈpendəns] n. 依赖性

equivalent conductance 摩尔电导

The sensitivity of the detector to an ion X (valency z and molecular concentration C) can be predict if its equivalent conductance (Λ_X) and that of the eluent ion E (Λ_E) are known. This depends from the difference ΔK between the equivalent conductances of ion X and that of E. ΔK can be calculated according to expression 5.2 knowing that the peak will be either positive or negative.

reciprocal [rɪˈsɪprəkl] n. 倒数

$$\Delta K = C(\Lambda_X - \Lambda_E) \qquad (5.2)$$

The conductance $G=1/R$ that corresponds to the reciprocal of the resistance R, is measured between two electrodes which are plunged into the conducting solution and across which is maintained a potential difference. G is expressed in Siemens (S). For a given ion, the conductance of the solution varies with the concentration of the electrolyte. This relationship is linear for very dilute solutions. The specific conductance (in S/mol) or conductivity k, permits the measure to be independent of the detection cell parameters:

$$k = GK_{cell} \qquad (5.3)$$

$K_{cell}=d/A$ represents the cell constant (area A and spacing d) be obtained by direct measurement, but is determined from a standard solution for which the conductivity k is known.

Finally the equivalent ionic conductance ($S \cdot m^2/mol$) represents the conductivity of an ion with a valence z, in an aqueous solution at 25℃, when the molar concentration C(mol/L) tends towards zero in water (Table 5.1).

$$\Lambda_0 = 1000k/C_z \qquad (5.4)$$

Table 5.1 Equivalent ionic conductivities of ions at infinite dilution in water at 25℃

Cations	Λ_o^+ (S · m^2 · mol^{-1})×10^{-4}	Anions	Λ_o^- (S · m^2 · mol^{-1})×10^{-4}
H^+	350①	OH^-	198
Na^+	50	F^-	54
K^+	74	Cl^-	76
NH_4^+	73	HCO_3^-	45
$1/2Ca^{2+}$	60	$H_2PO_4^-$	33

① 3500000 S · m^2/mol or 350 in S · cm^2/mol.

5.5 Ion suppressors

The mobile phase contains ions that create abackground conductivity, making it difficult to measure the conductivity due only to the analyte ions as they exit the column. To improve the signal to noise ratio, when using a conductivity detector, a device called a suppressor, designed to selectively remove the mobile phase ions is placed after the analytical column and before the detector. The principle consists to convert the mobile phase ions to a neutral form or replacing them by others of higher conductivity. Suppressor-based detection is more useful for anion analysis than for cation analysis.

neutral ['nju:trəl] *adj*. 中性的

The simplest model of a suppressor can be considered as column which contains a stationary phase having functional groups of opposing charge to those of the separating column. Such a chemical suppressor, which contains an anionic resin is associated to a cationic separation column. The mechanism of action can be described using the following example.

Suppose that a mixture containing the cations Na^+ and K^+ has been separated using a cationic column whose mobile phase contains dilute hydrochloric acid. In this acidic medium, at the outlet of the column, the Na^+ and the K^+ ions are accompanied by H^+ ions coming from the acid and Cl^- anions in order to assure the electroneutrality of the medium. After the separation column, the mobile phase flows through a second column which contains an anionic exchange resin whose mobile ion is OH^-. The Cl^- anions will be affixed on this column, thus displacing the OH^- ions that will react with H^+ ions in solution to give water. At the outlet of the suppressor, only (Na^+ OH^-) and (K^+ OH^-) species are found in water. The ions H^+ and Cl^- have effectively disappeared. As OH^- has a higher conductivity than Cl^-, detection of the Na^+ and K^+ ions is easier, because amplified (Figure 5.9).

hydrochloric acid 盐酸

electroneutrality 电中性 [ˈɪlektrəʊnju:trælɪtɪ]

neutralize ['nju:trəlaɪz] *vt*. 中和

In summary, for anion analyses (using a conductivity detector), ion suppressors neutralize the mobile phase, reducing its conductivity, while simultaneously increasing the sample's conductivity.

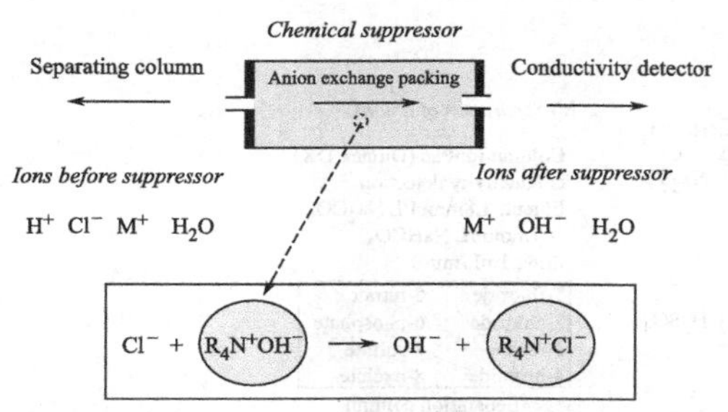

Figure 5.9 Chemical suppressor for an exchange cation column. For cation analysis, the mobile phase is often dilute HCl or HNO_3 solutions, which can be neutralized by an eluent suppressor that supplies OH^-. In this example, the anionic suppressor purges the mobile phase of H^+ ions and of almost all of the Cl^- ions, facilitating the detection of the cation M^+. The same principle holds for anion analysis. In this case, the mobile phase is often dilute NaOH or $NaHCO_3$, and the eluent suppressor supplies H^+ to neutralize the anion and retain or remove the Na^+

The limitation of this type of suppressor lies in its very large dead volume that reduces the separation efficiency, due to a remixing of the ions prior to their detection. The ions of the suppressor should be regenerated periodically and it should be used exclusively in the isocratic mode.

Other types of suppressors, having a high ionic capacity, have subsequently been developed. They are made of porous fibers or of micromembranes and possess very small dead volumes in the order of 30-50μL. That allows gradient elution with a negligible baseline drift. Figure 5.10(a) shows the passage of an anion A^-, in solution in a typical electrolyte used for anionic columns, through a suppressor with a cationic membrane.

Nowadays, continuous regenerated suppressors which make use of electrolytic reactions have been introduced for traces determinations. They behave either like a special column containing a resin which regenerates by electrolysis or like a membrane suppressor where the regenerating ions are produced in situ by electrolysis of water. Figure 5.10(c) illustrates the second procedure: it represents the passage of a cation, in a dilute hydrochloric acid solution, through a suppressor whose membrane is permeable to anions.

Alternately, if the problem consists in the separation of a mixture of anions on an anionic column (cationic material), with an eluent containing dilute sodium hydroxide, a membrane allowing the diffusion of cations will be chosen. At the cathode, the passage of hydronium ions towards the main flow of electrolyte will neutralize

purge
[pɜːdʒ]
vt.清除,(使)净化

principle
['prɪnsəpl]
n. 原理

regenerate
[rɪ'dʒenəreɪt]
vt. & *vi*. 回收(废热,废料等),使再生

exclusively
[ɪk'skluːsɪvlɪ]
adv. 唯一地,专门地,特定地,专有地,排外地

micromembrane
[maɪkrɒmemb'reɪn]
n. 微膜

order['ɔːdə(r)]
n. 量级

negligible
['neglɪdʒəbl]
adj. 可忽略的

electrolysis
[ɪlek'trɒləsɪs]
n. 电解

in situ[ɪn'saɪtuː]
原位

permeable
['pɜːrmɪəbl]
adj. 具渗透性的

OH^- ions. At the anode, Na^+ ions will migrate out of the mobile phase and will react with OH^- ions.

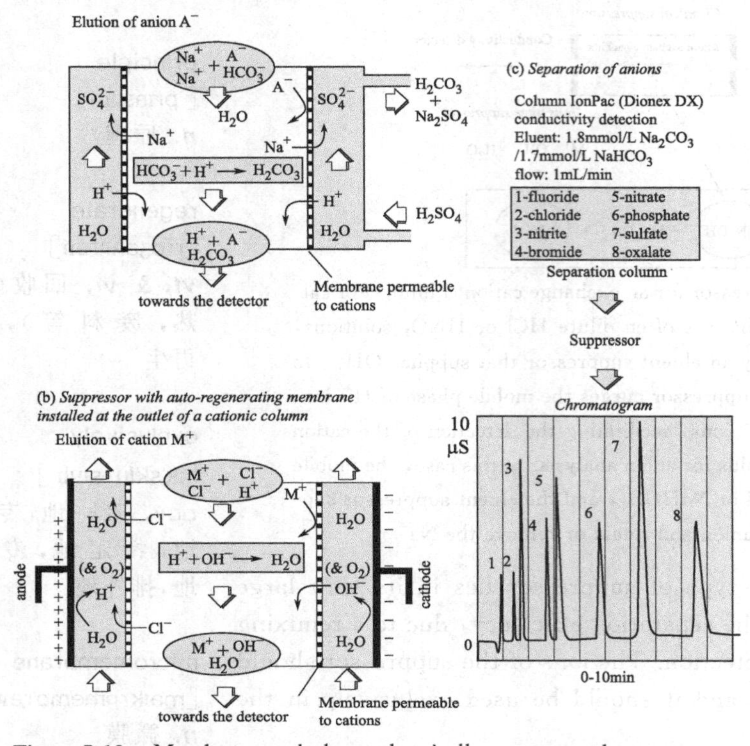

Figure 5.10 Membrane and electrochemically regenerated suppressors. There are two types of membranes, one type permeable to cations (H^+ and in this example Na^+), the other permeable to anions (OH^- and her Cl^-). (a) The microporous cationic is adapted to the elution of anions. Only cations can cross the membrane (corresponding to a polyanionic wall which keeps away the anions in the solution); (b) An anionic membrane suppressor placed, contrary to the preceding model, at the outlet of a cationic column. Ions are regenerated by the electrolysis of water. Note in both cases the counter flow circulation between the eluted phase and the solution of the post-column suppressor; (c) An example of a separation of inorganic cations (concentrations of the order of ppm) using a suppressor of this type

reliability [rɪˌlaɪəˈbɪləti] *n*. 可靠

5.6 Quantitative analysis by chromatography

standardize [ˈstændərdaɪz] *vt*. 使标准化

The significant development of chromatography in quantitative analysis is essentially due to its reliability and its use in standardized analyses. Trace and ultratrace analyses by chromatography are used, particularly the EPA methods for environmental analysis, although their costs are rather high. This type of analysis relies mainly on reproducibility of the separation and on the linear relationship between the injected mass of a compound onto the column and the area of the corresponding peak on the resultant chromatogram. This is an excellent comparative method used in many protocols, which, allied with software used for data treatment allow automation of all the calculations associated with these analyses.

The three most widely use methods are described below accompanied in their simplest formats.

5.6.1 Principle and basic relationship

In order to calculate the mass concentration of a compound appearing as a peak on a chromatogram, two basic conditions must be met. First, an authentic sample of the compound to be measured should be available, as a reference, to determine the detector sensitivity to this compound. Second, a software giving the heights or areas of the different eluting peaks of interest is also required. All of the quantitative methods in chromatography rely on these two principles. They are comparative but not absolute methods.

authentic [ɔːˈθentɪk] *adj*. 真正的

For a given tuning of the instrument, it is assumed that a linear relation exists for each peak of the chromatogram, over the entire concentration range, between its area and the quantity of the compound responsible for this peak in the injected sample. This applies for a given concentration range depending on the detector employed. This hypothesis is translated into the following equation:

hypothesis [haɪˈpɑːθəsɪs] *n*. 假设

$$m_i = K_i A_i \tag{5.5}$$

Where m_i is mass of the compound i injected on the column, K_i is the absolute response factor for compound i and A_i is the area of the eluting peak for compound i. The absolute response factor K_i (not to be confused with the partition coefficient), is not an intrinsic parameter of the compound since it depends upon the tuning of the chromatograph. To calculate the response factor K_i, according to expression 5.5, it is essential that both the area A_i and the mass m_i, of compound i injected on the column, are known. However, this mass is difficult to determine with precision since it relies simultaneously upon the syringe, upon the injector type (in GC), or upon the injection loop (in HPLC). This is why most chromatographic methods utilized for quantitative analyses, whether pre-programmed into an integrated recorder or in the multiplicity of available software, do not make use of the absolute response factors, K_i.

syringe [sɪˈrɪndʒ] *n*. 注射器

loop [luːp] *n*. 环

multiplicity [ˌmʌltɪˈplɪsətɪ] *n*. 多样性

furnish [ˈfɜːnɪʃ] *vt*. 提供

5.6.2 Areas of the peaks and data treatment software

To determine the areas of the peaks appropriate chromatographic software is used which also ensures not only the control and working of the chromatograph but also the data treatment to furnish a report corresponding to one of the pre- programmed methods of quantitative analysis.

The signal recovered by the detector is sampled by the analogue-digital converter (ADC) with a frequency of narrowest peaks in chromatograms obtained from GC with capillary columns. Each software package allows baseline correction, treatment of negative signals and all incorporate different methods to calculate peak areas (Figure 5.11).

frequency
[fri:kwənsi]
n. 频率

capillary
[kəpɪləri]
n. 毛细管，毛细血管，微管

baseline correction
基线校正

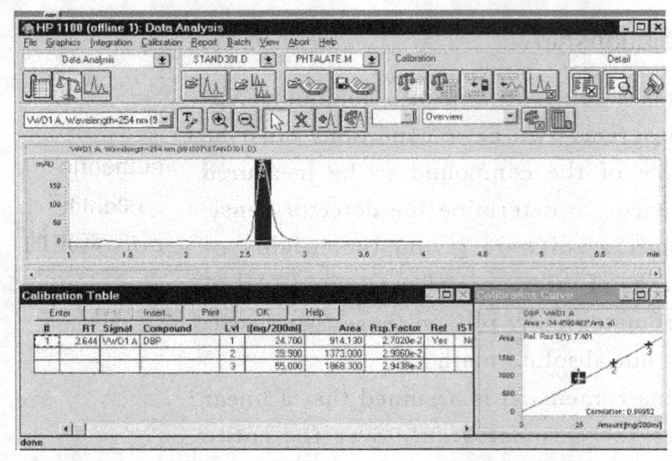

Figure 5.11 Quantitive analysis software for chromatography. Since the signal from the detector situated at the outlet of the column is analogical, an analogue-digital converter (ADC) is necessary. The stored chromatogram, digitized, serves as a basis for its exploitation by the software. Different zones of the screen can display the calibration curve, methodology, etc. (software Chemstation from Agilent Technologies)

The manual triangulation method and the "cut and weight" method (weight is considered proportional to area) are, of course, no longer employed. However, it is useful to remember that for a Gaussian eluting peak, the product of its width at half height by its full height, corresponds to approximately 94 per cent of the total area of the peak. In the same way, recorders with an integration system for measuring the peak areas are no longer used.

Gaussian
['gausɪən]
高斯的

5.6.3 External standard method

This method allows the measurement of the concentration (or percentage in mass) of one or more components that appear as resolved peaks on the chromatogram, even in the presence of other compounds yielding unresolved peaks. Easy to use, this method corresponds to the application of a principle common to many quantitative analysis techniques.

The modus operandi is based upon the comparison of two chromatograms obtained successively without changing the control settings of the chromatograph (Figure 5.12). The first chromatogram is

modus operandi (MO)
做法，手法

acquired from a solution (reference solution) of known concentration C_{ref} in a solvent. The standard and sample matrix should be as similar as possible. A volume V of this solution is injected. Analysis conditions must be identical. On the resulting chromatogram, the area A_{ref} of the corresponding peak is measured. The second chromatogram results from the injection of the same volume V of the sample in solution, containing an unknown concentration of the compound to be measured (conc. C_{unk}). The area of the corresponding peak is A_{unk}. Since an identical volume of both samples has been injected, the ratio of the areas is proportional to the ratio of concentrations which depend upon the masses injected ($m_i = C_i \cdot V$). Applied to the two chromatograms, expression (5.5) leads to relation (5.6), which characterizes this method:

single point calibration method
单点校正法

$$M_{ref} = C_{ref}\,V = KA_{ref},\ m_{sam} = KA_{unk}$$

$$C_{unk} = C_{ref}\frac{A_{unk}}{A_{ref}} \quad (5.6)$$

origin
[ˈɔːrɪdʒɪn]
n. 根源

The single point calibration method, as depicted in Figure 5.12, assumes that the calibration line goes through the origin. Precision will be improved if the concentrations of the reference solution and of the sample solution are similar.

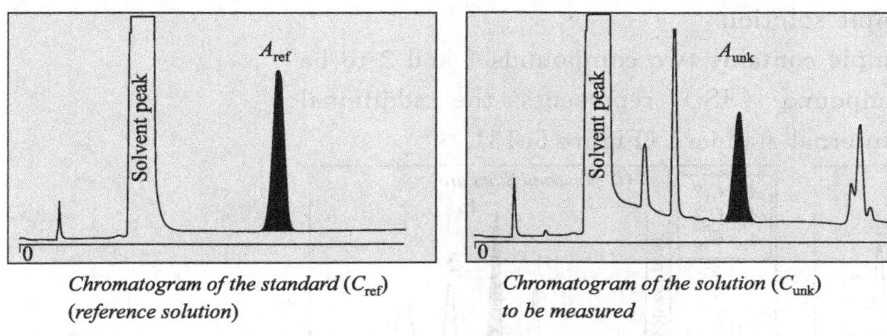

Chromatogram of the standard (C_{ref}) (reference solution)   *Chromatogram of the solution (C_{unk}) to be measured*

Figure 5.12 Analysis by the external standard method. The precision of this basic method is improved when several solutions of varying concentrations are used in order to create a calibration curve. For trace analyses by liquid chromatography it is sometimes advisable to replace the areas of the peaks by their heights as they are less sensitive to variations in the mobile phase flow rate

This technique, employing the absolute response factors, yields very reliable results with chromatographs equipped with an auto-sampler: a combination of a carousel sample holder and an automatic injector. This permits numerous measurements to be made without interruption, to the condition that no change in the apparatus tuning is made between injections.

carousel
[kærəˈsel]
n. 旋转木马

The reference solution periodically injected affords a control that can be used to compensate an eventual baseline drift during a sequence of programmed injections.

Precision can be also improved if several injections of the sample and the reference solutions are made, always using equal volumes. In a multilevel calibration, equal volumes of a series of standard solutions are injected. This allows to get a calibration curve of $A=f(C)$, obtained by a regression method (linear least-square or quadratic least-square). This leads to a more precise value for C_{unk} (Figure 5. 11).

linear least-square 线性最小二乘法

quadratic least-square 二次最小二乘法

This method, the only one adapted to gas samples, has the added advantage that nothing needs to be added to the sample solution, unlike the method described below.

5.6.4 Internal standard method

For trace analysis, it is preferable to use a method that relies on the relative response factor of each compound to be measured against a marker introduced as a reference. This means that any imprecision concerning the injected volumes, the principal constraint of the previous method, is compensated. As above, this more reliable method requires two chromatograms, one to calculate the relative response factors of the compounds of interest, and the other to analyse the sample.

constraint [kən'streɪnt] *n*. 限制，束缚，约束

The areas of the peaks to be quantified are compared with that of an internal standard (designated by IS), introduced at a known concentration within the sample solution.

Supposing that a sample contains two compounds 1 and 2 to be measured and that compound (IS) represents the additional compound for use as an internal standard (Figure 5. 13).

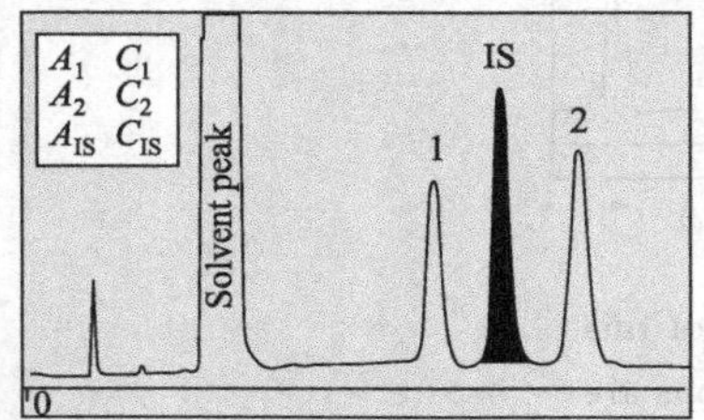

Chromatogram of the standards

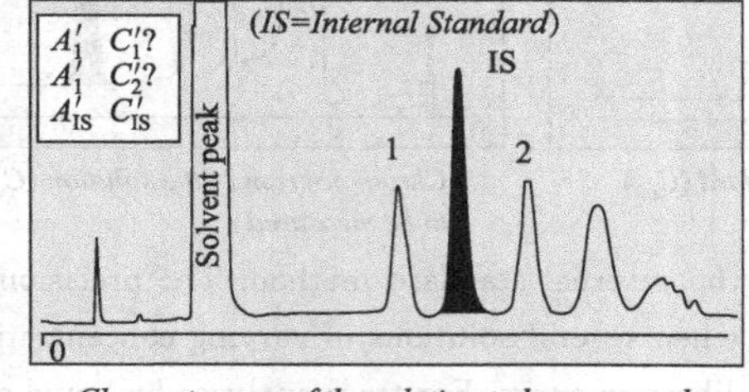

Chromatogram of the solution to be measured

Figure 5.13 Simple example illustrating the internal standard method

(1) Calculation of the relative response factors

A solution containing compound 1 at known concentration C_1, compound 2 at known concentration C_2 and the internal standard IS at known concentration C_{IS} is prepared then injected onto the chromatograph. A_1, A_2, A_{IS} will be the areas of the elution peaks in the chromatogram due to the three compounds. If m_1, m_2 and m_{IS} represent the real quantities introduced onto the column, of these three substances, then three relations of type 5. 5 can be derived:

$$m_1=K_1A_1$$
$$m_2=K_2A_2$$
$$m_{IS}=K_{IS}A_{IS}$$

$$\frac{m_1}{m_{IS}}=\frac{K_1A_1}{K_{IS}A_{IS}} \text{ and } \frac{m_2}{m_{IS}}=\frac{K_2A_2}{K_{IS}A_{IS}}$$

These ratios enable the calculation of the relative response factors of 1 and 2, against IS and designated by $K_{1/IS}$ and $K_{2/IS}$:

$$K_{1/IS}=\frac{K_1}{K_{IS}}=\frac{m_1A_{IS}}{m_{IS}A_1} \text{ and } K_{2/IS}=\frac{K_2}{K_{IS}}=\frac{m_2A_{IS}}{m_{IS}A_2}$$

Since the injected masses m_i are proportional to the corresponding mass concentrations C_i ($m_i = C_iV$), the above equations can be rewritten as follows:

$$K_{1/IS}=\frac{C_1A_{IS}}{C_{IS}A_1} \text{ and } K_{2/IS}=\frac{C_2A_{IS}}{C_{IS}A_2}$$

(2) Chromatogram of the sample-calculation of the concentrations

The second step of the analysis is to obtain a chromatogram for a given volume of a solution containing the sample to quantify and to which has been added a known quantity of internal standard IS. This will yield A'_1, A'_2 and A'_{IS} the areas of this new chromatogram being obtained under the same operating conditions as previously. If m'_1, m'_2 and m'_{IS} represent the quantities of 1, 2 and IS introduced into the column, then:

$$\frac{m'_1}{m'_{IS}}=K_{1/IS}\frac{A'_1}{A'_{IS}} \quad \text{and} \quad \frac{m'_2}{m'_{IS}}=K_{2/IS}\frac{A'_2}{A'_{IS}}$$

From the relative response factors calculated in the first experiment as well as from the known concentration of the internal standard within the sample, C'_{IS}, this leads to:

$$C'_1=C'_{IS}K_{1/IS}\frac{A'_1}{A'_{IS}} \text{ and } C'_2=C'_{IS}K_{2/IS}\frac{A'_2}{A'_{IS}}$$

Expanding to n components it is easy to calculate the mass concentration of the solute i using equation 5.7:

$$C'_i=C'_{IS}K_{i/IS}\frac{A'_i}{A'_{IS}} \qquad (5.7)$$

equally the percentage concentration of i can be expressed using equation (5.8);

$$x_i=(C'_i/\text{Mass of sample taken})\times 100\% \qquad (5.8)$$

This method becomes even more precise if several injections of the solution and of the sample are carried out. Often a same volume of an IS stock solution is spiked with all standards and samples.

In conclusion, this general and reproducible method demands nevertheless a good choice of internal standard, which should have the following characteristics:

- it must be stable, pure and not exist in the initial sample
- it must be measurable, giving an elution peak well resolved on the chromatogram

• its retention time must be close to that (or those) of the solute (s) to be quantified

• its concentration must be close to, or above that of the analytes to quantify to gain a linear response from the detector

• it must not interfere and co-elute with a sample component.

5.6.5 Internal normalization method

This method, also called "normalized to 100 percent" is used for mixtures for which each component is producing a peak on the chromatogram, in order to be able to make a complete assessment of the sample concerned. The solvent, if any, is typically ignored.

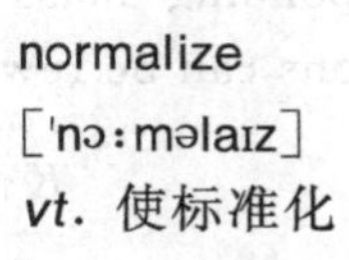

normalize [ˈnɔːməlaɪz] *vt.* 使标准化

Supposing that it is required to find the mass concentrations of three compounds 1, 2, 3 in a mixture (Figure 5.14). The analysis is again carried out in two steps.

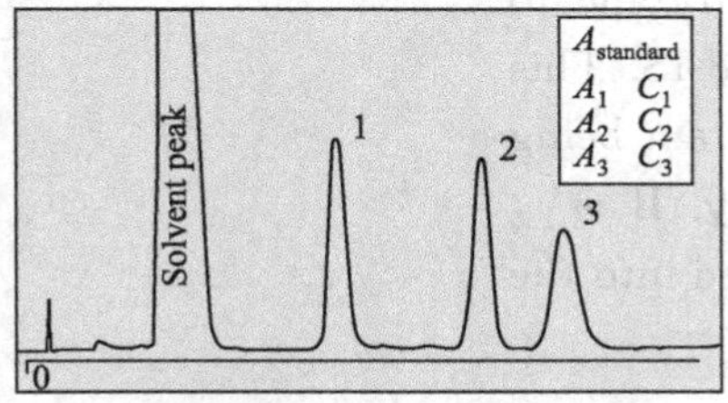

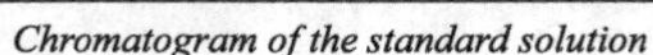

Chromatogram of the standard solution

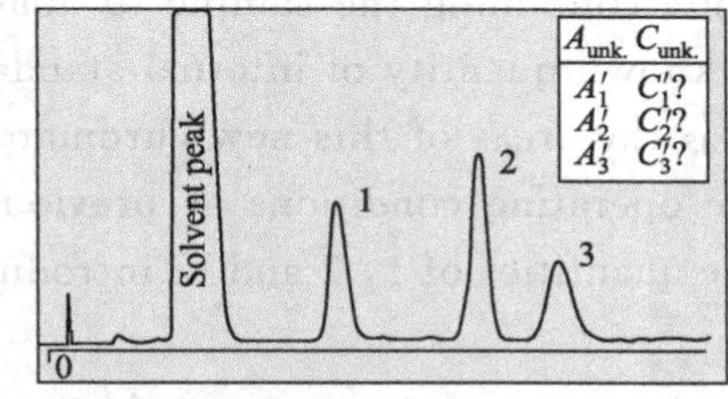

Chromatogram of sample solution

Figure 5.14 Analysis by internal normalization method. This method is commonly reported as the default for early integrators

(1) Calculation of the relative response factors

A standard solution containing the three compounds 1, 2 and 3 at known concentrations C_1, C_2 and C_3 is prepared. The chromatogram corresponding to the injection of a volume V of this standard solution shows three peaks of areas A_1, A_2 and A_3. These areas will be related to the masses m_1, m_2 and m_3 of the compounds in volume V, by three expressions of type 5.5.

One of the compounds 3, for example, is chosen as the substance for internal normalization. This compound 3 will serve to calculate the relative response factors $K_{1/3}$ and $K_{2/3}$ for compounds 1 and 2 with respect to 3. As previously deduced:

$$K_{1/3}=\frac{K_1}{K_3}=\frac{m_1A_3}{m_3A_1} \text{ and } K_{2/3}=\frac{K_2}{K_3}=\frac{m_2A_3}{m_3A_2}$$

Given that $m_i=C_iV$, then the following expressions for $K_{1/3}$ and $K_{2/3}$ are obtained:

$$K_{1/3}=\frac{C_1A_3}{C_3A_1} \text{ and } K_{2/3}=\frac{C_2A_3}{C_3A_2}$$

(2) Chromatogram of the sample-calculation of the concentrations

The next step consists to inject a sample of the mixture to be measured containing 1, 2 and 3. Labelling the elution peaks as A'_1, A'_2 and A'_3 will gain direct access to the percentage mass composition of the mixture represented by x_1, x_2 and x_3 via three expressions of the following form:

$$x_i = \frac{K_{i/3}A'_i}{K_{1/3}A'_1 + K_{2/3}A'_2 + A'_3} \times 100\% \tag{5.9}$$

The condition of normalization being that: $x_1 + x_2 + x_3 = 100$

If the procedure is extrapolated to n components normalized to the component j, a general expression for the response factor of a given compound can be obtained (equation 5.10).

$$K_{i/j} = \frac{C_i A_j}{C_j A_i} \tag{5.10}$$

It is also possible to determine $K_{i/j}$ by plotting a concentration-response curve for each of the solutes.

In a mixture containing n components, if A'_i designates the area of the elution peak of compound i, and if the internal reference is j, then the content of compound i will obey the following equation 5.11:

$$x_i = \frac{K_{i/j}A'_i}{\sum_{i=1}^{n} K_{i/j}A'_i} \times 100\% \tag{5.11}$$

Supposing that detector gives responses that are independent from the substance (i.e. relative response factors identical as for a TCD detector in GC), this method serves to give an estimation of relative concentrations. This was commonly reported as the default for early integrators (equation 5.12).

$$x_i = \frac{A'_i}{\sum_{i=1}^{n} A'_i} \times 100\% \tag{5.12}$$

Problems

5.1 0.604g of an undiluted stationary phase, comprising SO_3H groups is introduced to an erlenmeyer. 100mL of distilled water, is added and approximately 2g of NaCl, The liberated acidity is measured by sodium hydroxide 0.105mol/L in the presence of helianthin (methyl orange), as an indicator to identify the equivalence point (towards pH 4). If it is known that 25.4mL of the sodium hydroxide solution must be added to neutralise the acid liberated by the stationary phase, calculate its molar capacity in grams.

5.2 A mixture of proteins is separated on a column with a stationary phase of carboxymethylated cellulose. The internal diameter of the column is 0. 75cm and its length is 20cm. The dead volume is 3mL. The flow rate of the mobile phase is 1mL/min. The pH of the mobile phase is adjusted to 4. 8. Three peaks appear upon the chromatogram corresponding to the elution volumes V_1, V_2 and V_3 at 12mL, 18mL and 34mL respectively.

1. Does this arise from an anionic or cationic phase? Give reasons for your answer.

2. Why, when increasing the pH of the mobile phase, are the times of elution of the three compounds subject to modification? Predict whether these times are increased or decreased.

5.3 In measuring cyclosporin A (a treatment for skin and organ transplant rejection) by HPLC, according to a method derived from that of internal standard, the following procedure is employed.

Preparation of the samples: An extraction of 1 mL of blood plasma is made, to which is added 2mL of a mixture of water, and acetonitrile (80/20), containing 250ng of cyclosporin D as internal standard, which has a similar structure to cyclosporin A.

The 3mL of new mixture is passed on a disposable solid phase extraction column in order to separate the cyclosporins retained upon the sorbent. Following the rinsing and drying of the column, the cyclosporins are eluted with 1. 5mL of acetonitrile and are then concentrated to 200μL following evaporation of solvent. A fraction of this final solution is injected into the chromatograph.

1. What is the concentration factor of the original plasma following this treatment?

2. Can be rate of recuperation following the extraction step on the solid phase be deduced?

Sample: The standards are created with a plasma originally possessing no cyclosporin. Six different solutions are prepared by adding the necessary quantities of cyclosporin A to each in order to create solutions of 50ng/mL, 100ng/mL, 200ng/mL, 400ng/mL, 800ng/mL, and 1000ng/mL, 1mL from each of these solutions is subjected to the same extraction sequence following the addition of 250ng of cyclosporin D to each, as above.

ng/mL en cyclo. A	50	100	200	400	800	1000
Ratio of peaks heights cyclo. A/cyclo. D(R_h)	0. 25	0. 5	1. 02	2. 04	4. 05	5. 1

Column Supelco 75mm×4. 6mm. Silica gel 3μm, phase RP-8.

3. Determine the concentration in units of ng/mL of cyclosporin A in the blood plasma giving rise to the chromatogram reproduced below. This question should be attempted in two ways:

(a) By choosing a single point from a standard.

(b) By using the gradient which it is possible to draw from the data in table above (in both cases it will based upon the heights of the peaks).

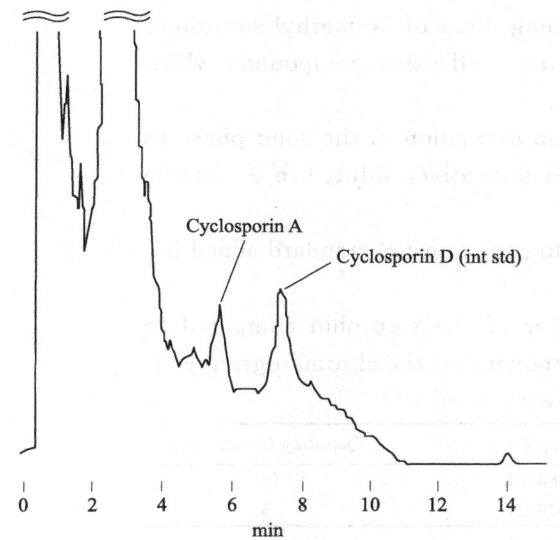

5.4 The method of internal normalization was chosen to determine the mass composition of a sample comprising a mixture of four esters of butanoic acid. To this end, a reference solution containing known % masses of these esters led to the following relative values of the response coefficients of the butanoates of methyl (ME), of ethyl (EE), and of propyl (PE), all three in ratio with butyl-butanoate (BE).

$K_{ME/BE}=0.919 \qquad K_{EE/BE}=0.913 \qquad K_{PE/BE}=1.06$

From the chromatogram of the sample under analysis, reproduced below, and the information given in the table, find the mass composition of this mixture (ignore the first peak at 0.68min.)

Peak No.	t_R	*compound*	*Area*/mV · min
1	0.68	—	0.1900
2	2.54	methyl ester(ME)	2.3401
2	3.47	ethyl ester(EE)	2.3590
3	5.57	propyl ester(PE)	4.0773
5	7.34	butyl ester(BE)	4.3207

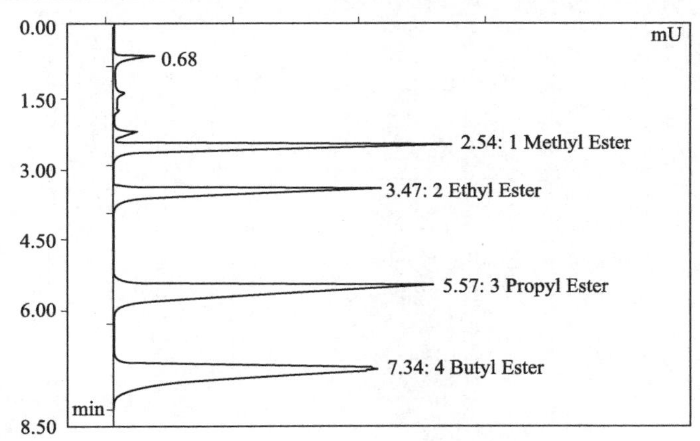

5.5 To measure serotonin (5-hydroxytryptamine), by the internal standard method, a 1mL aliquot of the unknown solution is added to 1mL of a solution containing 30ng of *N*-methyl-serotonin. This mixture is then treated to remove all other compounds which could interfere with the experiment.

The operation performed was an extraction in the solid phase to isolate the serotonin and its methyl derivative, diluted in a suitable medium.

1. Why is the compound forming the internal standard added before the extraction step?

2. Calculate the response factor of the serotonin compared to that of *N*-methyl-serotonin if it is known that the chromatogram yielded by the standards gave the following results.

Name	*Area*/μV · s	*Quantity injected*/ng
Serotonin	30885982	5
N-methyl-serotonin	30956727	5

3. From the chromatogram of the sample solution, find the concentration of serotonin in the original sample, if it is known that:

—area serotonin 2573832μV/s

—area *N*-methyl-serotonin 1719818μV/s

第5章 离子色谱法

离子色谱法（IC）是与高效液相色谱法（HPLC）具有许多相似性的一种分离技术，然而它在分离原理或检测模式等方面的许多新特点使其成为分离领域的一个分支。IC适用于离子与极性化合物的分离。流动相由含离子的水溶液组成，固定相则是离子交换树脂。除了利用吸光度或者荧光检测外，IC也利用待分离物质离子的电化学性质进行检测。IC最大的用途是对那些没有其他快速分离方法可用的阴离子进行分析。虽然IC是由于首次实现了对简单离子的分析而被大家熟知，然而其应用目前远不止于此。与毛细管电泳相比，IC的应用领域是分离各种各样的无机或有机物质，如复杂基质中的氨基酸、碳氢化合物、蛋白质或多肽等。

本章也会讲述色谱数据定量分析的主要方法。

5.1 离子色谱法基础

离子色谱分离技术涉及离子和极性化合物的分离。固定相中含有能与样品中待分析物质发生偶极相互作用的离子位点。高电荷密度的化合物，在固定相中停留的时间更长。与其他色谱法相比，这种交换过程要慢得多。对分子化合物而言，IC的分离机理与那些使用反相色谱柱的高效液相色谱相似。

有些HPLC色谱柱含有离子交换填料，但是它们起作用的方式却非常不同，因此它们不被看作离子色谱柱。因为它们需要使用不能被抑制器处理的浓缓冲液做流动相，从而不能兼容电导检测器。这些柱子使用更传统的HPLC检测方法（例如紫外或者荧光检测）。

离子色谱仪使用与HPLC相同的模块（图5.1），它们可以作为单独组件或集成模块使用。接触流动相的部件应由能承受洗脱液中酸或碱腐蚀的惰性材料组成。因为在含有大量离子的流动相中被测离子的浓度很低，样品中离子性物质的检测很困难。

样品中化合物的分离是在离子交换的基础上进行的，下面用两个经典的例子来说明。

如果分离的物质是阳离子（M^+），则使用能够交换阳离子的固定相色谱柱即阳离子柱，例如由含有磺酸盐（$—SO_3^-$）聚合物组成的固定相。因而这种固定相也就相当于聚阴离子。

同样如果分离的物质是阴离子（A^-），则选择一个能够交换阴离子的固定相色谱柱即阴离子柱，如由含有季铵基团聚合物组成的固定相。

为了理解分离机理，以含有季铵基团的阴离子柱为例，柱子已由碳酸氢根阴离子（如钠离子为抗衡离子）溶液组成的流动相达到平衡。固定相中所有的阳离子位点都已与流动相中的阴离子配对平衡（图 5.2）。

当样品中的阴离子 A^- 随流动相前进时，A^- 将会在流动相和固定相中发生由离子交换方程决定的一系列可逆分配平衡。

$$A_{MP}^- + [HCO_3]_{SP}^- \underset{2}{\overset{1}{\rightleftharpoons}} [HCO_3]_{MP}^- + A_{SP}^-$$

$$\frac{[A_{SP}^-][(HCO_3)_{MP}^-]}{[A_{MP}^-][(HCO_3)_{SP}^-]} = K_{equ} \tag{5.1}$$

箭头 1 指的是阴离子 A^- 在固定相上的吸附方向，箭头 2 则是指 A^- 返回流动相中沿色谱柱向下的移动方向。

K_{equ} 表示固定相中阳离子位点对两种阴离子的选择性差异。不同的阴离子其 K_{equ} 也不同，因此它们在柱子上的保留时间也不同。一种离子在柱子上的保留时间可以通过调节流动相的 pH 来控制。大多数仪器使用两个储存不同 pH 缓冲液流动相的储液槽和一个可编程泵，这种泵可以在分离的过程中调节流动相的 pH。

使用阳离子交换树脂可以得到类似的情况（如固定相是强酸性的 Polym-SO_3H）。

$$M_{MP}^+ + H_{SP}^+ \underset{2}{\overset{1}{\rightleftharpoons}} M_{SP}^+ + M_{MP}^+$$

这种极性物质保留在树脂中的交换现象称为固相萃取（图 5.3）。如果样品中含有离子 X 和 Y，并且 $K_Y > K_X$，那么 Y 离子在柱上的保留时间会大于 X 离子。

5.2 固定相

离子色谱可以细分为阳离子交换色谱和阴离子交换色谱。在阳离子交换色谱中，带正电的阳离子结合到带负电的固定相中；而在阴离子交换色谱中，则是带负电的阴离子结合到带正电的固定相中。柱子的填料主要由键合在惰性聚合颗粒上的反应层组成。固定相必须要满足一些实验要求，例如粒度分配窄、比表面积大、机械抗性强、酸性及碱性条件下稳定以及离子交换速度快等。

5.2.1 聚合物基材料

为了能承受柱子里的高压，必须使用足够坚硬的填料，最有名的固定相是由苯乙烯和二乙烯基苯形成的共聚物。这些填料是直径为 5～15μm 的球形颗粒（图 5.4），经过表面改性可以引入酸性或碱性官能团。

阳离子分离通常使用含有磺酸或羧酸基团的阳离子交换树脂。因此，使用浓硫酸来进攻共聚物树脂表面容易反应的芳香环以连接 SO_3H 官能团，得到一个强酸性的固定相并应用于阳离子交换过程。其中，阴离子是固定在高分子树脂上，而阳离子可以与流动相中的其他阳离子进行可逆交换。这些材料在一个广泛的 pH 范围内可以保持稳定，其交换容量大约为 mmol/g。

获得这些固定相的另一种方法是基于两种丙烯酸单体的共聚反应。根据所需固定相的性质，一个单体是阴离子（或阳离子），而为了确保固定相的亲水特征，另一个单体含有多羟基（图 5.5）。然而，由于它们的膨胀速度取决于流动相的组成，使这些树脂的使用有不便之处。所以它们一般用于中压色谱和一些生化应用。

从相同的共聚物开始可以合成阴离子交换树脂，首先通过结合氯甲基发生氯甲基化反应（Merryfield 树脂），然后根据固定相所需的碱性与仲胺或叔胺反应。

弱碱性固定相如 Polym-NMe_2 在与水接触时会产生一个弱离子相（Polym-NMe_2H)$^+$$OH^-$，尤其是当介质为碱性时。与此相反，在酸性介质中其活性表面发生强烈电离变为(polym-NMe_2H)$^+$$Cl^-$形式。这些树脂的交换容量随 pH 变化而变化。

5.2.2 硅胶基质材料

通过共价键合作用，多孔硅胶可以作为载体来键合带有磺酸基或季铵基的烷基苯，这种固定过程与 HPLC 中制备键合硅胶固定相的步骤类似。

这些固定相会使人从 IC 的性质联想到 HPLC 的性质。分离效果同时取决于离子系数和分配系数。与聚合物填料相比，二氧化硅填料通常表现出更高的分离效率。

5.2.3 树脂膜

用含有有机基团的单体制备得到了一种被称为“latex”的聚合物，使用该聚合物在一个憎水的载体材料表面生成一系列微珠（直径 0.1～0.2μm），形成一个厚度大约为 1～2μm 的连续类膜涂层。载体由直径约 25μm 的二氧化硅、玻璃或聚苯乙烯微球体制成（图 5.6)，它们能在固定相与流动相之间快速达到平衡。Latex 聚合物由两个不饱和单体如 1,3-丁二烯与马来酸或丙烯酸 2-羟乙酯的反应而来。

5.3 流动相

离子色谱的流动相通常使用 100％的有机或无机缓冲液来有效控制选择性，必要时加入少量甲醇或丙酮，用于溶解那些具有低解离程度的样品。根据固定相的类型，流动相中的反离子来自于酸（如高氯酸、苯甲酸、邻苯二甲酸、甲磺酸），或碱（对阴离子分析来说，最常见的是氢氧化钠和碳酸钠/碳酸氢钠）。

根据分离需要可调节流动相的 pH。洗脱液可以提前准备，但需记住一点：碱性溶液会吸收大气中的二氧化碳，这将会使保留时间发生变化。为了避免这些不便，可在泵和离子色谱仪（图 5.7）的注射器之间引入洗脱液发生器作为一个辅助部件。如果已知水流量和电流速率，那么浓度梯度就会受影响，这一措施很少用于离子色谱法中。

阴离子色谱图中的第一个峰是因为注射样品中的离子强度不同于洗脱液的离子强度而产生的。样品中的阴离子可取代那些吸附在柱填料上的阴离子（例如碳酸根离子/碳酸氢根离子或氢氧根离子）。被取代的阴离子随着流动相向前移动，当通过检测器时，色谱图上将会显示出一个向上的峰（图 5.8）。如果柱出口安装有抑制器（参见 5.5 节），同时流动相由碳酸盐组成，就会经常出现一个被称为“水浸”的负峰，该峰是由于在抑制流动相中形成的二氧化碳（以碳酸的形式存在）而产生的。

如果样品的离子强度大于洗脱液，那么将会出现一个正峰。这些峰给出了正在进行的色谱分析的保留时间。

5.4 电导检测器

除了基于紫外/可见辐射的吸收或荧光的光度检测器外，当流动相吸收光谱不明显时，可使用另一种基于电解质电导的检测模式。在色谱柱出口，两个微电极可以测量出流动相的电导率（电阻的倒数）。测量池的池体积（约 2μL）应足够小。该方法的难点在于从总信号中识别出来自于样品中的离子性物质产生的信号。为了做到能直接测量，流动相中的离子电荷必须尽可能低，因为电导率对温度有较高的依赖性（约 5%/℃），测量池还需要有误差小于 0.01℃的严格温控装置。

如果已知离子 X（价态为 z 和摩尔浓度为 C）和洗脱液的离子 E 的摩尔电导分别为 Λ_X 和 Λ_E，那么就可以预测检测器对离子 X 的灵敏度。这取决于离子 X 和离子 E 之间的 ΔK 值。如果已知二者同为正峰或负峰，ΔK 可以根据方程式(5.2) 来计算。

$$\Delta K = C(\Lambda_X - \Lambda_E) \tag{5.2}$$

电导 $G=1/R$，是电阻 R 的倒数，可以在导电溶液中的两个电极之间测量出并且在它们之间存在一个电位差。G 的单位为西门子（S）。对于特定的离子来说，溶液的电导随着电解质溶液的浓度变化。对稀溶液则呈线性关系。电导率（S/mol）或 k，可以不受检测池参数影响来测量。

$$k = GK_{cell} \tag{5.3}$$

$K_{cell}=d/A$ 是指通过直接测量（面积 A，间距 d）获得的电池常数，但实际上是通过测定已知电导率的标准溶液来得到的。

最后，当量离子电导（无限稀释摩尔电导率，$S \cdot m^2 \cdot mol^{-1}$）表示当摩尔浓度 C（$mol \cdot L^{-1}$）在水中趋于零时，在 25℃水溶液具有化合价为 z 的离子的电导率（表 5.1）。

$$\Lambda_0 = 1000k/C_z \tag{5.4}$$

5.5 离子抑制器

流动相中的离子可以产生背景电导率，当它们流出色谱柱时，使得待分析离子的电导率难以测量。当使用电导率检测器时，为了提高信噪比，可在分析柱之后、检测器之前安装一个抑制器，其目的是选择性地除去流动相中离子信号的干扰。它的原理是将流动相中的离子转换为中性形式或用其他有更高电导率的离子置换它们。相对阳离子分析而言，抑制型检测器对阴离子的分析更有用。

最简单的抑制器模型可以看作是一个含有固定相的柱子，它具有与分离柱相反电荷的官能团。由阴离子树脂组成的化学抑制器，应与阳离子分离柱相连接。其作用机理可以使用下列示例进行说明。

假设用一根阳离子色谱柱分离一个含有 Na^+ 和 K^+ 的混合溶液，稀盐酸溶液做流动相。在酸性介质中，流动相中的 H^+ 伴随着 Na^+ 和 K^+ 同时到达柱子的出口处，同时为确保流动相呈电中性，溶液中含有 Cl^-。在分离柱之后流动相经过第二根柱子（即抑制器），它是迁移离子为 OH^- 的阴离子交换树脂柱。流动相中的阴离子 Cl^- 会吸附在柱子上并置换出 OH^-，这样 OH^- 将中和溶液中 H^+ 生成水分子。因此在抑制器的出口处，H^+ 和 Cl^- 被有效地除去，在水中只检测到（Na^+OH^-）和（K^+OH^-）（图 5.9）。因为 OH^- 比 Cl^- 具有更高的电导率，由于放大作用，Na^+ 和 K^+ 的检测变得更容易（图 5.9）。

总之，对于阴离子的分析（使用电导率检测器），离子抑制器可中和流动相，降低其导电性，同时增加了样品的电导率。

由于在检测之前会有离子的再混合，这种抑制器的缺点在于它的死体积非常大，会降低分离效率。抑制器应定期再生，且使用等度洗脱。

具有高离子容量的其他类型抑制器也先后被研发了出来。它们是由微膜或多孔纤维组成，在 30～50μL 范围内具有非常小的死体积，这样就可以忽略基线漂移的影响，使用梯度洗脱。图 5.10(a) 为在阴离子柱的典型电解质溶液中阴离子 A^- 通过阳离子膜抑制器时的过程示意。

如今，利用电解反应的连续再生抑制器已被用于痕量测定。它们作用方式或者像可以通过电解再生的特殊树脂柱，或者是通过在线电解水产生再生离子的膜抑制器。图 5.10(b) 为第二种情况：在稀盐酸溶液中，阳离子通过可渗透阴离子的膜抑制器时的过程示意。

同样地，如果使用稀氢氧化钠做洗脱剂，使用阴离子柱（阳离子材料）来分离阴离子混合物，需要选择允许阳离子扩散的膜。在阴极，随电解质流动的水合氢离子将中和 OH^-。在阳极，Na^+ 会迁移出流动相与 OH^- 反应。

5.6 色谱定量分析

色谱定量分析的发展本质上是由于它性能可靠和在标准化分析中的应用。通过色谱可以进行痕量和超痕量分析，特别是用 EPA 方法进行环境分析时，尽管它们的成本相当高。这种类型的分析主要由于分离的重现性和色谱柱上分析物质的进样量与其在色谱图中相应峰面积之间的线性关系。这是一种优良的比较方法，结合数据处理软件，可以让与这些分析相关的计算实现自动化。

三种使用最广泛的方法简单介绍如下。

5.6.1 原理与基本关系

为了计算色谱图上出峰的化合物的质量浓度，必须满足两个基本条件。首先，被测化合物的真实样品是可用的，以作为参考来确定检测器对这一化合物的灵敏度。第二，还需要一个软件来计算不同洗脱峰的峰高或峰面积。所有的色谱定量方法依赖于这两个原则。它们是相对方法，而不是绝对方法。

对于给定的仪器调谐条件，假设色谱图上的每个峰面积与对应该峰的化合物的量在整个浓度范围内存在线性关系。根据所使用的检测器不同，该假设适用于一定浓度范围。这种假设可以表示为下面的公式：

$$m_i = K_i A_i \tag{5.5}$$

式中，m_i 是化合物 i 在色谱柱中的进样质量；K_i 为化合物 i 的绝对响应因子；A_i 为化合物 i 的峰面积。

绝对响应因子 K_i（不要与分配系数混淆）不是化合物的固有参数，因为它依赖于色谱仪的调谐条件。计算响应因子时，根据表达式(5.5)，需要知道峰面积 A_i 以及注射到色谱柱上化合物 i 的质量 m_i。然而，因为 m_i 同时依赖于注射器、进样器类型（GC）以及进样环（HPLC），该质量难以精确地测定。这就是不管是预先编程的整合记录仪，还是各种可以使用的软件，大多数色谱分析都不利用绝对响应因子 K_i 进行定量分析的原因。

5.6.2 峰面积与数据处理软件

为了测定峰面积的大小，要使用合适的色谱软件，这不仅确保了色谱仪的控制和正常工作，还可以根据预编程的定量分析方法对数据进行处理生成报告文件。

检测器检测的信号以气相色谱毛细管柱得到的色谱图上最窄色谱峰的频率通过模数转换器（ADC）进行采集。每个软件包含基线校正、负峰的处理以及引入不同的方法来计算峰面积（图 5.11）。

当然，我们不再采用手动三角测量法和“剪切和称重”的方法（重量与面积成正比）。然而需知道，对一高斯洗脱峰，其半峰高处的宽度等于它全部高度的一半时，相当于大约94%的总峰面积。同样，带有集成系统的记录仪也不再用于测量峰面积。

5.6.3 外标法

该法允许对色谱图上可分辨峰的一种或多种组分的浓度或质量分数进行测量，甚至是其他未出峰的化合物。这种方法易于使用，与很多定量分析技术的应用原理类似。

这种方法是基于在不改变色谱条件时连续得到的两个色谱图间的比较（图 5.12）。第一个色谱图来自于一个已知浓度为 C_{ref} 的标准溶液（参比溶液）。标准品和样品基质应该是尽可能地相似，进样体积为 V，分析条件必须相同，所得色谱图中测量对应物质的峰面积为 A_{ref}。第二个色谱图是相同体积 V 的待测样品溶液，含有待测化合物（浓度为 C_{unk}），相应峰面积为 A_{unk}。由于两个样品的进样体积相同，那么峰面积之比与浓度成正比，而浓度又取决于进样质量（$m_i=C_iV$）。将式(5.5) 应用到这两个色谱图，则可以推导出式(5.6)。

$$M_{ref}=C_{ref}V=KA_{ref}, m_{sam}=KA_{unk}$$

$$C_{unk}=C_{ref}\frac{A_{unk}}{A_{ref}} \tag{5.6}$$

如图 5.12 所示的单点校准方法假定校准曲线经过原点。如果参比溶液和样品溶液的浓度相近，精确度还将进一步提高。

该方法中，使用带有自动进样装置（一个圆盘样品架和一个自动进样器组合在一起）的色谱仪，采用绝对响应因子会得到更可靠的结果。这能使测量过程连续进行而不会因为每次进样而停止。

在一系列程序进样过程中，周期性地进样参比溶液会抵消最终的基线漂移影响。

如果重复进样等体积的样品溶液和参比溶液，精度也会提高。在一个多级校正实验中，注入一系列等体积的标准溶液后，通过回归方法（线性最小二乘法或二次最小二乘法）得到 $A=f(C)$ 校准曲线。这也会使 C_{unk}（图 5.11）的值更加精确。

该方法是适用于气体样品分析的唯一方法，有很多其他优点，它不像下文中要讲的方法那样需要在样品溶液中加入其他物质。

5.6.4 内标法

对于痕量分析，建议优先使用待测化合物与参比物质的相对校正因子进行计算的方法。这意味着如进样体积、上述方法的缺点等任何不精确的因素都将会抵消。如上所述，这种更可靠的方法需要使用两个色谱图，一个色谱图用来计算目标化合物的相对响应因子，另外一个是分析样品的色谱图。

在样品溶液中加入一定已知浓度的内标物（表示为 IS），将待定量物质的峰面积与内标物的峰面积进行对比。

假设一个样品含有两个待测化合物 1 和 2，则 IS 是加入样品中的内标物（图 5.13）。

(1) 计算相对响应因子

制备一个化合物1的浓度为C_1，化合物2的浓度为C_2和内标物IS的浓度为C_{IS}的溶液进行色谱分析。A_1、A_2、A_{IS}分别是这三种化合物在色谱图上的峰面积。m_1、m_2和m_{IS}代表这三种物质在柱子上的真实质量，可以推导出三个如式(5.5) 所示的关系式：

$$m_1 = K_1 A_1$$

$$m_2 = K_2 A_2$$

$$m_{IS} = K_{IS} A_{IS}$$

$$\frac{m_1}{m_{IS}} = \frac{K_1 A_1}{K_{IS} A_{IS}} , \quad \frac{m_2}{m_{IS}} = \frac{K_2 A_2}{K_{IS} A_{IS}}$$

这些比值可以用来计算化合物1和2相对于内标物IS的相对响应因子，分别用$K_{1/IS}$和$K_{2/IS}$表示：

$$K_{1/IS} = \frac{K_1}{K_{IS}} = \frac{m_1 A_{IS}}{m_{IS} A_1} \quad , \quad K_{2/IS} = \frac{K_2}{K_{IS}} = \frac{m_2 A_{IS}}{m_{IS} A_2}$$

由于进样质量正比于相应物质的质量浓度C_i（$m_i = C_i V$），上述方程可以改写为：

$$K_{1/IS} = \frac{C_1 A_{IS}}{C_{IS} A_1} \quad , \quad K_{2/IS} = \frac{C_2 A_{IS}}{C_{IS} A_2}$$

(2) 计算样品浓度的色谱图

分析第二步是获得样品溶液的色谱图，其中含一定量内标物和一定体积的待测物质。在与上面相同的操作条件下得到的新色谱图产生A'_1、A'_2和A'_{IS}三个峰面积。如果m'_1、m'_2和m'_{IS}分别表示化合物1、2和内标物IS的质量，则得到：

$$\frac{m'_1}{m'_{IS}} = K_{1/IS} \frac{A'_1}{A'_{IS}} \quad , \quad \frac{m'_2}{m'_{IS}} = K_{2/IS} \frac{A'_2}{A'_{IS}}$$

从第一个实验得到的相对响应因子以及样品中内标物的已知浓度，可推出：

$$C'_1 = C'_{IS} K_{1/IS} \frac{A'_1}{A'_{IS}} \quad , \quad C'_2 = C'_{IS} K_{2/IS} \frac{A'_2}{A'_{IS}}$$

当溶质数目增大到n个组分后，使用式(5.7) 很容易可以计算出溶质i的质量浓度：

$$C'_i = C'_{IS} K_{i/IS} \frac{A'_i}{A'_{IS}} \tag{5.7}$$

物质i的百分含量可以使用式(5.8) 计算出来：

$$x_i = (C'_i / \text{分析样品的量}) \times 100\% \tag{5.8}$$

如果用此方法对标液和样品进行多次进样分析，结果会更精确。通常所有的标准溶液和样品溶液中都会加入相同体积的内标物储备液。

总之，这个通用的和可重复的方法要求必须选择一个好的内标物，它应该满足以下特点：

- 必须是稳定的，纯的，而且在初始样品中不存在；
- 必须是可测量的，在色谱图上可以给出有很好分辨率的色谱峰；
- 其保留时间必须接近待定量的溶质（S）；
- 其浓度必须接近或高于待测物质，并能从检测器获得线性响应；
- 必须不干扰且不与样品组分共同洗脱。

5.6.5 内标归一化法

为了对分析样品进行全面评价，选用一种称为"100%归一化法"的方法，其适用于那些每个组分都在色谱图上产生色谱峰的混合物。溶剂峰通常忽略不计。

假设需要测定混合物中三种化合物 1，2，3 的质量浓度，那么分析过程同样是两步。

(1) 计算相对校正因子

制备含有三种已知浓度为 C_1、C_2 和 C_3 的化合物 1、2 和 3 的标准溶液，进样量为 V 的标准溶液的色谱图上给出 A_1、A_2 和 A_3 三个峰面积。根据式(5.5) 峰面积的大小与体积为 V 的溶液中化合物的质量 m_1、m_2 和 m_3 相关。

例如：选择化合物 3 作为内标归一化的物质。该化合物将用于计算化合物 1 和 2 的相对响应因子 $K_{1/3}$ 和 $K_{2/3}$。正如前面的推导：

$$K_{1/3}=\frac{K_1}{K_3}=\frac{m_1A_3}{m_3A_1}\quad,\quad K_{2/3}=\frac{K_2}{K_3}=\frac{m_2A_3}{m_3A_2}$$

鉴于 $m_i=C_iV$，则 $K_{1/3}$ 和 $K_{2/3}$ 的表达式为：

$$K_{1/3}=\frac{C_1A_3}{C_3A_1}\quad,\quad K_{2/3}=\frac{C_2A_3}{C_3A_2}$$

(2) 计算样品浓度的色谱图

接下来的步骤是注射含有待测物 1、2 和 3 的混合物样品。色谱图中标记为 A'_1、A'_2 和 A'_3 的峰面积，可以直接用来计算该混合物中 3 种组分的质量分数为 x_1、x_2 和 x_3，表示如下：

$$x_i=\frac{K_{i/3}A'_i}{K_{1/3}A'_1+K_{2/3}A'_2+A'_3}\times 100\% \tag{5.9}$$

归一化的前提条件是：$x_1+x_2+x_3=100$

如果该方法外推到一个 n 组分的样品对化合物 j 进行归一化，那么对于给定的化合物 i 的相对响应因子的一般表达式为：

$$K_{i/j}=\frac{C_iA_j}{C_jA_i} \tag{5.10}$$

另外，也可以通过为每个溶质绘制一个浓度-响应因子曲线来确定 $K_{i/j}$。

在含有 n 个组分的混合物中，如果 A'_i 表示化合物 i 的峰面积，且内标物为 j，那么化合物 i 的含量可以用方程式(5.11) 表示：

$$x_i=\frac{K_{i/j}A'_i}{\sum\limits_{i=1}^{n}K_{i/j}A'_i}\times 100\% \tag{5.11}$$

假定检测器响应信号独立于被测物质（如 GC 的 TCD 检测器，其相对响应因子均相同)，该方法可用于估算化合物 i 的相对浓度。对早期的积分仪来说，这通常是默认设置[方程式(5.12)]。

$$x_i=\frac{A_i}{n}\times 100\% \tag{5.12}$$

Chapter 6

Size Exclusion Chromatography

Size exclusion chromatography (SEC) is a method by which molecules can be separated according to their size in solution, thus relating indirectly to their molecular masses. To achieve this, stationary phases contain pores through which compounds are able to diffuse to a certain extent. Although the efficiency of separation can never attain that observed with HPLC, SEC has become an irreplaceable tool to separate natural macromolecules in order to study the distribution of synthetic polymer masses. Though the separation of compounds according to their sizes is not the most efficient process for small and medium molecules, this approach remains very useful in industry where the products are most often mixtures of compounds of very different masses. The instrumentation is comparable to that used in HPLC.

irreplaceable
[ɪrpleɪsəbl]
adj. 不能替代的，不能调换的

6.1 Principle of SEC

Size exclusion chromatography (SEC) is based upon the ability of the sample molecules to penetrate into the highly porous "bead"-like structure of the stationary phase (Figure 6.1). Separation arises only as a result of the different degrees of penetration. Molecules of comparatively smaller weight are slowed in their progression in the column because they can enter into the stagnant mobile phase within the pores of the packing. This method is referred to as gel filtration chromatography (GFC) when the stationary phase is hydrophilic (the mobile phase being aqueous) and as gel permeation chromatography (GPC) when the stationary phase is hydrophobic (the mobile phase being a non-aqueous system).

The total volume V_M of the mobile phase in the column can be considered as the sum of two values $V_M = V_I + V_P$: the interstitial volume V_I (external to the pores) and the volume of the pores V_P. V_I, called the void volume, represents the volume of mobile phase necessary to transport a large molecule assumed to be excluded from

assumed
[əsju:md]
adj. 假定的，假装的

the pores and V_M is the volume accessible to a small molecule that can enter all the pores of the packing (volume V_S).

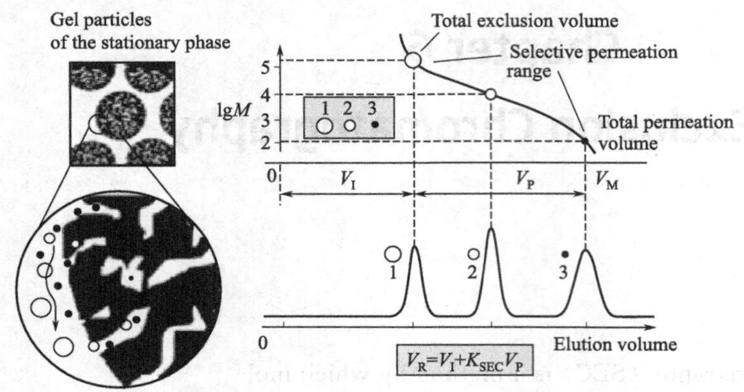

Figure 6.1 Migration across a stationary phase packing. Left, illustrative description of separation in SEC by a porous packing according to the size of the pores. The non-porous part of the bead, called the backbone, is inaccessible to the sample molecules. Right, a chromatogram displaying the separation of three species (1, 2, 3) of different sizes. The large molecules (excluded) 1 are the first to arrive followed by medium sized molecules (partial access) 2, and finally by the smallest (full access) 3. The elution volumes V_R are located between V_I for $K_{SEC}=0$ and V_M for $K_{SEC}=1$

description
[dɪˈskrɪpʃn]
n. 描述，描写，类型，说明书

inaccessible
[ɪnækˈsesɪbl]
adj. 难达到的，难接近的，难见到的

The general expression giving the retention (or elution) volumes V_R are therefore

$$V_R=(V_I+K_{SEC}V_P)+KV_s \quad (6.1)$$

K_{SEC}, the diffusion coefficient, represents the degree of penetration of a species dissolved in the volume V_P ($0<K_{SEC}<1$). Ideal SFC retention is only governed by the continuous exchange of solute molecules between the void volume and the stagnant mobile phase within the pores. If these conditions are attained, Nernst coefficient K is equal to zero and retention volumes V_R are comprised between V_I and V_M, giving:

$$V_R=V_I+K_{SEC}V_P \quad (6.2)$$

For the majority of modern packing materials, V_I and V_P are both of the order of 40 percent of the volume of the empty column.

When V_R/V_M is greater than 1, the behavior of the compound on the column no longer follows rigidly the mechanism of size exclusion but as in HPLC it undertakes physicochemical interactions with the support ($K>0$). Each stationary phase is adapted to a separation range expressed in terms of two masses; a higher mass and a lower mass, above and below which there is no obtainable separation. Molecules whose diameter is larger than those of the greatest pores are excluded from the stationary phase ($K_{SEC}=0$). This explains the expression size exclusion for this situation. They flow through the column without being retained or separated. They appear as a single peak in the chromatogram at the position V_I (Figure 6.1). In contrast, the retention volume of the very small molecules is represented by $V_R = V_M$ (in this case, $K_{SEC} = 1$). The larger the part of sample molecules in the pores, the larger the retardation. To increase this range of separation, which is fixed by the difference between these two volumes, two or three columns can be put in a series, solvents of low viscosity are used and column are occasionally warmed. The diffusion coefficients K_{SEC} are independent of the temperature.

retardation
[ri:tɑ:ˈdeɪʃn]
n. 延迟

coefficient
[kəʊɪˈfɪʃnt]
n.[数]系数，率

6.2 Stationary and mobile phases

SEC stationary phases are constituted of reticulated organic polymers or minerals that are used as porous rigid or semi-rigid beads (3 to 20μm). Pores diameters are within the 4-200nm range. These packings, usually called gels, must withstand the pressure at the head of the column and the temperature until about 100℃ in order to allow their utilization for various applications.

A reduction in the diameter of the particles - a gauge for efficiency - reduces the interstitial passages, rendering migration more difficult for the large, excluded molecules. For this reason, it is preferable to increase the size of the particles and to compensate by using a longer column. Standard columns have a length of 30cm, (with an internal diameter of 6.5mm). Their efficiency N can attain 10^5 plates/m.

interstitial
[ɪntəˈstɪʃl]
adj. 间质的，空隙的；*n*. 填隙原子

SEC is sub-divided into two techniques:

(1) gel permeation chromatography (GPC).

The material packing most often used is a copolymer styrene-divinylbenzene (PS-DVB) with an organic mobile phase such as tetrahydrofuran, a good solvent for most polymers. Trichloromethane as well as hot trichlorobenzene are also employed to dissolve synthetic polymers that are not soluble in other solvents (Figure 6.2). GPC is mainly used in chemical analysis.

(2) gel filtration chromatography (GFC).

aqueous
[ˈeɪkwɪəs]
adj.水的，水般的

For aqueous SEC, stationary phases must be hydrophilic. Some are based upon a PS-DVB base particle that is made biocompatible with a hydrophilic coating containing hydroxyl or sulfonic groups. Others are based upon polyvinyl alcohols either pure or copolymerized with polyglyceromethacrylates or vinyl polyacetates. These packings are called gels since they swell on contact with aqueous mobile phases. Porous silica gels containing hydrophilic surfaces are equally employed (presence of glyceropropyl groups) [$\equiv Si(CH_2)_3OCH_2CH(OH)CH_2OH$]. Adsorption phenomena ($K > 0$) are weak, even for

adsorption
[ædˈsɔ:pʃən]
n.[化学][物]吸附(作用)

the small molecules. GFC is mainly used to separate bio-polymers (e.g. polysaccharides) or other water-soluble biological macromolecules (e.g. proteins).

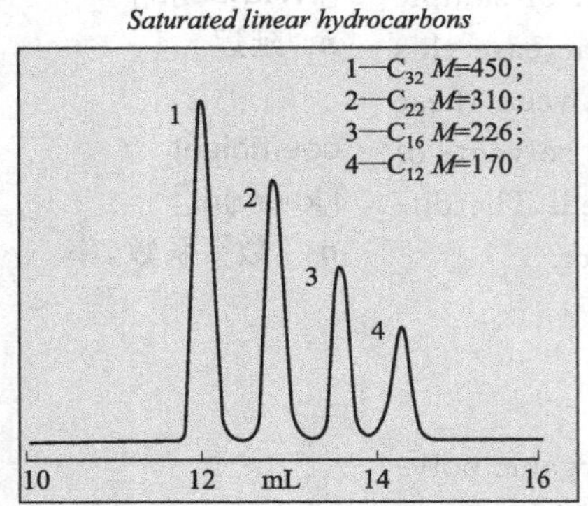

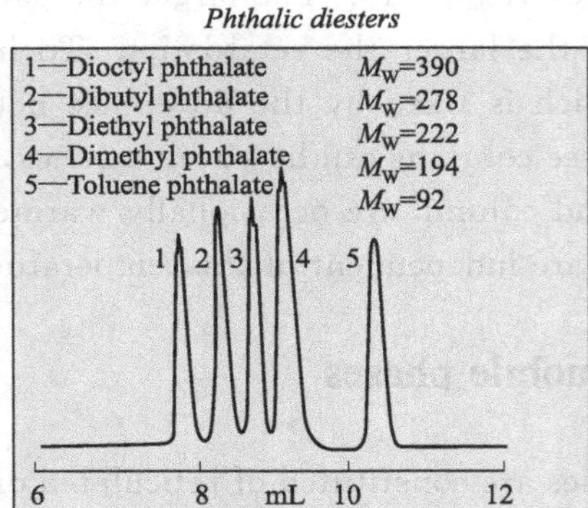

Conditions:Column, PLgel 5μm, Pores,5nm, Dimensions, 300mm×7.5mm.
Eluent: tetrahydrofurane. Flow-rate: 1mL/min.

Figure 6.2 GPC. When the stationary phase has small pores, organic compounds of small to medium molecular weight can be separated

The term gel filtration should not be confused with the current procedure of filtration which would create the reverse effect. The larger molecules have a greater difficulty than the smaller ones in crossing the filter.

6.3 Calibration curves

For a given solvent, each stationary phase is described by a calibration curve made with isomolecular standards of known masses, M: polystyrenes in THF, polyoxyethylenes, pullulanes or polyethyleneglycols, etc. (Figure 6.3 and Table 6.1). The curves representing lg$M$ as a function of the elution volume have a sigmoidal shape. However, by combining stationary phases of differing porosities, manufacturers can provide mixed columns for which the calibration curves are linear over a broader range of masses. These curves are fairly indicative since size and mass are not mutually dependent parameters when passing from one polymer to another.

pullulan [ˈpʌljulən] *n.* 支链淀粉

manufacturer [mænjuˈfæktʃrəz] *n.* 制造商，厂商

permeation [pəmɪˈeɪʃən] *n.* 渗透，[化学]渗透作用

Table 6.1 Permeation range (Da) of three gels for various standard compounds

Standard	G2000:12.5nm pores	G3000:25nm pores	G4000:45nm pores
Globuar protein	5000-100 000	10 000-500 000	20 000-7 000 000
Dextran	1000-30 000	2000-70 000	4000-500 000
Polyethylene glycol	500-15 000	1000-35 000	2000-250 000

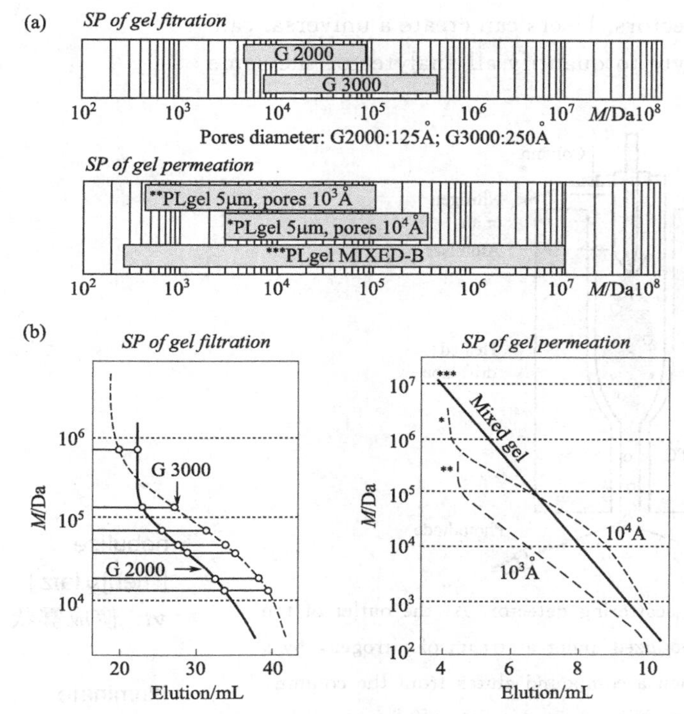

Figure 6.3 Characteristics of stationary phases used for gel filtration (GFC) and GPC columns. (a) Graphs indicating the mass ranges for two phases in gel filtration and for three phases in gel permeation; (b) Calibration curves [$\lg M = f(V)$] for these different phases obtained with proteins for gel filtration and with polystyrene standards for the others (known molecular weights). A weak slope from the linear section reveals a better resolution between neighboring masses. This is the case when the pores are regular dimension. The curves $\lg M = f(K)$, more rarely studied, reveal the same aspect. (reproduced courtesy of Tosohaas and Polymer Lab.). To avoid protein aggregate formation, denaturing compounds are sometimes introduced to the aqueous mobile phase ($1Å = 10^{-10m}$)

aggregate
[ˈægrɪgeɪt]
n. 集合体; *adj*. 聚的,集合的

6.4 Instrumentation

Instrumentation is similar to that employed in HPLC excepting the columns which are bigger. To increase the resolution, it is customary to use two or three columns, with different porosities, in series.

The most commonly used detector is the refractive index detector, considered as universal since for polymers a variation of the refractive index is at first approximation independent of the molecular mass. As this detector is not very sensitive, other detectors are sometimes added to it. They are based upon the light absorption (UV detector) or fluorescence or light scattering (Figure 6.4). This last detector provides a more uniform response to structurally similar analytes than light absorption detectors. Users can create a universal calibration set from a single analyte to quantify all analytes of the same class.

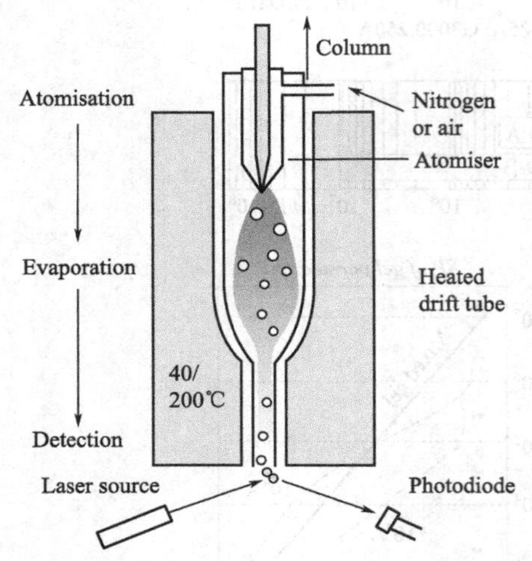

Figure 6.4 Evaporative light scattering detector. At the outlet of the column the mobile phase is nebulized under a stream of nitrogen, by a specially atomizing device. When a compound elutes from the column, the droplets under evaporation give a suspension of fine particles. Illuminated by a laser source, they scatter the light from a lamp via the Tyndall effect (what happens is comparable to the diffusion by fog through a car headlight beam). The signal detected by a photo-diode is proportional to the concentration of the compound illuminated. Irrespective of the substance, the response factors are very close. This detector is only useful for sample components that cannot be vaporized in the heated section of this detector.

nebulize
[ˈnebjʊˌlaɪz]
vt. 使成雾状

illuminate
[ɪˈljuːmɪneɪt]
vt. 照明，启发

6.5 Applications of SEC

The main applications are found in the analysis of synthetic polymers or biopolymers because this method permits the separation of nominal masses ranging from 200 to more than 10^7 Da. The absence of a chemical interaction with the stationary phase, associated with a rapid elution time and the possibility to recover all of the analytes, convey many advantages.

The choice of stationary phase best adapted to a given separation is made by consultation of the calibration curve of various columns. The column of choice is that which presents a linear range over the masses of the compounds formed in the sample (Figure 6.5). The calibration must be carried out with the same type of standards, because macromolecules can have a variety of forms, from pellet like to thread like (Table 6.1).

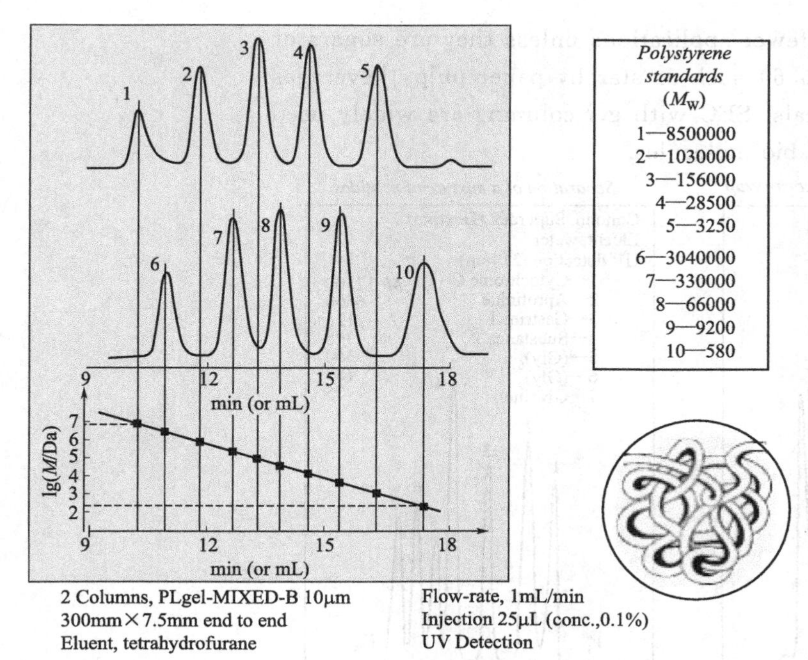

Figure 6.5 Determination of molecular weight in a SEC experiment. The column can be calibrated by using two complementary mixtures of polystyrene standards. Log molecular weight versus retention volume plot. This curve is linear over a wide range of masses due to the use of a mixed stationary phase. Bottom right, conformation supposed for a lipophilic rod or random coil polymer in an aqueous solution. The resulting volume is called hydrodynamic volume of the macromolecule

complementary [kɒmplɪ'mentrɪ] *adj*. 补足的，补充的

hydrodynamic ['haɪdrəʊdaɪ'næmɪk] *adj*. 水力的，流体动力学的

6.5.1 Molecular weight distribution analysis

Synthetic polymers differ from small molecules in that they cannot be characterized by a single molecular weight. Even in its pure state, a polymer corresponds to a distribution of macromolecules of different weights. SEC chromatograms and integrated peak areas allows the determination of the molecular weight distribution as well as the most probable mass and the mean mass. This assumes that the mass and the molecular volumes are directly related. Furthermore, the calibration of the column must be done with standards of the same family and with the same mobile phase flow rate.

6.5.2 Other analysis

SEC, which allows high-speed separations, is used for control analysis of samples with unknown compositions. These samples usually contain polymers and small molecules, which is often the case for numerous commercial or industrial products, such as in the biodegradation of polymers.

For typical organic compounds, which can readily be analyzed by

HPLC or GC, there are fewer applications unless they are sugars or polysaccharides (Figure 6.6) such as starch, paper pulp, beverages and certain pharmaceuticals. SEC with gel columns are widely used for aqueous separation of bio-molecules.

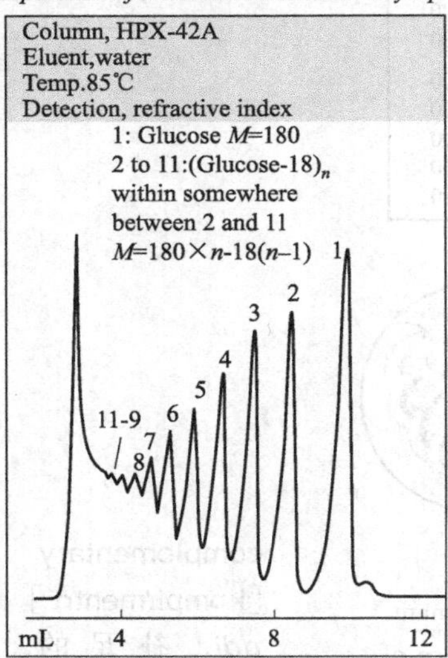

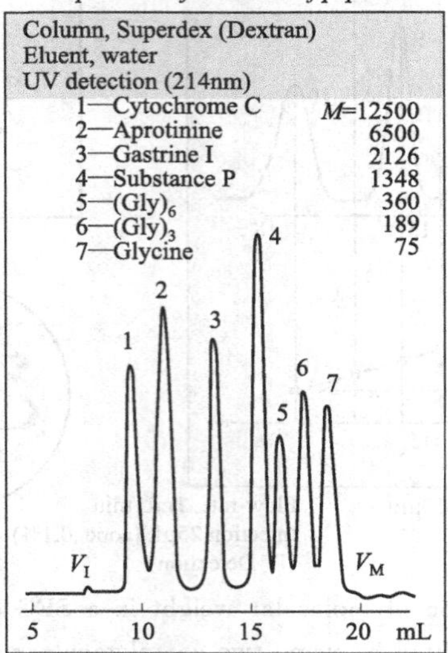

Figure 6.6 Gel filtration chromatogram. Left, separation of a mixture of glucose oligomers ranging from a single unit (glucose $M=180$), up to 20 units (M around 3000). Right, separation of a mixture of diverse proteins and glycine oligomers.

oligomer [ˈɔlɪgəmə] *n*. 低聚物,低聚体

Problems

6.1 Occasionally in SEC values of $K>1$ are observed. How might this phenomenon be accounted for?

6.2 Gel permeation chromatography is to be used to separate a mixture of four polystyrene standards of molecular mass: 9200, 76000, 1.1×10^6 and 3×10^6 daltons. Three columns are available for this exercise. They are prepacked with gel with the following fractionation ranges for molecular weights:

A: 70000 to 4×10^5 daltons

B: 10^5 to 1.2×10^6 daltons

C: 10^6 to 4×10^6 daltons

How might these four polymers be separated in a single operation if it is permitted to use two of the above columns end to end?

6.3 A solution in THF of a set of polystyrene standards of known molecular mass was injected onto a column whose stationary phase is effective for the range 400-3000 daltons. The flow rate of the mobile phase (THF) is 1mL/min. The chromatogram below was obtained.

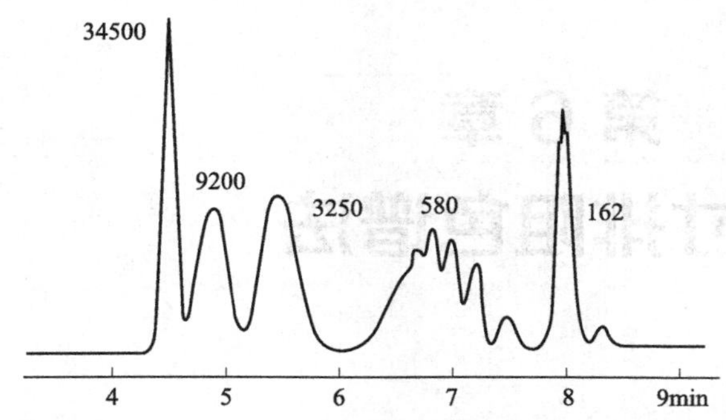

1. What is the total exclusion volume for the column (i. e. the interstitial volume), and what is the volume of the pores (the intra-particle volume)?

2. Calculate the diffusion coefficient K_{SEC} for the polystyrene standard whose relative molecular mass is 3250 daltons.

第6章
尺寸排阻色谱法

尺寸排阻色谱（SEC）是一种根据溶液中分子尺寸进行分离的方法，因此该方法间接地与它们的分子量有关。为了达到这一目的，固定相含有能使化合物在一定程度内扩散的微孔。虽然分离效率不能像HPLC那么高效，但SEC已成为分离天然大分子化合物不可替代的工具，以研究合成聚合物的分子量分布。对于中、小尺寸化合物，根据它们的尺寸进行分离并不是最有效的方法，但是由于工业产品通常是分子量差异较大的化合物的混合物，这种方法在工业中仍然是非常有用的，其经常用于分离不同分子量化合物的混合物。SEC的仪器装置与HPLC的装置类似。

6.1 SEC的原理

SEC的原理是基于样品分子在固定相中高度多孔的“珠”式结构中的渗透能力不同（图6.1）。分离仅仅是不同渗透程度的结果。分子量较小的分子可以进入填料孔内的停滞流动相中从而导致其在色谱柱运行过程中相对被减速。当固定相亲水（流动相为水）时，称为凝胶过滤色谱法（GFC）；而当固定相疏水（流动相是一种非水体系）时，则称为凝胶渗透色谱（GPC）。

色谱柱中流动相的总体积 V_M 是两部分体积的和：$V_M=V_I+V_P$，即孔隙体积 V_I（孔外面的）和孔自身的体积 V_P 之和。

V_I 称为孔隙体积，表示用来输送排除孔外的大分子时所需的流动相体积，V_M 是用来输送可以进入填料所有孔的小分子时的体积（V_S）。

保留（或洗脱）体积 V_R 的一般表达式是

$$V_R=(V_I+K_{SEC}V_P)+KV_s \tag{6.1}$$

K_{SEC} 为扩散系数，表示溶解在体积 V_P 中某物质在色谱柱中的渗透程度（$0<K_{SEC}<1$）。理想SEC的保留只受溶质分子在孔隙体积和孔内的静止流动相之间不断交换控制。如果达到这些条件，能斯特系数 K 等于零，保留体积 V_R 在 V_I 和 V_M 之间，则有：

$$V_R=V_I+K_{SEC}V_P \tag{6.2}$$

对于大多数现代填充材料，V_I 和 V_P 均在空柱体积的40%范围内。

当 V_R/V_M 远大于1时，化合物在柱上的行为不再严格遵循尺寸排阻的机制，而是像高效液相色谱一样，遵循与载体的物理化学相互作用（$K>0$）。

每个固定相的适用分离范围依据两个质量：上限分析质量和下限分析质量，高于或低于这个范围都得不到分离。分子直径大于填料中最大孔的分子被排除在固定相外（$K_{SEC}=0$），这也是尺寸排阻的意思所在。它们流过色谱柱而不被保留或分离。它们在色谱图 V_I 的位置出现一个单峰（图 6.1）。与此相反，非常小的分子的保留体积表示为 $V_R=V_M$（在此情况下，$K_{SEC}=1$）。在孔中的样品分子越多，延迟越大。这两个体积的差值决定了分离范围，为了增加这个分离范围，可以串联两个或三个色谱柱、使用低黏度的溶剂和偶尔加热色谱柱。扩散系数 K_{SEC} 与温度无关。

6.2 固定相和流动相

SEC 固定相由多孔刚性或半刚性（3～20μm）的网状有机聚合物或矿物组成，内孔的直径为 4～200nm。为了适应各种应用，这些通常称为凝胶的填料，要求必须经得起柱头的高压和最高大约 100℃的温度。

粒子直径是分离效率的衡量标准，减小粒径，减小了间隙通道，使排除在孔外的大分子的迁移更加困难。出于这个原因，建议增大颗粒尺寸，使用较长的柱子来进行补偿。标准柱子长度为 30cm（内径为 6.5mm），其理论塔板数（N）可达到 10^5 块/米。

SEC 细分为两种技术。

（1）凝胶渗透色谱（GPC）

最常使用的材料是苯乙烯-二乙烯基苯（PS-DVB）的共聚物，使用有机流动相，如大多数聚合物的良好溶剂四氢呋喃。三氯甲烷以及热的三氯苯也可以用来溶解不溶于其他溶剂的合成聚合物（图 6.2）。GPC 主要用于化学分析。

（2）凝胶过滤色谱（GFC）

对于水性 SEC，固定相必须是亲水性的。有些是基于 PS-DVB 颗粒，内含羟基和磺酸基的亲水性涂层而使其具有生物相容性。其他是基于纯的聚乙烯醇颗粒，或者是乙烯醇与甲基丙烯酸丙三酯或乙酸乙烯酯共聚的颗粒（图 6.3）。因为它们与水性流动相接触时会膨胀，这些填料被称为凝胶。同样也使用含有亲水表面的多孔硅胶凝胶（存在丙三醇基团）[$\equiv Si(CH_2)_3OCH_2CH(OH)CH_2OH$]）。即使针对小分子，吸附现象（$K>0$）也较弱。GFC 主要用于分离生物聚合物（如糖）或其他水溶性生物大分子（如蛋白质）。

不应把凝胶过滤与当前的过滤操作相混淆，这将产生相反的效果。较大的分子比较小的分子更难穿过滤纸。

6.3 校正曲线

对于一个给定的溶剂，每个固定相可用已知质量的等分子量标准物质制作的校准曲线来定义，M 指溶解在四氢呋喃（THF）中的聚苯乙烯、聚氧乙烯类化合物、支链淀粉或聚乙二醇等的分子量（图 6.3 和表 6.1）。图 6.3 表明 $\lg M$ 作为洗脱体积的函数曲线为 S 形。然而，通过结合不同孔隙率的固定相，制造商可以提供在一个较宽的质量范围内校正曲线为线性的混合柱。这些曲线相当明确地指出，当从一种聚合物传递到另一种聚合物时，分子大小和质量是两个无关的参数。

6.4 仪器

SEC 的仪器装置除了色谱柱更大以外，其余部分与高效液相色谱的仪器类似。为了提高分离度，通常将两个或三个不同孔隙度的色谱柱串联在一起。

最常使用的检测器是示差折光检测器，由于对聚合物来说，折射率的变化首先是近似独立于分子量的，故它被看作通用型的检测器。由于该检测器不是很灵敏，有时也会和其他检测器一起使用。这类检测器基于光吸收（紫外检测器）或荧光或光散射原理（图 6.4）。光散射检测器比光吸收检测器对结构类似物可提供更加一致的结果。用户可以创建一个从单一分析物到定量同类所有分析物的通用的校准方法。

6.5 SEC 的应用

SEC 方法允许分离的分子量从 200Da 到超过 10^7Da，主要用于合成聚合物或生物聚合物的分析。与固定相没有化学相互作用、快速洗脱和回收所有分析物的可能性，都是 SEC 的优势。

通过对各色谱柱的校正曲线进行校正，可以选择对某一特定分离物质最适用的固定相。选择样品中所含化合物的质量均在该色谱柱的线性范围之内（图 6.5）。校准必须用同一类型的标准物质进行，因为大分子可以有从球状到线状等多种形式（表 6.1）。

6.5.1 分子量分布分析

合成聚合物不同于小分子的地方在于它们不能表示为单一的分子量。即使在它们纯的状态，聚合物仍相当于不同质量大分子的分布。SEC 色谱图和积分峰面积可以测量分子量分布，以及最可能质量和平均质量，这是在质量和分子体积是直接相关的假设基础上的推定。此外，色谱柱的校准必须在相同标准体系和相同的流动相流速下进行。

6.5.2 其他分析

SEC 允许高速分离，可用于控制分析未知组成的样品。这些样品含有聚合物和小分子，通常情况下为许多商业或者工业产品，如可生物降解的聚合物。

因为典型的有机化合物可以很容易地通过 HPLC 或 GC 进行分析，所以 SEC 的应用越来越少，除非它们是糖或多糖（图 6.6），例如淀粉、纸浆、饮料和某些特定药物。使用凝胶柱的 SEC 技术广泛用于生物分子水溶液的分离。

Chapter 7

Capillary Electrophoresis and Electrochromatography

Capillary electrophoresis (CE) is a very sensitive separation method which has been developed largely through the experience acquired from high performance liquid chromatography, thin layer chromatography and from electrophoretic methods. Bio-molecules and compounds of lower molecular mass, difficult to study by HPLC or classical methods of electromigration on slab gel, become separable by CE. From now on, high performance capillary electrophoresis (HPCE), suited to automation, has replaced the traditional electrophoretic techniques on gels. CE, officially recognized by several regulatory agencies, the FDA among them, included in the pharmacopoeia, can be used for quantitative analysis. Capillary electrochromatography (CEC) is a relatively new hybrid separation method that couples the high separation efficiency of CE with HPLC.

FDA
abbr.(美国)食品和药物管理局(Food And Drug Administration)

pharmacopoeia
[ˌfɑːməkəˈpiə]
n. 药典

7.1 From zone electrophoresis to capillary electrophoresis

Capillary electrophoresis corresponds to an adaptation of the more general electrophoresis methodology. This reparative technique is based upon the differential migration of the species, whether or not they carry an overall electric charge, present in the sample solution, under the effect of an electric field and when supported by an appropriate medium.

reparative
[ˈrepərətɪv]
adj. 弥补的

In the classic semi-manual electrophoretic technique used mainly in bio-analysis, a small slab or strip of plastic material covered by a porous substance (a gel) is impregnated with an electrolyte buffer. The two extremities of the covered gel system are dipped into two independent reservoirs containing the same electrolyte and linked to the electrodes of a continuous voltage supply (Figure 7.1). The sample is deposited in the form of a transverse band, which is cooled and then bedded between two isolating plates. Under the effect of several parameters that act jointly-voltage (500V or more in the case of small molecules), charge, size, shape, temperature and viscosity-the

electrolyte
[ɪˈlektrəlaɪt]
n. 电解液,电解质

hydrated species migrate from one end to the other, generally towards the electrode or pole opposite sign on a time scale that can vary from a few minutes to over an hour. Each compound is differentiated by its mobility but the absence of a measurable solvent front, compared to TLC, requires that the distance of migration is estimated in relation to that of a substance used as an internal marker. The detection of species, following migration, through a contact process, are transferred onto a membrane where they are derivatized with specific reagents as silver salts, or Coomassie Blue (Figure 7.1). Markers containing radioactive isotopes (^{32}P or ^{3}H) can also be used to give traces that can be exploited in the same manner as in TLC.

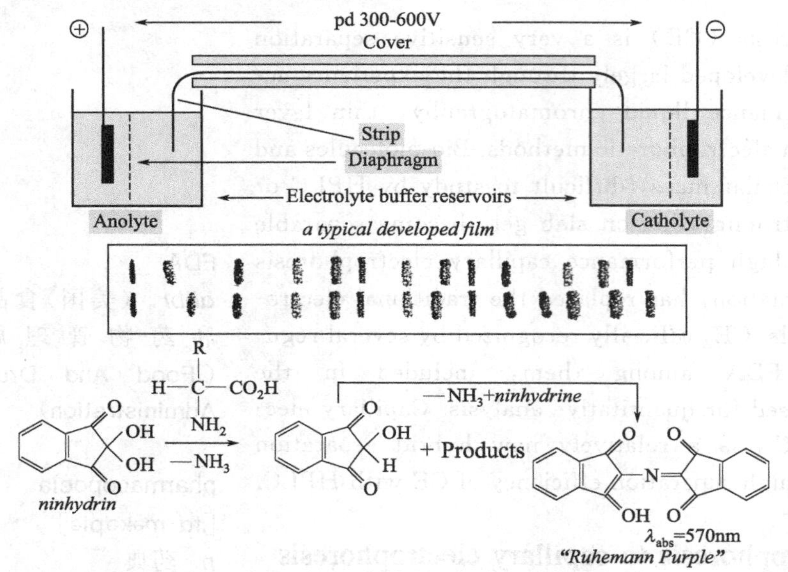

Figure 7.1 Schematic of a horizontal electrophoresis apparatus. Top, each compartment is separated by a diaphragm in order to avoid contamination of the electrolyte by secondary products which are formed during contact with the electrodes. This technique is operated either at constant current or at constant voltage or at constant power. The flat gel is between two plates of glass. Middle, aspect of a standard strip support after revelation. Bottom, ninhydrin is often used to reveal proteins or amino acids. This reagent is transformed on contact with an amino acid that it degrades, to an unstable compound intermediate which reacts in turn with a second molecule of ninhydrin yielding "Ruhemann purple" This reagent is one of several compounds existing (e. g. Fluorescamine) to determine position of particular species following migration on the strip. Horizontal gels offer many advantages for nucleic acid separation and remain a widely-used tool for the molecular biologist

In capillary electrophoresis, the covered gel slab of the classic technique is replaced by a narrow-bore open-ended fused silica capillary (15-100μm diameter). The capillary of length L, varying between 20 and 80cm, is filled with an electrolyte aqueous buffer so-

hydrated [ˈhaɪdreɪtɪd] *adj*. 含水的，与水结合的

mobility [məʊˈbɪlətɪ] *n*. 流动性，流动

Coomassie Blue 考马斯亮蓝

strip[strɪp] *n*. 带，条状

diaphragm [ˈdaɪəfræm] *n*. 隔膜

anolyte[ˈænəlaɪt] *n*. 阳极电解液

catholyte[ˈkæθəlaɪt] *n*. 阴极电解液

ninhydrin [nɪnˈhaɪdrɪn] 茚三酮

amino [əˈmiːnəʊ] *adj*. 氨基的

Ruhemann Purple 罗曼紫

fluorescamine [fluːəˈreskəmiːn] *n*. 荧光胺

nucleic acid [nuˌklɪkˈæsɪd] 核酸

slab[slæb] *n*.平板

lution as the two reservoirs (Figure 7.2). For a better control of the migration time, it is advisable to place the capillary in a thermostatic oven. A small inner diameter of the capillary reduces Joule heating to negligible levels and allows the use of high electric fields for very rapid separations (the voltage applied to the electrodes can attain 30kV).

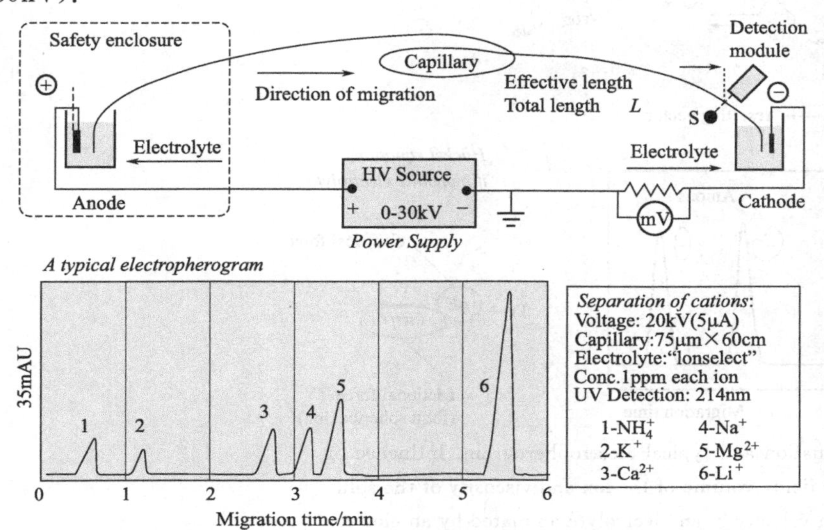

Figure 7.2 Schematic of a capillary electrophoresis instrument. The electrolyte is an aqueous ionic solution, filtered, degassed and containing a variety of additives. A small volume of the sample is injected at the positive end of the capillary. For a voltage, greater than 5-600V/cm (30 kV if $L = 50$cm) special insulation becomes necessary to avoid shorts and arcs of current in a humid atmosphere. For safety reasons, one electrode is usually at ground. The effective distance of migration l is around 10cm shorter than the length L of the capillary. Below is a typical electropherogram corresponding to the separation of several canons (ordinate given in units of milli-absorbance). The electrolyte is a commercially available mixture. The distortion of the peaks is due to electrodispersion. This phenomenon is caused by the differences in conductivity, and hence field, in each zone. The peaks would be symmetrical if their speeds were closer together

electropherogram
电泳图(谱)

absorbance
[əb'sɔ:bəns]
n. 吸光度,吸收率

distortion
[dɪ'stɔ:ʃən]
n. 变形,失真

electrodispersion
[elektrədɪs'pɜ:ʃn]
n. 电分散作用

A detector is placed near the cathode compartment, at a distance from the head of the capillary. The signal obtained provides the baseline for the electropherogram (Figure 7.2) which yields information about the composition of the sample. The only species detected are those which are directed towards the cathode.

7.2 Electrophoretic mobility and electro-osmotic flow

Particles suspended in a liquid, like solvated molecules, can carry electric charges. The sign and size of the charge depend upon

solvated
['sɒlveɪtɪd]
adj. 溶剂化的

the nature of the species, that of the electrolytic medium as well as on the pH (Figure 7. 3). This net charge originates from the fixation to the particle surface, of ions contained in the buffer electrolyte.

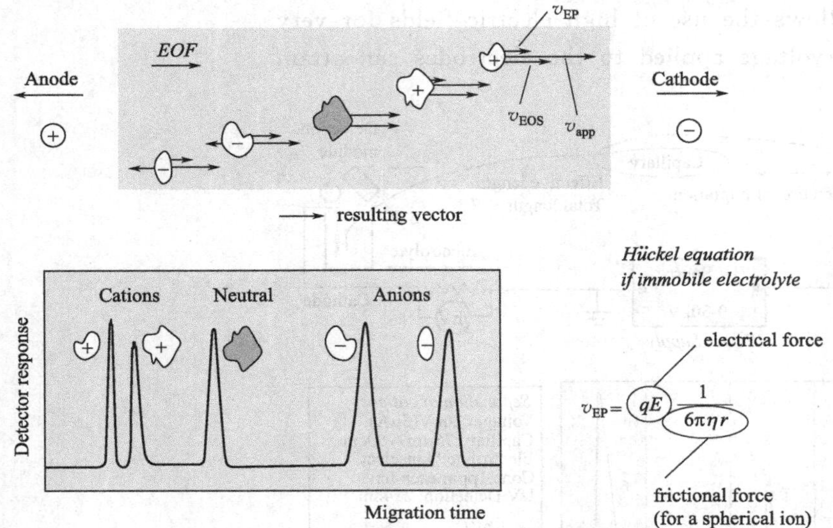

Figure 7.3 Huckel equation and typical electropherogram. Influence of the net charge, electric field, volume of the ion and viscosity of the solution upon the migration velocity in an electrolyte animated by an electroosmotic flow. Higher charge and smaller size confer greater mobility, whereas lower charge and larger size confer lower mobility. The separation depends approximately upon the charge-to-size ratio of each species. Uncharged species move at the same velocity as the electroosmotic flow. The smaller anions arrive last since they would normally go towards the anode. At low pH, electroosmotic flow is weak and anions may never reach the detector

Under the effects of various phenomena or simultaneous actions such as temperature, viscosity, voltage, these particles will have migration velocities that will be faster, the smaller they upon the charge-to-size are and the greater their charge. The separation depends on ratio of the hydrated analyte ions.

For each ion between the electric particle of charge q, limiting migration velocity v_{EP} results from the equilibrium force F, which is exerted in the electric field E acting on the and the forces resulting from the solution viscosity η. Neutral species are poorly separated except if an ionic agent is added to the electrolyte in order to associate with them.

Hückel proposed an equation taking account of the influence of these factors upon the electrophoretic velocity v_{EP} of an ion considered as a sphere of radius r.

The charged species present in the sample are submitted to two principal effects, which are their individual electrophoretic mobility

analyte
[ænə'lɪt]
n.（被）分析物

equilibrium
[iːkwɪ'lɪbrɪəm]
n. 平衡

radius
['reɪdɪəs]
n. 半径

and the more global electroosmotic flow.

7.2.1 Electrophoretic mobility-electromigration

A compound bearing an electric charge move within the electrolyte assumed to be immobile at a velocity v_{EP} (m/s) which depends upon the conditions of the experiment and of its own electrophoretic mobility μ_{EP}. For a given ion and medium, μ_{EP} is a constant which is characteristic of that ion. This parameter is defined from the electrophoretic migration velocity of the compound and the exerted electric field E, using expression (7.1):

$$\mu_{EP}=\frac{v_{EP}}{E}=v_{EP}\frac{L}{V} \qquad (7.1)$$

In the above equation, L designates the total length of the capillary (m) and V the voltage applied across the extremities of the capillary. A positive or negative sign is assigned to the electrophoretic mobility, according to the cationic or anionic nature of the species; μ_{EP} ($m^2/V \cdot s$) is equal to zero for a globally neutral species. μ_{EP} can be obtained from an electropherogram by calculating v_{EP} of the compound in a known electric field E (V/m), taking into account of the velocity of the electrolyte (cf. Section 7.2.2). This parameter depends not only on the charge carried by the species, but also on its diameter and on the viscosity of the electrolyte.

7.2.2 Electro-osmotic mobility-electro-osmotic flow

The second factor that controls the migration of the solute is the *electro-osmotic flow* (*EOF*). This flow corresponds to the bulk migration of the electrolyte through the capillary. In gel electrophoresis, this flow is small, but in capillary electrophoresis it becomes more important because of the internal wall of the capillary. It is characterized by the electro-osmotic mobility μ_{EOS}, as defined by a relation similar to (7.1). v_{EOS} represents the velocity of the electro-osmotic flow.

$$\mu_{EOS}=\frac{v_{EOS}}{E}=v_{EOS}\times\frac{L}{V} \qquad (7.2)$$

In order to calculate μ_{EOS}, the electro-osmotic flow velocity v_{EOS} must first be determined. This corresponds to the velocity in the electrolyte of a species without any global charge. This is done by measuring the migration time t_{nm} for a neutral marker to migrate over the distance l of the capillary ($v_{EOS}=l/t_{nm}$).

As a neutral marker, an organic molecule that is non-polar at the pH of the electrolyte used and easily detected by absorption in the near UV, is selected (e.g. acetone, mesityl oxide or benzyl alcohol).

The electro-osmotic flow has several origins in relation with the

global
['gləubəl]
adj. 总体的，球形的

bear
[beə(r)] *vt*. 具有

exerted
[ɪg'zɜːtɪd]
adj. 外露的

neutral
['njuːtrəl]
adj.中性的

EOF (electro-osmotic flow)
电渗流

gel [dʒel]
n. 凝胶，胶体

mesityl oxide
异亚丙基丙酮

benzyl alcohol
[化]苯甲醇，苄醇

internal wall of the capillary. A capillary made of fused silica not having been subjected to any particular treatment bears on its inner surface numerous silanoates (Si—OH) which become ionized to silanoates (Si—O^-) when the pH of the electrolyte rises above 3.

Under these conditions, a fixed polyanionic layer is formed that attracts cations present in the electrolyte. These cations arrange themselves into two layers of which one is attached to the internal wall while the other remains slightly mobile (Figure 7.4). Between these two layers a potential difference (Zeta potential) develops, the value of which depends upon the concentration of the electrolyte and the pH. When the electric field is applied, H_3O^+ and ions being solvated by water molecules migrate towards the cathode. In capillary electrophoresis instruments, the net effect of the electro-osmotic flow is to impose for all charged species present in the sample the direction oriented from the anode towards the cathode. The anions progress in an anti-electro osmotic fashion.

cation
['kætaɪən]
n. 阳离子

Zeta potential
['ziːtəpə'tenʃl]
n. 电动电势

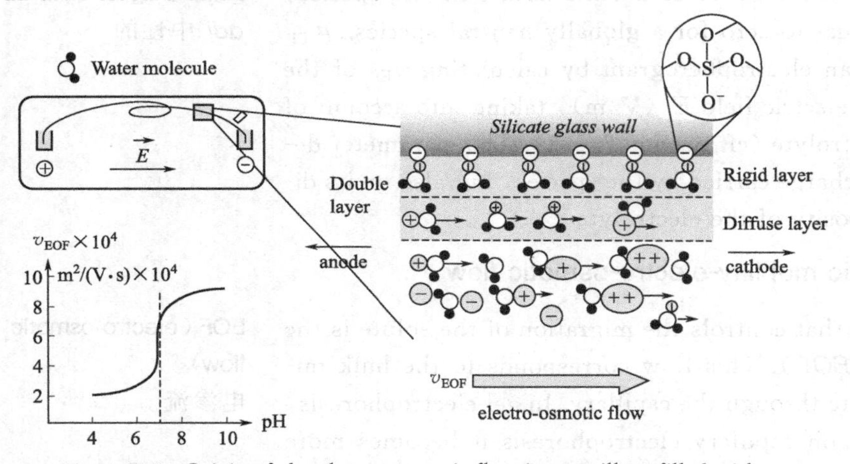

Figure 7.4 Origin of the electro-osmotic flow in a capillary filled with an electrolyte. Model of the double layer. If the inner wall has not been treated (polyanionic layer of a silica or glass capillary) then a pumping effect arises from the anodic to the cathodic compartment: this is the electro-osmotic flow which is reliant upon the potential which exists on the inner surface of the wall. If the wall is coated with a non polar film (e. g. octadecyl) then this flow no longer exists. The electro-osmotic flow is proportional to the thickness of the double cationic layer attached to the wall. It is reduced if the concentration of the buffer electrolyte increases. v_{EOS} is pH dependant: between pH 7 and 8 v_{EOS} can increase by as much as 35 percent

Usually, a capillary with a negative inner surface generates a linear displacement of the electrolyte (an electro-osmotic flow) directed towards the cathode. In contrast, if a surfactant is added, such as tetra-alkyl ammonium, capable of reversing the polarity of the inner wall, then the electro-osmotic flow is directed towards the anode

alkyl
['ælkɪl]
n. 烷基，烃基；adj. 烷基的，烃基的

(Figure 7.5). By treating the wall with an alkylsilane to make it hydrophobic, proteins become separable, otherwise they tend to adsorb at the surface of bare fused silica.

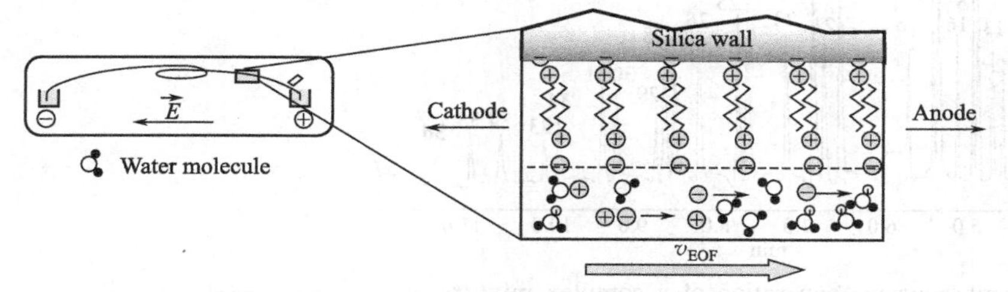

Figure 7.5 Effect of a cationic surfactant upon the silica inner wall. As the migrations must always be made in the direction where the detector is, the voltage polarity of the instrument must therefore be reversed in order for anionic species to progress naturally in the same direction as the electrolyte, that is, toward the detector

7.2.3 Apparent mobility

According to the previous arguments, each ion has an apparent migration velocity v_{app} which depends on both the electrophoretic velocity and the electro-osmotic flow (relation 7.3):

$$v_{app}=v_{EP}+v_{EOS} \tag{7.3}$$

v_{app} is easily calculated from the electropherogram. Using l the effective length of the capillary, and t_m, the migration time, v_{app} is given by the expression:

$$v_{app}=l/t_m \tag{7.4}$$

The apparent electrophoretic mobility μ_{app} is defined by an expression analogous to (7.1) or (7.2), such that:

$$\mu_{app}=\frac{v_{app}}{E}=v_{app}\frac{L}{V} \tag{7.5}$$

and, by consequence

$$\mu_{app}=\frac{lL}{t_m V} \tag{7.6}$$

By combining the electro-osmotic flow and the apparent mobility it is possible to calculate the true electrophoretic mobility of the species carrying charges. From relation (7.3) it can be written:

$$\mu_{EP}=\mu_{app}-\mu_{EOS}$$

or alternatively $$\mu_{EP}=\frac{Ll}{V}\times\left(\frac{1}{t_m}-\frac{1}{t_{nm}}\right) \tag{7.7}$$

HPCE is commonly used for the separation of anionic species. Generally, the polarity of the instrument is reversed in order to change the location of the detector from the cathode end to the anode end of the capillary. As a result, the electro-osmotic flow direction is also reversed. Only anions whose μ_{EP} values are greater than those of μ_{EOS} will be detected (Figure 7.6).

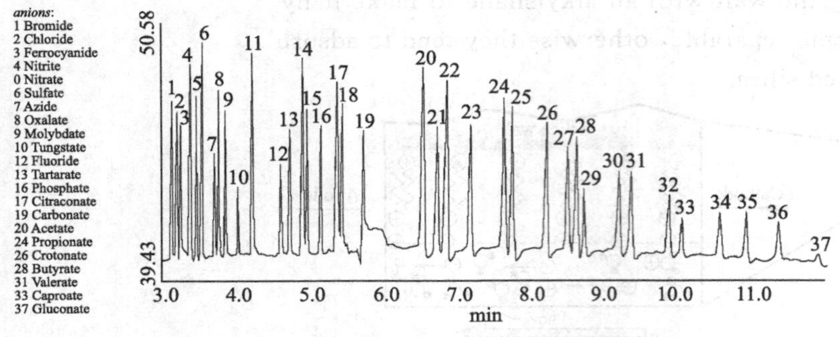

Figure 7.6 An electropherogram. Separation of a complex mixture of anions. The lack of peak symmetry (see also Figure 7.2), is due to variations in the electric field along the whole length of the capillary caused by the dynamic ionic composition

7.3 Instrumentation

To introduce a micro-volume of sample, which must not exceed 1 percent of the effective length of the capillary in order to protect the resolution, two procedures are used:

- Hydrostatic injection. This action is achieved by dipping the end of the capillary into the solution (also an electrolyte) containing the sample while creating a slight vacuum at the opposite end. The procedure can be improved by exerting a pressure of around 50mbar upon the sample solution.

hydrostatic [ˌhaɪdrəˈstætɪk] *adj*. 静水力学的，流体静力学的

- Electro-migration injection. This technique, used in gel electrophoresis, is achieved by putting the sample at a potential of appropriate polarity compared to the other extremity and briefly dipping the capillary into it. In contrast to the previous procedure this mode of injection induces a discriminatory action upon the compounds present, which leads to a non-representative composition of the sample.

The hydrostatic injection method is less precise than in HPLC because injection loops do not exist for volumes between 5-50nL. The quantity entering the capillary is dependent upon many of the parameters that appear in the well-known Poiseuille expression which gives the flow rate F in a tube (radius r, length L) for a liquid having a dynamic viscosity η (expression 7.8). The application of this formula results in an approximate value for what might be termed the "entering flow rate" in the capillary.

formula [ˈfɔːmjʊlə] *n*. 公式，准则

$$F=\Delta p\ \frac{\pi r^4}{8\eta L} \qquad (7.8)$$

This relationship reveals that if the radius of a capillary is doubled, then the volume entering will be 16 times greater. This volume is also proportional to the difference in pressure Δp.

Indirect UV detection is used for analytes—as inorganic ions—which do not absorb UV radiation. It involves using a buffer in the capillary such as a chromate or a phthalate which have high absorption coefficients under these experimental conditions. As analytes move past the detector the amount of light passing through the capillary increases as UV absorbing buffer is excluded, leading to negative peaks (Figure 7.7 and 7.8).

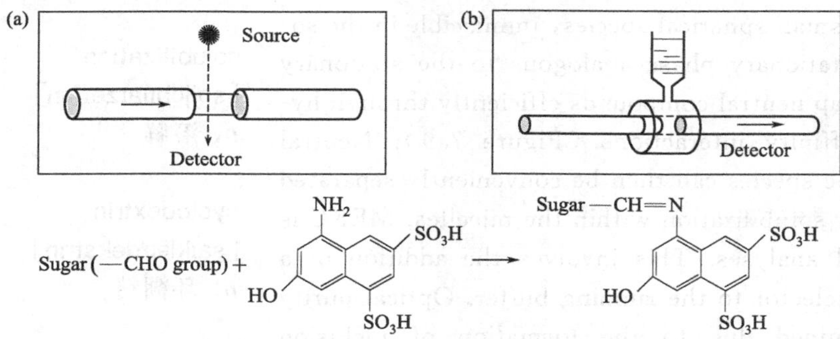

Figure 7.7 UV detection and post-separation derivatization. Analytes passing the light source absorb UV radiation resulting in a decrease in light reaching the detector. (a) The measuring cell comprising the capillary itself. (b) Post-derivatization can be undertaken by cutting the capillary just prior to the detector and by introducing an appropriate reagent. Below, example of a reaction to transform a sugar into a fluorescing derivative

derivatization [dəˌrɪvəˈtɪzeɪʃən] n. 衍生(作用),衍生化

chromate [ˈkrəumɪt] n. 铬酸盐

phthalate [ˈθæleɪt] n. 邻苯二甲酸盐[酯]

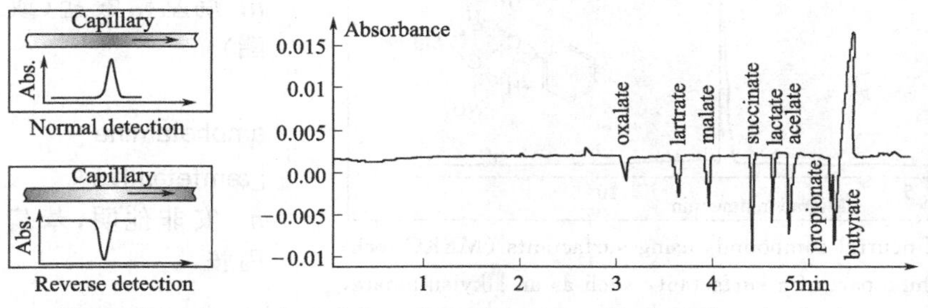

Figure 7.8 Separation of the principal organic acids in white wine by indirect UV detection

7.4 Electrophoretic techniques

7.4.1 Capillary zone electrophoresis (CZE)

This mode of electrophoresis, in which the electrolyte migrates through the capillary, is the most widely used. In this mode, samples are applied as a narrow band that is surrounded by the electrolyte buffer. This electrolyte can be, depending upon the application, acidic (phosphate or citrate) or basic (borate) or an amphoteric substance (a molecule possessing both an acidic and an alkaline function). The

phosphate [ˈfɒsfeɪt] n. 磷酸盐

citrate [ˈsɪtreɪt] n. 柠檬酸盐

borate [bɔːreɪt] n. 硼酸盐

electro-osmotic flow increases with the pH of the liquid phase or can be rendered inexistent. This procedure is also called, in contrast to CGE, free solution electrophoresis.

7.4.2 Micellar electrokinetic capillary chromatography (MEKC)

In this variant of CZE, adapted to the separation of neutral or polar molecules, a cationic or anionic surfactant, e.g. sodium dodecylsulfate, is added in excess to the background electrolyte to form charged micelles. These small spherical species, immiscible in the solution, form a pseudo-stationary phase analogous to the stationary phase in HPLC. They trap neutral compounds efficiently through hydrophilic/hydrophobic affinity interactions (Figure 7.9). Neutral molecules as well as ionic species can then be conveniently separated as a direct result of their solubilization within the micelles. MEKC is useful to perform chiral analyses. This involves the addition of a cyclodextrin as a chiral selector to the running buffer. Optical purity analysis can be determined due to the formation of inclusion complexes between a hydrophobic portion of the solute and the cavity of the cyclodextrin, having different stabilities (Figure 7.9).

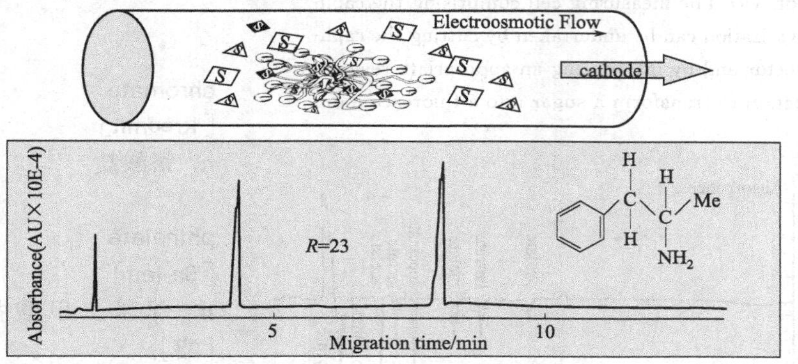

Figure 7.9 Separation of neutral compounds using surfactants (MEKC technique). Above, the lipophilic part of a surfactant, such as an alkylsulfonate, can bind moderately to certain molecules of the substrate S. The negatively charged micelles are directed towards the cathode by the electro-osmotic flow. Below, electropherogram of a racemic sample of an amphetamine in the presence of a cyclodextrin. The electrolyte, NaH_2PO_4 25mmol (pH2.5), contains 5 per cent of gamma-cyclodextrin polysulfate. At this pH the electro-osmotic flow is negligible. The hydrogenosulphate ions of the cyclodextrins being directed towards the anode collect the amphetamine molecules retained in their hydrophobic cavities. The two enantiomers, in equal quantity, form two diastereomer complexes which migrate at different velocities. For this experiment the anode is on the detector side

7.4.3 Capillary gel electrophoresis (CGE)

This technique represents the transposition of the classic agarose

immiscible [ɪ'mɪsəbəl] *adj*. 不能混合的，不融合的

solubilization [sɒljʊbɪlaɪ'zeɪʃən] *n*. 溶解

cyclodextrin [saɪkləʊ'dekstrɪn] *n*. 环糊精

lipophilic [lɪpə'fɪlɪk] *adj*. 亲脂性的

alkylsulfonate [ɔːlkɪl'sʌlfɒneɪt] *n*. 烷基磺酸盐（或酯）

amphetamine [æm'fetəmiːn] *n*. 安非他明，苯基丙胺

gamma-cyclodextrin polysulfate γ-环糊精聚硫酸盐

hydrogenosulphate *n*. 硫酸氢盐

diastereomer [daɪə'stɪərɪəʊmə] *n*. 非对映体，非对映异构体

or polyacrylamide gel (PAG) electrophoresis into a capillary. The capillary is filled with an electrolyte impregnated into a gel. This produces a filtration effect, which decreases the electro-osmotic flow and minimizes convection and diffusion phenomena. Fragile oligonucleotides can be separated in this way.

CGE has been adapted to DNA sequencers. Special instruments fully automated (from sample loading to data analysis) have been designed with multiple capillaries that can simultaneously analyze many samples through a fluorescent-based detection. The modified technique is known as capillary array electrophoresis for the separation of the DNA or RNA fragments based on their size.

In classic zone electrophoresis, the support on which the migration takes place, can contain a polyacrylamide gel impregnated with the electrolyte. If the latter contains sodium dodecylsulfate (SDS), then the electrophoresis is named SDS-PAGE, after its acronym. Separations are improved by a filtering phenomenon that superimposes to the electric forces.

7.4.4 Capillary isoelectric focusing (CIEF)

This technique consists of creating a stable and linear pH gradient in a surface-treated capillary, which has been filled with a solution that contains ampholytes.

Ampholyte or zwitterions. An ampholyte is a molecule that contains both acidic and basic groups. At a particular pH, known as the isoelectric point, the charge on the groups is balanced and the molecule is neutral. If an ampholyte is placed in a pH gradient and electrophoresed it will migrate to the point at which it is uncharged and then stops moving.

At the anode side, the capillary is plunged into H_3PO_4 solution while on the cathode side, it is dipped into a NaOH solution. When an electric field is applied to the extremities of a capillary filled with a mixture of solutes and ampholytes, a pH gradient appears. Each of the charged analytes (generally, proteins) migrates through the medium until it reaches the region where the pH has the same value as its isoelectric point (at its pI, the component net charge is zero). Then, by maintaining the electric field and using a hydrostatic pressure, these separated species migrate towards the detector. In order to ensure the high performance of analysis, standards of pI (pI markers) are needed. The high resolutions obtained with this procedure allow especially the separation of peptides whose pI may differ by only 0.02 pH units.

oligonucleotide [ɒlɪgəʊ'nju:klɪətaɪd] *n.* 低(聚)核苷酸，寡核苷酸

SDS 十二烷基硫酸钠

superimpose [su:pərɪm'pəʊz] *vt.* 添加，重叠，附加

ampholyte [æm'fəʊli:t] *n.* 两性电解质

isoelectric point 等电点

dip [dɪp] *vt.* 浸，泡，蘸，舀取，把伸入；*n.* 下沉，下降，倾斜，浸渍

extremity [ɪk'streməti] *n.* 末端，极限

7.5 Performance of CE

The performance of CE, for the separation of biopolymers is comparable to that of HPLC. The success of a separation relies on the choice of an appropriate buffer medium adapted to the analysis. Only a very small quantity of sample is required and the reagent consumption or solvent is negligible (Figure 7.10). The sensitivity is very high: by using laser induced fluorescence (LIF) detection a few thousands of molecules can be observed.

LIF 激光诱导荧光

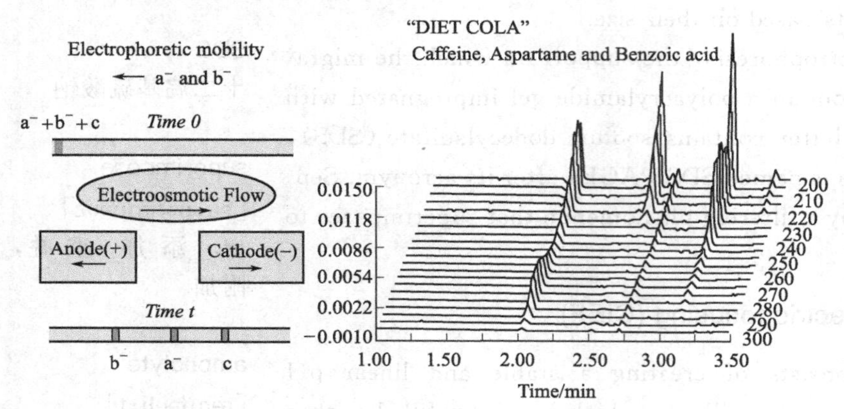

Figure 7.10 Electro-kinetic separation of three species a^-, b^- and c. Left, The electro-osmotic flow carrying all of the charged or neutral species along with it, is directed towards the cathode. The negative species though attracted by the positive pole cannot overcome the electro-osmotic flow and are therefore displaced towards the cathode. Right, separation of caffeine c and of the anions of aspartame a^- and of benzoate b^- from a sample of "DIET COLA". Presentation in the form of a 3D electropherogram

aspartame [ˌəˈspɑːteɪm] *n*. 阿斯巴甜,天冬甜素

benzoate [ˈbenzəʊeɪt] *n*. 苯甲酸盐

However, the reproducibility of the analyses is less certain than in HPLC because there are many subtle factors that can affect the precision of the injected volume. For hydrostatic introduction, relative standard deviation is of the order of 2 per cent, if an internal standard is used. When the efficiency *N* of the separations is sufficiently large, it becomes possible to separate isotopes of an element (Figure 7.11).

HPCE in its 'lab-on-a-chip' version is at the frontier of classic chemistry and is particularly interesting to molecular biologists for the analysis of degradation products of proteins.

degradation [ˌdegrəˈdeɪʃən] *n*. 退化,降格,降级

The separation parameters (efficiency, retention factor, selectivity, resolution) are accessible from electropherograms by using either the same formulae as in chromatography, or those specific to

formulae [ˈfɔːmjəliː] *n*. 公式(formula 的复数)

HPCE.

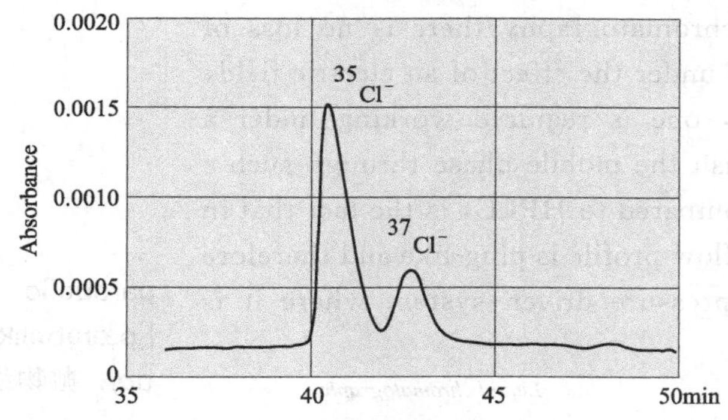

Figure 7.11 Separation of chlorine isotopes. When the efficiency is very high isotopes of the same element can be separated which as in this example leads to two clearly defined peaks. Conditions: capillary of 75μm/47cm. $V=20\text{kV}$, $T=25℃$, electrolyte: chromate/borate pH 9.2.

dispersion
[dɪ'spɜːʃn]
n. 散布，离差，驱散

The theoretical efficiency of a separation N^- as high as one million plates/metre in a column of length, L^- can be calculated from its effective length l and from the diffusion coefficient D (cm^2/s). This latest parameter is linked to the dispersion σ and to the migration time t_m via Einstein's law ($\sigma^2=2Dt_m$) Expression 7.9 is attained (or 7.10 if t_m is eliminated by using expression 7.6). These expressions may be use also to calculate D if N is experimentally available.

$$N=\frac{l^2}{2Dt_m} \tag{7.9}$$

$$N=\frac{\mu_{app}}{2D}\cdot V\cdot\frac{l}{L} \tag{7.10}$$

Relation 7.10 reveals that the efficiency grows up with the voltage applied. The macromolecule, for which the diffusion factors are slighter than those of small molecules, inevitably lead to better separations (Figure 7.12). Finally in HPCE, the resolution R between two peaks can be calculated by using the efficiency N, the difference in the speed of migration of the two solutes (Δv) and their average speed $\bar{v}$ (7.9):

$$R=\frac{\sqrt{N}}{4}\times\frac{\Delta v}{v} \tag{7.11}$$

7.6 Capillary electrochromatography

Capillary electrochromatography (CEC) is a separation method that borrows from CZE and HPLC, by associating the electromigration of ions with their partitioning between the phases, typical to chromatography. Instead of a hydraulic pump, it is the high voltage power supply of the CE instrument that generates a flow through the packed bed capillary. In electrochromatography there is no loss of charge: each H_3O^+ ion migrates under the effect of an electric field, while in liquid chromatography, one is required working under a pressure close to 1000 bars to push the mobile phase through such a capillary column. The benefit, compared to HPLC, is the fact that in an electrically driven system the flow profile is plug-like and therefore much more efficient than in a pressure driven system where it is parabolic (Figure 7.12).

parabolic [ˌpærəˈbɒlɪk] adj. 抛物线的

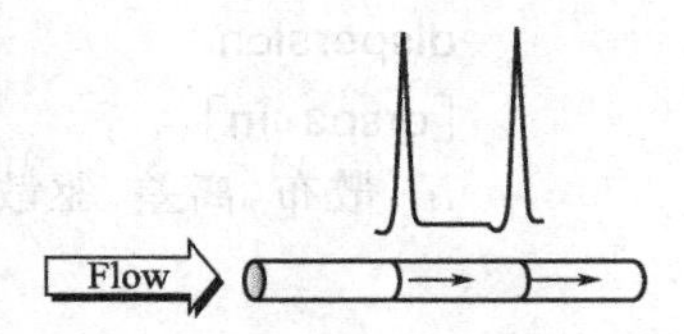

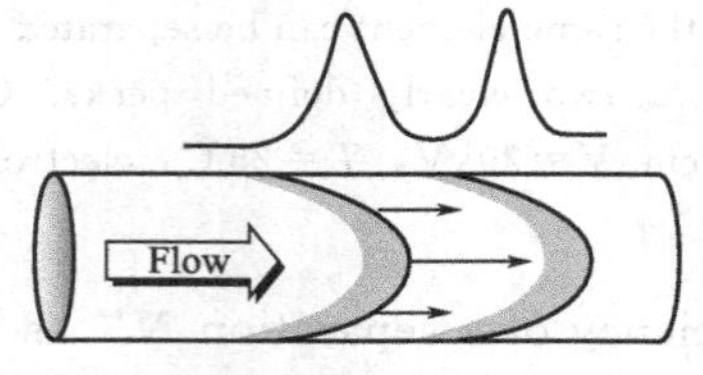

Figure 7.12 Comparison of the conjugated effects of diffusion and of the mode of progression of the mobile phase in HPCE and in HPLC. Diffusion which increases as the square of the diameter of the tube is smaller in electrophoresis than in liquid chromatography, especially if the sample mixture comprises compounds of greater molecular mass. Elsewhere in CE, the electrolyte is drawn along by the internal wall, while it is pushed in HPLC and in EC. Finally, the profile of the hydrodynamic flow is almost perfectly flat and not parabolic as in HPLC

The method requires a high-purity fused capillary filled with standard HPLC packing materials constituted of particles that can be of very small diameters 1-3μm or of continuous bed (monoliths) since there is no back pressure.

monolith [ˈmɒnəlɪθ] n. 整体材料

Since the surface areas of micro-particulate silica based packing materials are much greater than that of the capillary walls, a significant part of the electroosmotic flow is generated by the remaining surface silanol groups of the stationary phase (Figure 7.13). The packing behaves in a fashion analogous to an untreated internal capillary wall. This does not apply in the case of totally base deactivated materials, such as RP-18 silica which have minimal levels of residual silanol groups.

deactivate [diːˈæktɪveɪt] vt. 使无效,失效

silanol [ˈsɪleɪnɒl] n. 硅烷醇

The stationary phase acts also with varying success as a retaining material. Nonaqueous mobile phases (for hydrophilic analytes) and untreated silica packings may also be used. The separation factor is very high but a change of composition of the mobile phase or of the organic modifiers will affect the electro-osmotic flow and then the selectivity (Figure 7.14)

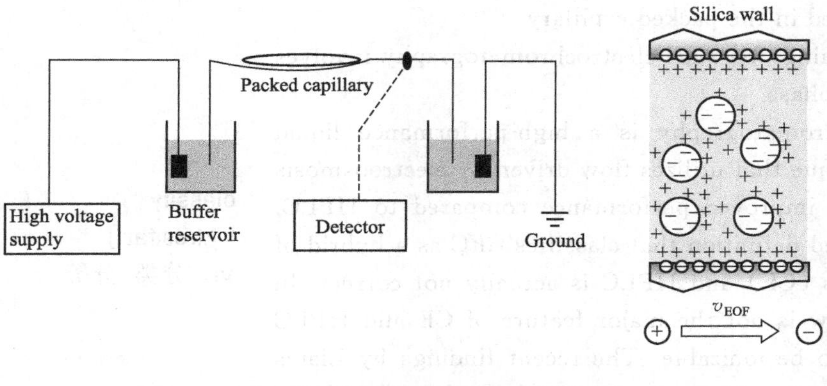

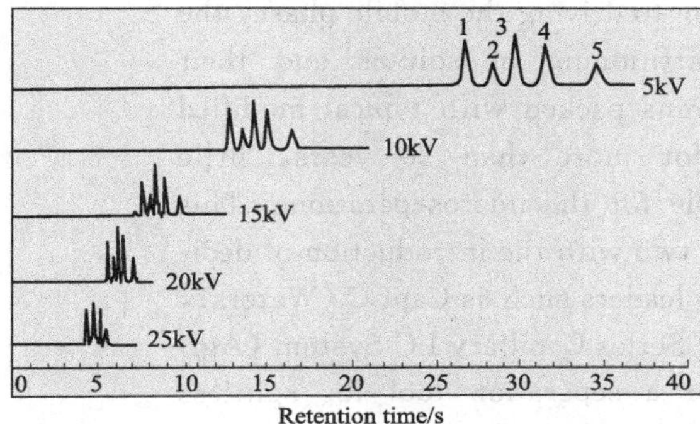

Figure 7.13 Schematic of an electrochromatographic installation and the representation of a small part of the packed capillary. The different attempts to separate a mixture of neutral compounds by varying the voltage, verifies to a first approximation, the relation existing between voltage and velocity of electro-osmotic flow

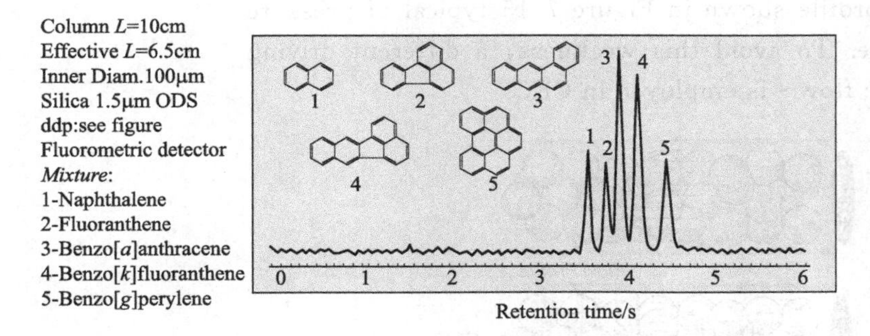

Figure 7.14 Separation of aromatic hydrocarbons by electrochromatography. In this example, the use of a very short column and a high voltage enable the separation to be made in few seconds (for $V=28$kV of a mixture of non-ionized compounds), with a good resolution. An appreciable efficiency is attained with this procedure

Some problems remain, such as:

• the speed of separation is limited by the upper limit of the electro-osmotic flow velocity

• the precise control of the volume of sample introduced into the capillary

• the heat generated in the packed capillary

• the band broadening, because electrochromatography involves the use of a stationary phase.

Capillary electrochromatography is a high-performance liquid phase separation technique that utilizes flow driven by electroosmosis to achieve significantly improved performance compared to HPLC. The frequently published definition that classifies CEC as a hybrid of capillary electrophoresis (CE) and HPLC is actually not correct. In fact, electroosmotic flow is not the major feature of CE and HPLC packings do not need to be ionizable. The recent findings by Liapis and Grimes indicate that, in addition to driving the mobile phase, the electric field also affects the partitioning of solutes and their retention. Although capillary columns packed with typical modified silica beads have been known for more than 20 years, little commercial equipment was available for the microseparations. This has changed during the last year or two with the introduction of dedicated microsystems by the industry leaders such as CapLC (Waters), UltiMate (LC Packings), and 1100 Series Capillary LC System (Agilent) that answered the need for a separation tool for splitless coupling with high resolution mass spectrometric detectors.

classify
['klæsɪfaɪ]
vt. 分类,分等

Capillary μ-HPLC is currently the simplest quick and easy way to clean up, separate, and transfer samples to a mass spectrometer, the feature valued most by researchers in the life sciences. However, the peak broadening of the μ-HPLC separations is considerably affected by the parabolic profile shown in Figure 7.15 typical of pressure driven flow in a tube. To avoid this weakness, a different driving force - electroosmotic flow - is employed in CEC.

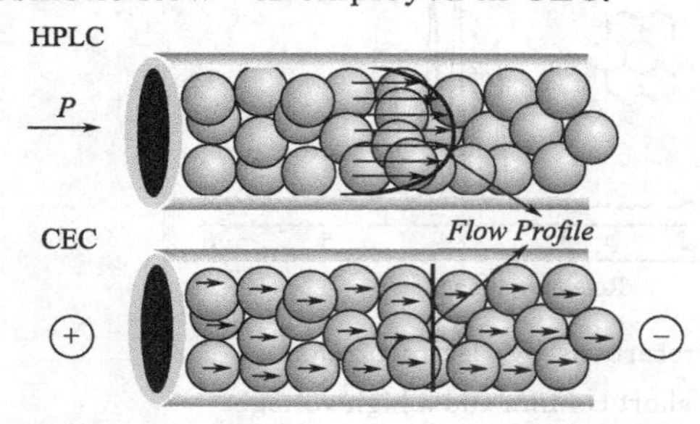

Figure 7.15 Flow profiles of pressure and electroosmotically driven flow in a packed capillary

Robson et al. in their excellent review mention that Wiedemann has noted the effect of electroosmosis more than 150 years ago. Cikalo at al. defines electroosmosis as the movement of liquid relative to a stationary charged surface under an applied electric field. According to this definition, ionizable functionalities that are in contact with the liquid phase are required to achieve the electroosmotic flow. Obviously, this condition is easily met within fused silica capillaries the surface of which is lined with a number of ionizable silanol groups. These functionalities dissociate to negatively charged Si-O^- anions attached to the wall surface and protons H^+ that are mobile. The layer of negatively charged functionalities repels from their close proximity anions present in the surrounding liquid while it attracts cations to maintain the balance of charges. This leads to a formation of a layered structure close to the solid surface rich in cations. This structure consists of a fixed Stern layer adjacent to the surface covered by the diffuse layer. A plane of shear is established between these two layers. The electrostatic potential at this boundary is called zata potential. The double-layer structure is schematically shown in Figure 7.16. Table 7.1 exemplifies actual thickness of the double-layer in buffer solutions with varying ionic strength.

functionality [fʌŋkʃə'nælətɪ] *n.* 功能

ionic strength 离子强度

Also, comparison of other parameters (pressure, efficiency) for capillary columns operated in pressurized and electrically driven flow is given in Tables 7.2 and 7.3, respectively.

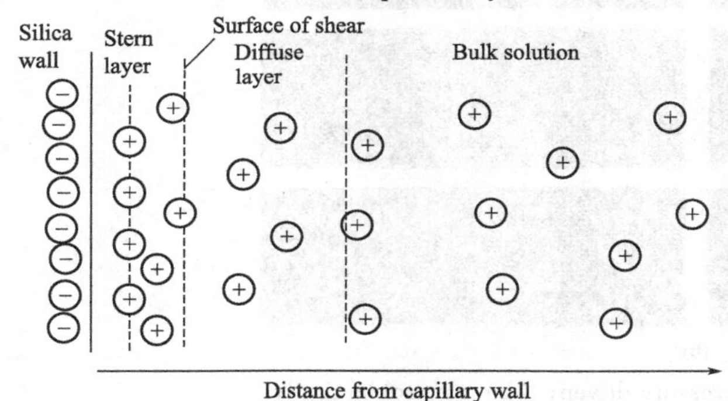

Figure 7.16 Scheme of double-layer structure at a fused silica capillary wall

Table 7.1 Effect of buffer concentration *c* on thickness of the electrical double layer

c/(mol/L)	δ/nm
0.1	1.0
0.01	3.1
0.001	10.0

Table 7.2 Comparison of parameters for capillary columns operated in pressurized and electrically driven flow

flow①	Pressurized flow		Electroosmotic flow	
Packing size/μm	3	1.5	3	1.5
Column length/cm	66	18	35	11
Elution time/min	33	n.a.	18	6
Pressure/MPa	40	120②	0	0

① Column lengths, elution times, and back pressures are given for a capillary column affording 50000 plates at a mobile phase velocity of 2mm/s.

② The back pressure exceeds capabilities of commercial instrumentation (typically 40MPa).

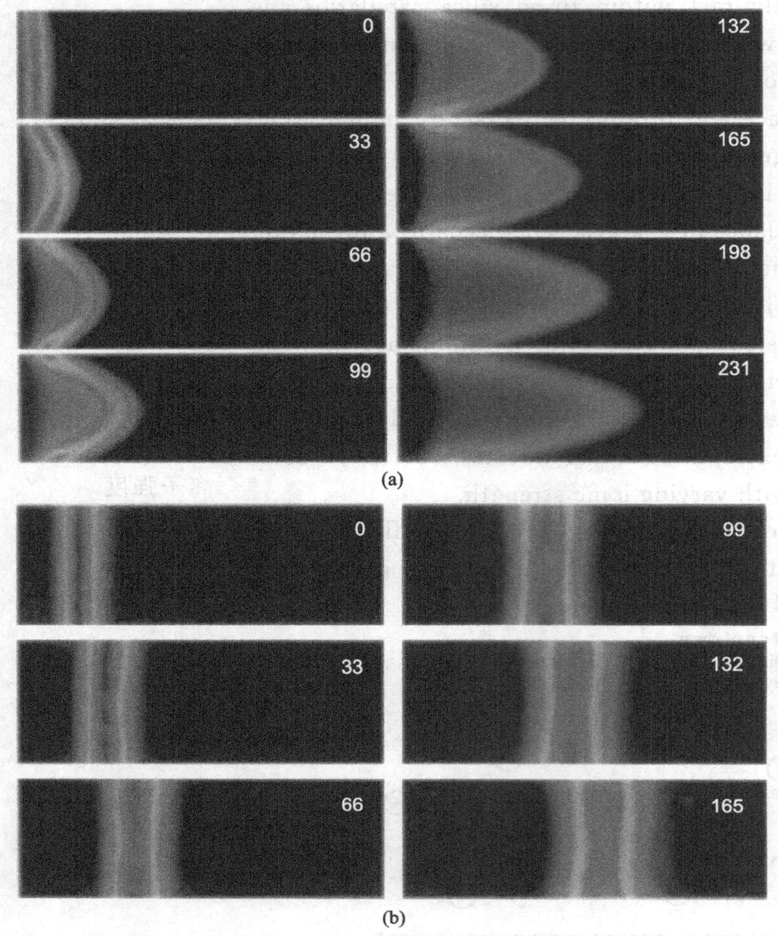

Figure 7.17 Images of: (a) pressure-driven; (b) electrokinetically driven flow. Conditions: (a) flow through an open 100μm (i. d.) fused-sillica capillary using a caged fluorescein dextran dye and pressure differential of 5cm of H_2O per 60cm of column length; viewed region 100μm by 200μm; (b) flow through an open 75μm (i. d.) fused-silica capillary using a caged rhodamine dye; applied field 200V/cm, viewed region 75μm by 188μm. The frames are milliseconds as measured from the uncaging event

After applying voltage at the ends of a capillary, the cations in the diffuse layer migrate to the cathode. While moving, these ions drag along molecules of solvating liquid (most often water) thus producing a net flow of liquid. This phenomenon is called electroosmotic flow. Since the ionized surface functionalities are located along the entire surface and each of them contributes to the flow, the overall flow profile should be flat. Indeed, this has been demonstrated in several studies and is demonstrated in Figure 7.17. Unlike HPLC, this plug-like flow profile results in reduced peak broadening and much higher column efficiencies can be achieved.

broad
[brɔːd]
v. 扩宽,扩展

Table 7.3 Comparison of efficiencies for capillary columns packed with silica particles operated using pressurized and electrically driven flow.

d_p[1]/μm	Pressurized flow HPLC		Electroosmotic flow CEC	
	L[2]/cm	Plates/column	L/cm	Plates/column
5	50	45,000	50	90,000
3	30[3]	50,000	50	150,000
1.5	15[3]	33,000	50	210,000

① Particle diameter.

② Column lengths.

③ Column length is detected by the pressure limit of commercial instrumentation (typically 40MPa).

Problems

7.1 An electrophoresis analysis in free solution (CZE), calls for the use of a capillary of 32cm and with effective length of separation 24.5cm. The applied voltage is 30kV. Under the conditions of the experiment the peak of a neutral maker appeared upon the electropherogram at 3min.

1. Calculate the electrophoretic mobility, μ_{EP}, of a compound whose migration time is 2.5min. Give the answer in precise units.

2. Calculate the diffusion coefficient under these conditions for this compound, remembering that the calculated plate number, $N=80000$.

7.2 The apparatus for an experiment using a fused silica capillary is set up. The total capillary length $L=1$m while the effective capillary length $l=90$cm. The applied voltage is 30kV. The detector is located towards the cathode and the electrolyte is a buffer of pH 5.

In a standard solution, a compound has a migration time of $t_m=10$min.

1. Sketch a diagram of the apparatus described.

2. From the information given, can it be deduced whether the net charge carried by the compound is positive or negative?

3. Calculate the apparent electrophoretic mobility μ_{EP} of the compound.

4. If a small molecule not carrying a charge has a migration time $t_m=5$min, deduce the value of the electro-osmotic flux μ_{EOS}.

5. Calculate the electrophoretic mobility of the compound μ_{EP}.

6. What is the sign of the net charge carried by the compound?

7. What would happen if the fused silica capillary was coated by trimethyl-chlorosilane?

8. Supposing that the pI of the compound was 4, what would be the sign of its net charge if the pH of the electrolyte was lowered

to 3?

9. Calculate the number of theoretical plates if the diffusion coefficient of the solute is $D = 2\times10^{-5}\,cm^2/s$.

10. Form the relationship between the efficiency N and the diffusion D, explain why small molecules have poorer separations than larger molecules, and why for the smaller variant these separations are much better when the capillary is narrower.

7.3 The table below lists a series of proteins along with their perspective isoelectric points (pI). By completing the table with positive (+) or negative (−) signs or zeros (0) indicate for each protein whether the net charge will be positive, zero or negative at the three values of pH specified.

Protein	pI	pH=3	pH=7.4	pH=10
Insulin	5.4			
Pepsin	1			
Cytochrome C	10			
Haemoglobin	7.1			
Albumin (serum)	4.8			

第 7 章
毛细管电泳和电色谱

毛细管电泳（CE）是一种非常灵敏的分离方法，它主要由高效液相色谱（HPLC）、薄层色谱（TLC）及电泳技术发展而来。一些生物分子和低分子量的化合物难以通过高效液相色谱或经典的平板凝胶的电迁移来研究，往往可以通过毛细管电泳来分离。HPCE 因适宜于自动化已经取代了传统的凝胶电泳技术。它已得到 FDA 等几个官方监管机构的认可，应用于包括药典的定量分析方法中。毛细管电色谱（CEC）是一种相对新颖的混合分离方法，它同时兼顾了 CE 与 HPLC 二者的高分离效率。

7.1 从区带电泳到毛细管电泳

毛细管电泳与很多通用的电泳方法具有一致性。这种改进后的技术是基于在电场的作用下，在适宜的介质中，无论样品溶液中的各组分整体是否携带电荷，各组分都会有不同的迁移速度。

在传统的主要用于生物分析的半手工电泳技术中，是将一个小的铺有多孔物质（凝胶）材料的塑料平板浸泡在电解质缓冲溶液中。覆盖有凝胶的两个末端浸入含有相同电解质的两个独立的库槽内，同时与提供连续电压的电极相连。样品以横向带的形式沉积，冷却后保留在两隔离板之间。在几个参数的影响下，如共同电压（500V，或者对小分子而言更大的电压）、电荷、大小、形状、温度和黏度，水合组分从一端迁移到另一端，一般向与它电性相反的电极移动，时间可能是几分钟到一个多小时。每种化合物根据其迁移率不同来分离，但与薄层色谱法相比，没有可供测量的展开剂前沿，组分迁移的距离通过它相对于一种内部标记物移动的距离来估算。迁移之后组分的检测通过接触法，将各组分转移到一张薄膜上，与专有试剂银盐或考马斯亮蓝发生衍生反应（图 7.1）。常用的标记物含有可用于示踪的放射性同位素（^{32}P 或 ^{3}H），这在 TLC 中也经常使用。

在毛细管电泳中，传统技术中铺有凝胶的平板被窄孔径、开放式的熔融石英毛细管（直径 15～100μm）所取代。在长度 20～80cm 间的毛细管中充满了和两个电极槽相同的电解质缓冲溶液（图 7.2）。为了更好地控制迁移时间，最好将毛细管放在恒温箱中，较小的毛细管内径可使焦耳热小到可以忽略，并可采用高电场实现非常快速的分离（电极电压可以达到 30kV）。

检测器靠近阴极槽，与毛细管的进口端有一定距离。检测到的信号给出电泳谱图的基线（图 7.2），由此可以得到样品的成分信息。样品中只有那些向阴极移动的组分才能被检测。

7.2 电泳淌度和电渗流

像溶剂化分子一样，微粒悬浮在液体中会携带电荷。所带电荷的多少和电性由微粒的性质、电解质液体介质以及 pH 值（图 7.3）所决定。微粒所带的净电荷取决于粒子表面吸附的电荷以及缓冲液中含有的离子。

在诸如温度、黏度、电压等因素的单独或共同作用下，微粒的电荷/尺寸比越小，所带电荷越大，则微粒的迁移速度将越快。分离取决于水合离子的比例。

对于电场中每个带电量为 q 的颗粒而言，有限的迁移速度 v_{EP} 来源于平衡力 F，即电场 E 所施加的电场力以及由于溶液黏度 η 所产生的阻力的平衡。中性物质分离效果不佳，除非添加一种离子试剂到电解质中与中性物质发生相互作用。

休克尔考虑到这些因素对一个半径为 r 的球形离子的电泳速度 v_{EP} 的影响，提出了一个方程。

样品中的带电组分主要受两个因素影响，即它们各自的电泳淌度和整个溶液的电渗流。

7.2.1 电泳淌度-电迁移

假定电解质溶液中带有一定电荷的某化合物以 v_{EP}（m/s）的速度移动，移动的速度取决于实验条件及其自身的电泳淌度 μ_{EP}。对于给定的离子和介质，μ_{EP} 是一个该离子特有的常数，此参数由该化合物的电泳迁移速度和所施加的电场 E 通过式(7.1) 得出。

$$\mu_{EP}=\frac{v_{EP}}{E}=v_{EP}\frac{L}{V} \tag{7.1}$$

在上述方程中，L 是毛细管的总长度，m；V 是施加在毛细管两端的电压。电泳淌度的符号是正或负依据该组分是阳离子或阴离子来确定；对所有电中性组分，μ_{EP} [$m^2/(V\cdot s)$] 等于 0。通过电泳可以计算化合物在已知电场 E(V/m) 中的 v_{EP}，同时考虑到电解质的电渗速度，就可以得到 μ_{EP}（参见第 7.2.2 节）。此参数不仅取决于组分所带的电荷，还取决于该组分的直径及电解质的黏度。

7.2.2 电渗淌度-电渗流

电渗流是控制溶质迁移的第二个因素。这种流动是电解质溶液在毛细管中的整体移动。在凝胶电泳中，电渗流很小，但在毛细管电泳中，由于毛细管的内壁（有双电层）而使其变得非常重要。电渗流的强弱用电渗淌度 μ_{EOS} 表示，μ_{EOS} 通过类似于式(7.1) 的关系来确定。v_{EOS} 表示电渗流的速度。

$$\mu_{EOS}=\frac{v_{EOS}}{E}=v_{EOS}\frac{L}{V} \tag{7.2}$$

为了计算 μ_{EOS}，必须先求得电渗流速度 v_{EOS}。它相当于电解质溶液中没带电荷的组分的迁移速度。通过测量某电中性标记物在毛细管中迁移距离 l 所用时间 t_{nm} 而求得（$v_{EOS}=l/t_{nm}$）。

只有在电解质缓冲液 pH 下无极性，且易于在近紫外吸收检测到的有机分子，才能选作电中性的标记物（如丙酮、异亚丙基丙酮或苯甲醇）。

电渗流的形成与毛细管的内壁有关，一根没有经过任何特殊处理的熔融石英毛细管，其内表面含有大量的硅羟基（Si—OH），当管内的电解质缓冲液的 pH 大于 3，硅羟基会电离为带负电的硅氧基（$Si—O^-$）。在这种条件下，毛细管内壁会成为一个聚阴离子层来吸引电解质中的阳离子。这些阳离子会分为两层，其中一层附着到毛细管内壁，而另一层仍可稍微移动（图 7.4）。这两层之间的电位差（ζ 电势），其大小取决于电解液的浓度和 pH。在外加电场的作用下，H_3O^+和其他水合离子向负极移动。在毛细管电泳仪中，电渗流的净效应是强制样品中所有的带电组分定向地从正极向负极移动。而样品中的阴离子的移动方向则与电渗流方向相反。

通常情况下，带负电的毛细管内壁会产生电解质朝向阴极线性移动的电渗流。相反，如果内壁覆盖了表面活性剂，如四烷基铵，将改变内壁的电性，则电渗流将流向阳极（图 7.5）。用烷基硅烷处理毛细管内壁使其疏水，可用于分离蛋白质，否则蛋白质会吸附在裸露的熔融石英毛细管内壁上，不能被分离。

7.2.3 表观淌度

如前所述，每个离子的表观迁移速度 v_{app} 取决于电泳速度和电渗流［式(7.3)］，从电泳易求得 v_{app}，它与毛细管有效长度 l、迁移时间 t_m 的关系符合式(7.4)：

$$v_{app}=v_{EP}+v_{EOS} \tag{7.3}$$

$$v_{app}=l/t_m \tag{7.4}$$

表观淌度μ_{app}定义的数学表达式类似式(7.1) 或式(7.2)，为：

$$\mu_{app}=\frac{v_{app}}{E}=v_{app}\frac{L}{V} \tag{7.5}$$

因此：

$$\mu_{app}=\frac{lL}{t_m V} \tag{7.6}$$

结合电渗流和表观淌度就可能计算出带电组分真正的电泳淌度。式(7.3) 也可以写成：

$$\mu_{EP}=\mu_{app}-\mu_{EOS}$$

或

$$\mu_{EP}=\frac{Ll}{V}\times\left(\frac{1}{t_m}-\frac{1}{t_{nm}}\right) \tag{7.7}$$

HPCE 常用于分离阴离子组分，改变仪器的正负电极可以将检测器的位置从毛细管的阴极端变为阳极端。这样的话电渗流的方向也就颠倒过来了。这时只有那些其μ_{EP}远大于μ_{EOS}的阴离子才会被检测到（图 7.6）。

7.3 仪器

为了保证分离度，样品体积不得超过毛细管有效长度的 1%，要将如此微量的样品注入毛细管，采用两种方法。

• 流体力学方式　此方法是将毛细管的一端浸入含有样品的溶液（也是电解质）中，从而在毛细管另一端形成轻微的真空，依靠毛细管两端的压力差将液体导入。在样品溶液上施加 50mbar 左右的压力可以改进此法。

• 电迁移方式　相比其他简单地将毛细管插入样品池来进样的方法（流体力学方式），

凝胶电泳中的进样是在样品端施加一定电压来实现的。与流体力学进样法相比，电动进样会导致进样歧视，从而导致注入的物质不能代表样品的组成。

都采用流体力学方式进样时，电泳的准确度比 HPLC 的差。这是因为没有体积在 5～50nL 之间的注射器。进入毛细管内的样品体积取决于许多参数，可通过著名的泊肃叶（Poiseuille）方程求出。即运动黏度为η的液体在半径为r、长为L的管子中流动时的流速的计算公式［式(7.8)］。由该式会得到一个近似值，可认为是毛细管的“进样流速”。

$$F=\Delta p\ \frac{\pi r^4}{8\ \eta L} \tag{7.8}$$

上式表明，如果毛细管的半径翻倍，则进样体积将扩大 16 倍。该体积还与压力差Δp成正比。间接紫外检测用于检测无机离子等不吸收紫外线的分析物。在毛细管中使用含铬酸盐或邻苯二甲酸盐等物质的缓冲溶液，在实验条件下这些物质具有较高的吸光系数。当分析物经过检测器时，因为吸收紫外线的缓冲液被排除在外，所以穿过毛细管的光量将增加，从而导致负峰的产生（图 7.7 和图 7.8）。

7.4 电泳技术

7.4.1 毛细管区带电泳（CZE）

CZE 是应用最广泛的电泳方法。样品被加在电解质缓冲溶液中，在电场的作用下移动通过毛细管。根据具体的应用，这种电解质可以是酸性（磷酸或柠檬酸盐）或碱性（硼酸）或两性物质（分子同时具有酸性和碱性官能团）。电渗流随液相的 pH 值或增加或不受其影响。与 CGE 相对比，这种方法也被称为自由溶液电泳。

7.4.2 胶束电动毛细管色谱（MEKC）

MEKC 是改良了的 CZE，适用于分离中性或极性分子，在背景电解质缓冲溶液中加入一定量的阳离子或阴离子表面活性剂，如十二烷基硫酸钠，以形成带电荷的胶束。这些微小的球形胶束，与液体不混溶，这样就形成了类似于 HPLC 中的固定相的一种假固定相。它们通过亲水/疏水亲和相互作用（图 7.9）有效地捕集保留溶液中的中性化合物。溶液中的中性分子以及离子组分，因为它们在胶束中溶解能力不同从而很方便地被分离。MEKC 也可分离手性分析物。这主要涉及将环糊精作为手性选择剂加入到缓冲液中。疏水部分与环糊精内腔作用形式不同稳定性的包合物的包合作用来进行。

7.4.3 毛细管凝胶电泳（CGE）

CGE 将传统的凝胶电泳中的琼脂糖或聚丙烯酰胺凝胶（PAG）引入毛细管电泳。毛细管中填充了浸满电解质的凝胶。这会产生一个过滤效应，从而降低电渗流，减少对流扩散现象。可用本方法分离寡核苷酸。

CGE 已经应用于 DNA 测序，采用多根毛细管且可同时分析多个样品的带荧光检测器的全自动仪器（从进样到数据分析）已经开发出来了，即所谓的毛细管阵列电泳，该技术可基于分子大小的不同来分离核酸或核糖核酸片段。

在传统的区带电泳中，组分迁移的介质可以是浸满电解质的聚丙烯酰胺凝胶。如果含有十二烷基硫酸钠，则可以称为 SDS-PAGE（即十二烷基硫酸钠-聚丙烯酰胺凝胶电泳）。由于凝胶的过滤作用，增强了电场的分离作用。

7.4.4 毛细管等电聚焦（CIEF）

本方法是在表面处理过的毛细管内进行的，含有两性电解质的溶液可在毛细管中形成稳定、线性的 pH 梯度。

两性电解质是含有酸性和碱性官能团的分子。在特定的 pH 值下，即所谓的等电点，官能团上电荷是平衡的，分子呈电中性。如果某两性电解质放入具有 pH 梯度的环境中电泳，那么它将迁移到使它不带电的等电点，然后停止移动。

在阳极一端，毛细管插入 H_3PO_4 溶液，而阴极端则浸入 NaOH 溶液。当电场加到一个充满溶质和两性电解质混合物的毛细管两端时，毛细管中就形成了 pH 梯度。每个带电分析物（通常指蛋白质）通过介质迁移直到到达某区域，该区域的 pH 与它的等电点 pI 相同（在等电点，该组分的净电荷为零）。然后，保持电场并使用静水压力，使这些分离的组分迁移到检测器。为了确保高效分析，需要有 pI 的标准物（即 pI 标记物）。本方法的高分离度使得 pI 仅相差 0.02pH 的多肽分离成为可能。

7.5 CE 的效率

CE 分离生物大分子的效率可媲美高效液相色谱法（HPLC）。分离的成功依赖于选择合适的缓冲介质。分析所需的样品量极少，消耗的试剂或溶剂可以忽略不计（图 7.10）。灵敏度非常高：采用激光诱导荧光（LIF）检测可以检测数千个分子。

然而，分析的重现性逊于高效液相色谱法，因为有许多微妙的因素会影响进样体积的精确度。对于流体力学进样，如果使用内标，相对标准偏差为 2%。当分离效率 N 足够大时，可能可以用 CE 分离元素的同位素（图 7.11）。

高效毛细管电泳（HPCE）的“芯片实验室”技术处于传统化学的研究前沿，而且特别吸引那些要分析蛋白质降解产物的分子生物学家的兴趣。

分离参数（效率、保留因子、选择性、分离度）可以使用与色谱中共用或 HPCE 中特有的公式，从电泳谱图中得到。

分离的理论效率塔板数 N 高达 1 百万块/m，N 可以通过有效长度 l 和扩散系数 D（cm^2/s）求得［式(7.9)］。最后一个参数（即扩散系数）通过 Einstein 方程与区带方差 σ 和迁移时间 t_m 相关联（$\sigma^2=2Dt_m$）。这样就得到式(7.9）或式(7.10）［将式(7.6）代入式(7.9)，消掉 t_m，即得］。如果由实验得到分离的理论效率 N，则也可通过这些公式求得 D。

$$N=\frac{l^2}{2Dt_m} \tag{7.9}$$

$$N=\frac{\mu_{app}}{2D}V\frac{l}{L} \tag{7.10}$$

式(7.10）表明，效率 N 与电压成正比。由于大分子的扩散因子比小分子的要小，所以二者可以得到很好的分离（图 7.12）。

最后，在 HPCE 中，两峰之间的分离度 R 可通过效率 N、两组分移动速度的差（Δv）和它们的平均速度 $\bar{v}$，通过式(7.11）算出。

$$R=\frac{\sqrt{N}}{4}\times\frac{\Delta v}{v} \tag{7.11}$$

7.6 毛细管电色谱

毛细管电色谱（CEC）是借鉴了CZE和HPLC而发展起来的分离方法，它将离子的电迁移与它们在固定相和流动相之间具有不同的分配系数相结合，从而实现各组分间的分离。与色谱相比，它没有液压泵来驱动流体，而是由CE仪的高压电源产生的电渗流来驱动流体通过毛细管填充柱。在电色谱中没有电荷损失，每个H_3O^+在电场的作用下迁移，而在液相色谱中，让流动相通过一根相同的毛细管柱需要近1000bar的压力。与HPLC相比，在电驱动系统里，流体的截面是塞状流，因而比压力驱动的系统要高效得多，后者的流体截面是抛物线（图7.12）。

CEC需要一根高纯度的熔融毛细管，管中填充了标准HPLC颗粒填充材料或连续床（整体柱），前者填充颗粒直径可以非常小，至1～3μm，而后者没有背压。

由于二氧化硅材质的微粒填充材料的比表面积比毛细管壁的比表面积要大很多，所以电渗流主要是由固定相表面的硅羟基产生的（图7.13）。这种填料的作用类似于未处理的毛细管内壁。在未经处理内壁的毛细管管内填充完全碱性失活的材料（如残留少量硅羟基的RP-18硅胶）不是很合适。

固定相作为一种保留材料，也可以灵活多变。也可能采用非水的流动相（对亲水性分析物）和未经处理的二氧化硅填料。分离因子非常高，但流动相组成或有机改性剂的改变将影响电渗流，从而影响选择性（图7.14）。

在传统的液相色谱中，塔板高度可以用修正的van Deemter塔板高度方程表示。不过仍然会存在一些问题，例如：

- 分离的速度受限于电渗流速度的上限；
- 精确控制导入毛细管的样品体积；
- 在填充毛细管中产生的热；
- 区带增宽，因为电色谱使用了固定相。

毛细管电色谱是一种高效的液相分离技术，它利用电渗驱动流体获得了较HPLC更高的效率。经常将CEC定义为综合了毛细管电泳（CE）和高效液相色谱（HPLC）优点的一种分离方法，实际上这并不正确。事实上，电渗流并不是毛细管电泳的主要特征，且高效液相色谱填料不需要是可电离的。Liapis和Grimes最近的研究表明，除了驱动流体移动，电场还会影响溶质的分配和保留。

虽然采用改性硅胶颗粒填充的毛细管柱已有20多年的历史，但是直到现在，因为涉及标准尺寸的柱系统，微分离的商业仪器还不多。在过去的一两年里，行业领导者们推出许多新的微流系统，如CapLC（美国Waters公司）、UltiMate（美国LC Packings公司）和1100系列毛细管液相色谱系统（美国安捷伦公司），以应对市场对不分流串联高分离度的质谱检测器的分离仪器的需求。毛细管μ-HPLC是目前最简单、快速、简便的净化、分离和转移样品到质谱仪的方法，这些已由生命科学领域的研究人员所证实。然而，如图7.15所示，μ-HPLC分离的峰增宽很大程度上是由于在传统方法中，采用压力驱动管中流体形成抛物线形的流体截面所导致的。为避免这个弱点，CEC采用了不同的驱动力——电渗流。

Robson 等在文章中提到，早在 150 多年前 Wiedemann 就已注意到电渗的影响。Cikalo 等将电渗定义为在外加电场作用下，液体相对于带电的固体表面发生的位移。依照这个定义，为了得到电渗流，与液相接触的固体必须要有电离的功能。显然，这个条件在熔融石英毛细管内很容易满足，因为其表面带有大量能电离的硅羟基。这些硅羟基电离为附着在管壁上的带负电荷的硅氧基阴离子（Si—O^-）和移动的 H^+。带负电荷的管壁排斥与它接触的液体中的阴离子，但吸引阳离子以保持电荷平衡。结果在靠近固体的表面形成了富含阳离子的层状结构。该层状结构由一个固定的 Stern 层及其外面的扩散层组成。Stern 层和扩散层之间形成一个剪切面。剪切面所具有的电势称为 ζ 电势。双电层结构如图 7.16 所示。表 7.1 为在不同离子强度的缓冲溶液中双电层的实际厚度。表 7.2 和表 7.3 分别比较了在压力流和电驱动流下操作的毛细管柱的其他参数（压力、柱效）。

在毛细管两端施加电压后，扩散层中的阳离子向阴极迁移。在移动时，这些离子会带动液体分子（通常是水）一起移动，从而产生了液体的净流动。这种现象称为电渗流。由于整个管壁上都有离子化的官能团，而它们每一个都对电渗流的形成做出了贡献，总体流体流动截面应该是平坦的（图 7.17）。事实上，好几个研究已经证明了这点。

如图 7.17 所示。与高效液相色谱法相比，此种塞状流体截面既减小峰增宽，又可获得更高的柱效。

Chapter 8

Sample Preparation

For all chemical analysis, the analyte, which refers to the species to be measured, must be in a sufficient quantity and suitable form for the instrument used. The majority of samples require a specific pretreatment. This preliminary stage, which follows the so-called sampling procedure, has long been the "bottleneck" in the analytical process. It is often a critical step in a chemical analysis, because it has an influence upon the end result. Sample preparation is therefore an essential step in analysis, as important as the measurement itself, greatly influencing its reliability, its accuracy and its cost. Since trace analyses (less than 1mg/L of analyte in a crude sample), now represent at least half of all analysis being performed, then powerful methods of extraction are required. The particular mode of enrichment is determined by the instrument or the methodology used. This domain is currently very actively studied and benefits from recent knowledge in robotics to shorten the time of preparation. Moreover, the large number of analyses of very small samples has stimulated the development of state-of-the-art equipment and up-to-date procedures, some of which are described in this chapter, dedicated to samples intended for chromatographic analysis.

8.1 The need for sample pretreatment

In order to start an analysis, a sample representative of the batch under study is required. This contains the species being studied called the analyte, mixed, in general, with a variable number of other compounds which constitute together what is called the matrix. When measuring traces of an analyte within other compounds thousands of times more abundant, there is a fear that interferences between analyte and other compounds may occur. Even the best analytical techniques cannot remedy problems generated by poor sample pretreatment. This makes essential a meticulous sample preparation. Since this supplementary work is often tedious and time consuming (Figure

abundant
[ə'bʌndənt]
adj. 丰富的，盛产

remedy
['remɪdɪ]
vt. 补救，纠正；
n. 补救，治疗

meticulous
[mɪ'tɪkjələs]
adj. 谨小慎微的，过度重视细节的

8.1) methodologies involving adsorption, extraction or precipitation methods have been developed to make it easier and when possible automatic.

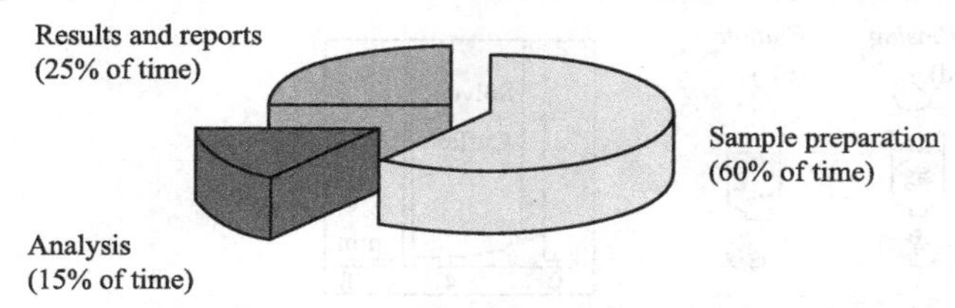

Figure 8.1 Statistics displaying the proportion of the time spent in each stage of a chromatographic analysis. Sample preparation generally represents a large fraction of the total time dedicated to the complete analysis

8.2 Solid phase extraction (SPE)

SPE is currently used for the purification or concentration of a crude extract prior to the measurement of one or several of its constituents. SPE is a clean analytical procedure which has drastically changed the classical approaches of solvent extraction as liquid/liquid extraction in a separating funnel, comparatively slower and that use relatively large amounts of liquid organic solvents.

SPE operates in a small open plastic cartridge, similar to the barrel of a syringe, which contains an appropriate solid sorbent (e.g. a cartridge of 100 to 300mg of a RP-18 phase type) upon which the process consists of passing a known volume of the liquid sample (Figure 8.2). Two methods can be used:

The currently most used method comprises an initial step in which the compound of interest is retained by the sorbent-containing column, followed by the elimination, through rinsing, of the greatest possible number of undesired substances for the subsequent measurement. The analyte is then recovered in a small volume of an appropriate solvent as a solution strongly enriched. This extraction procedure allows not only isolation of the analyte but also its preconcentration, which is particularly useful in traces analysis.

The second method, less frequently used, though very simple, consists of passing the sample through the column in order that the undesired compounds are retained while the analytes pass freely. This approach is attractive because it eliminates the elution (desorption step). On the other hand, the analyte solution is purified, but not concentrated (enrichment factor of 1).

The solid sorbents closely resemble to that of the solid stationary phases of liquid chromatography. Bonded silica gels (of reverse phase polarity) with a particle size of between 4 and 100μm, allow a

drastically
[ˈdræstɪkəlɪ]
adv. 彻底地，激烈地

separating funnel
分液漏斗

elimination
[ɪˌlɪmɪˈneɪʃən]
n. 除去

preconcentration
[ˈpriːkɒnsenˈtreɪʃən]
n. 预富集，预浓缩

percolation of faster flow rate. Other adsorbents, containing graphite or co-polymers such as styrene- divinylbenzene of large functional group bonded surfaces, are more stable in acidic solutions.

percolation [pɜːkəleɪʃən] *n*. 过滤，浸透

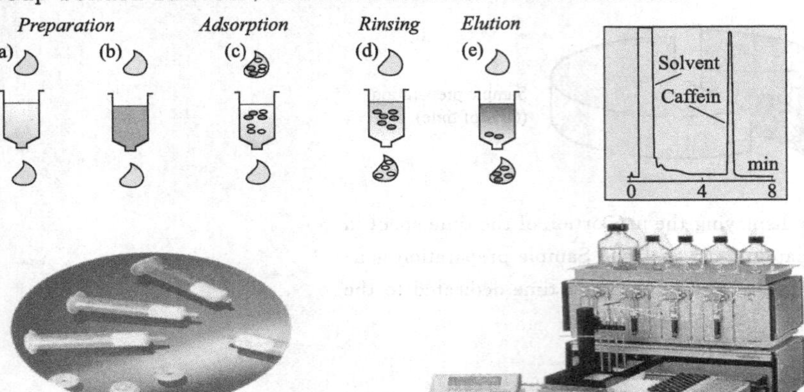

Figure 8.2 Solid phase extraction. The separation of an analyte from the matrix. In this example, the analyte is the only compound retained on the cartridge. Note the following steps: (a), (b) activation and rinsing of the sorbent prior to use; (c) introduction of a known volume of the sample; (d) interference elimination by rinsing; (e) desorption of the analyte by percolation. On the right, chromatogram (GC) obtained following treatment by this sequence of 1mL of coffee. Below, on the left, SPE cartridges with reservoirs of 1 to 3mL; on the right, automated sample preparation

cartridge [ˈkɑːtrɪdʒ] *n*. 卡盒，墨盒，胶卷盒

There are also extraction discs (0.5mm thickness and 25 to 90mm in diameter) used as a kind of chemical filters in the same way as filter paper fit in the funnel of a vacuum flask. The discs are made of bonded phase silica particles trapped in a porous Teflon or glass fiber matrix. They retain traces of organic compounds present in large volumes of aqueous solutions, something that is not possible with the cartridges described above. The analyte is recovered by percolating a solvent through the filter. Trace analysis is made easier because the enrichment factor can be as high as 100.

trapped [træpt] *adj*. 收集的；*vt*. 圈住，诱捕

percolating [ˈpɜːkəleɪtɪŋ] *n*. 渗透；*adj*. 渗透的

Non-polar compounds are isolated from a polar matrix with a cartridge or a disc of C-18 derivatized silica gel. Polar compounds are isolated from a non-polar matrix with a cartridge filled with a normal phase. Charged species can be isolated on an ion exchange phase. These three situations are well adapted to analysis by HPLC.

isolated [ˈaɪsəˌleɪtɪd] *adj*. 隔离的；*v*. 使隔离（isolate 的过去式和过去分词），使绝缘

Two particular points require attention. During the extraction step the break- through volume must not be reached since beyond this amount the compound to be extracted will no longer be trapped by the cartridge. Elsewhere, during the percolation step, the recovery of the compound absorbed is rarely complete. If the measuring technique calls upon detection by mass spectrometry, this quantification problem can be resolved by introducing, prior to extraction, of a given volume of the sample, a known quantity of this same compound labelled with a stable isotope. The behavior during this sequence of the two forms of the compound being the same, the comparison of their respective signals will permit the establishment of their ratios and consequently the concentration of the unknown. This is the equivalent of the method of internal standards using a standard for which the relative response coefficient will be 1.

A current research theme concerns the preparation of adsorbents having a marked affinity for a particular type of compound. For this, polymers are synthesized which possess recognition sites in the form of surface deformations, appropriate for trapping molecular targets. The principle resembles that of immuno-extraction, which is even more selective.

In conventional liquid/liquid extraction, the analyte is diluted in a large volume of solvent. Consequently, when the extracted solution is concentrated, the non-volatile impurities are contained in the solvent. This frequently makes the method unacceptable unless ultra-pure solvents are used. For example, if 1mg of analyte is mixed with 1mL of 99.9 percent pure solvent, this latter brings a mass of added contaminants equal to that of the analyte.

8.3 Immunoaffinity extraction

Retention of the analytes with classic SPE cartridges is based upon hydrophobic interactions. This method can lack selectivity if the matrix in which the target analyte is, contains interferent compounds of which greater concentration is much that are likely to interfere. A lot of progresses have been made in developing new types of cartridges based on immunoadsorbers, able of an efficient enrichment of a single analyte from complex environmental matrices. This molecular recognition is based upon the interactions of antigen/antibody type.

By fixing through covalent bonding an antibody on an adapted solid support, it will result in a sorbent that will fix the corresponding antigen, as a molecularly imprinted polymer (Figure 8.3). Immunoaffinity techniques, now used extensively in pharmaceutical and biotechnology applications, can be adapted to extract small organic molecules of environmental samples. There exist extraction cartridges containing antibodies of herbicides or polycyclic aromatic hydrocarbons (HPA) adapted to the extraction of these compounds by a rapid percolation.

quantification
[kwɒntɪfɪ'keɪʃn]
n. 定量,量化

labelled
['leɪbəld]
v. 贴上标签

isotope
['aɪsətəʊp]
n. 同位素

deformation
[diːfɔː'meɪʃn]
n. 变形

immunoadsorber
免疫吸附器

antigen
['æntɪdʒən]
n. [免疫] 抗原

antibody
['æntɪbɒdɪ]
n. [免疫] 抗体

covalent bonding
共价键

support
[sə'pɔːt]
n. 载体

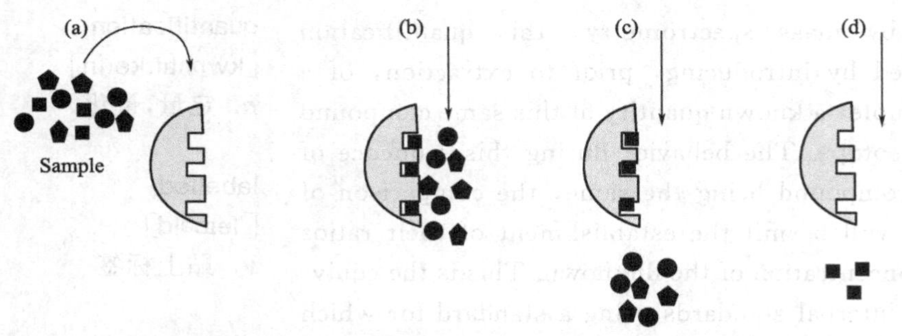

Figure 8.3 Principle of immunoaffinity extraction. The different steps of a classic procedure on a cartridge containing the immobilized antibodies adapted, on this illustration, to the square molecules. After the fixing step (b), rinsing (c), then elution (d) of the analyte is done with an organic

8.4 Microextraction procedures

8.4.1 Solid-phase microextraction (SPME)

integrate [ˈɪntɪgreɪt] *vt.* 使……完整；使……成整体

SPME integrates sampling, extraction, concentration and sample introduction into a single solvent-free step. This technique uses a 1 to 2cm long segment of porous fused silica fibre coated with an appropriate adsorbent material. This fibre is held in a syringe-like device, protected by the steel needle (Figure 8.4). The device' s plunger is depressed to expose the fibre to the sample then retracted at the end of the sampling time, the fibre may be introduced into the aqueous solution or in the space situated above it surface (headspace). This process does not need solvents. The sample analytes are directly extracted and concentrated on the extraction fibre. Then the fibre is introduced into a particular injector of a GC, where, through the effect of the temperature, the compounds of interest are desorbed and then separated. The SPME technique can also be routinely used in combination with HPLC, HPCE and MS. Sensitivity is usually low-to-mid per trillion, but this procedure is rather expensive despite the

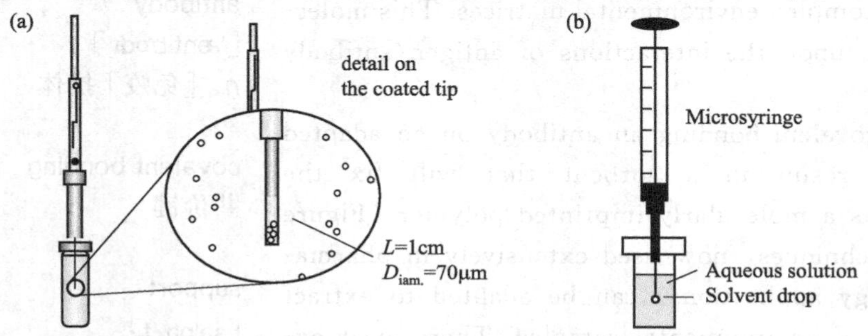

Figure 8.4 Micro-extraction procedures. (a) Micro-extraction using an adsorption fibre; (b) Micro-extraction with a single drop of organic solvent. Advantages of this technique are: ultra small volume of sample, low chemical consumption and fast analysis

same fibre can be used around 50 times.

8.4.2 Liquid-phase microextraction (LPME)

This technique also called single-drop microextraction, operates in a two-phase mode. Analytes are extracted from 1 to 4mL aqueous samples into 25-50μL of an organic very pure solvent, as hexane. It uses a microdrop of solvent at the tip of a conventional microsyringe, rather a fragile silica fibre as above. The microdrop, formed at the point of the syringe needle, is immersed over several minutes into the aqueous sample solution containing the products to be extracted (Figure 8. 4b). LPME combines extraction and concentration in one step. Subsequent analysis is performed by GC. Quantification is possible, provided operating conditions are held constant through all of the calibration procedure. The method has been applied to the analysis of chemical solvents in foods and pharmaceuticals, natural plant components, environmental volatile and semi-volatile contaminants in water.

fragile
['frædʒaɪl]
adj. 脆的,易碎的

The purification of small quantities of aqueous solutions traditionally performed using a separating funnel (liquid-liquid extraction) can be carried out advantageously by a small column containing a porous chemically inert material which is strongly water adsorbent. The aqueous sample is first absorbed into the column (the size must be such that all of the solution should be absorbed). At this stage, the water diffuses into the inert material and behaves as an immobile film of aqueous phase that permits the neutralization and washing the aqueous phase with a non-miscible solvent.

diffuse
[dɪfju:s]
vt. 扩散,传播

8.5 Gas extraction on a cartridge or a disc

In order to measure low concentrations of molecular compounds present in gaseous samples, for example a polluted atmosphere, the best method consists of trapping them by passing a known volume of the sample through a short tube of absorbing material or through a disposable SPE cartridge (Figure 8. 5).

The composition of the trap-column is variable: molecular sieves, graphite-based carbon black, organic polymers. A trap can be composed of a series of several adsorbents. A pump rate is used to deliver a pre-programmed volume of gas, in relation to the capacity of the trap (flow rate of 0. 1 to 1L/min).

sieve
[sɪv]
vt. 筛,滤;*n*. 筛子,滤网

The compounds absorbed are recovered either through extraction by means of solvents (often using carbon disulfide), or by thermal desorption in a carrier gas which avoids analyte dilution and the introduction of artefacts.

artefact
['ɑ:tɪfækt]
n. 人工制品,加工品

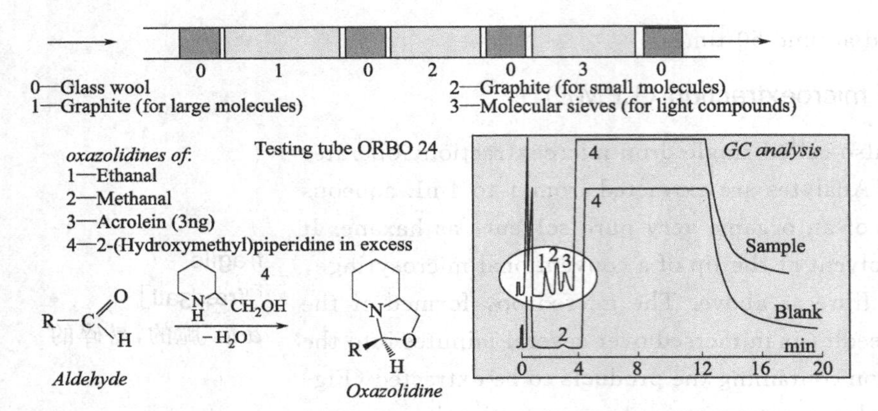

Figure 8.5 Gas extraction. Also called the dynamic purge-and-trap method. Principle of a gas/solid extraction column. A chemical transformation used to detect an aldehyde by derivativization

derivativization [dəˈrɪvəˌtɪzeɪʃən] *n*. 衍生(作用)

This technique can be adapted to gas phase chromatography. The trap is inserted into a special oven and heated with a temperature gradient able to attain 350℃ in a few seconds. The desorbed compounds are directly introduced into the GC injector. As an alternative to these special ovens there exists extraction tubes of a diameter sufficiently small that can be directly introduced into a modified GC injector.

The recovery of the compounds is considered to be satisfactory when it attains 60 percent, but it is often almost quantitative with this device.

subsequent [ˈsʌbsɪkwənt] *adj*. 后来的，随后的

The same extraction tube filled with several layers of adsorbents enables the retention of a broad range of molecules with different polarities. Each layer protects the next, which is more active. The first layer serves to purge the sample of heavy compounds. Lighter compounds and gases are adsorbed by the subsequent layers. To regenerate the trap, the sweeping gas circuit is reversed, so that the high molecular weight compounds retained at the entrance of the tube, are eliminated first (Figure 8.5).

regenerate [rɪˈdʒenəreɪt] *vt*. & *vi*. 使再生，回收(废热，废料等)

A new range of problems appear, affecting more than 50 per cent of the samples treated for trace analyses: absorption by the walls of the containers, action of atmospheric oxygen, evaporation of the analytes, etc. As a consequence, those methods are favored which lead to an enhanced enrichment with a short extraction time.

The same principles are applied for developing detector tubes, which contain, along with the adsorbent, a reagent that causes a specific derivatization of the analyte. The flexibility of this method is due to the wide choice of adsorbent phases that permit selectivity of extraction. When molecules are unstable on the adsorbent, they can be transformed to a specific derivative by a reagent mixed with the adsorbent.

8.6 Headspace

This relatively simple technique can provide sensitivity similar to the above dynamic purge-and-trap analysis. Headspace is a sampling device used in tandem with a GC installation (Figure 8. 6). It is reserved for the analysis of volatile compounds present in a matrix, which cannot be itself directly analyzed by chromatography. This is generally the case as most samples are composed of a wide variety of compounds that differ in molecular weight, polarity, and volatility. Basic principles of headspace are illustrated by the two following variations.

static mode
静态模式

Static mode: the sample (solid or liquid matrix) is placed in a glass vial capped with a septum such that the sample occupies only a part of the vial's volume. After thermodynamic equilibrium between the phases present (1/2 to 1h), a sample of the vapour is taken. Under these conditions, the quantity of each volatile compound present in the headspace (volume above the liquid) will be proportional to its concentration in the matrix. After calibration (using methods of internal or external standards), it is possible to match the real concentrations in the sample with those of the vapours injected in the gas chromatograph (Figure 8. 6).

be proportional to
与……成比例

calibration
[kælɪ'breɪʃn]
n. 校准,刻度,标度

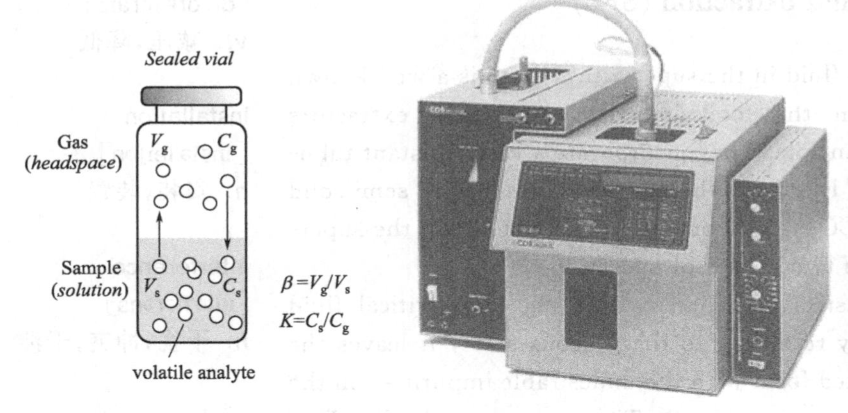

Figure 8.6 Static headspace. Sample vial in a state of equilibrium. The state of equilibrium is under the control of two factors: $K=C_s/C_g$ and $\beta=V_g/V_s$. In addition to routine single extraction (one sampling per vial), modes of multiple headspace extraction are incorporated in this model G1888

Dynamic mode: instead of working in a closed environment, a carrier gas such as helium is either passed over the surface of the sample or bubbled through it in order to carry the volatile components toward a trap where they are adsorbed and concentrated (Figure 8. 7). The procedure continues with a thermal desorption of the trap using reversed gas flow in respect of the injection in the chromatograph. This "purge-and-trap" principle is semi- quantitative and delivers a sample without residue.

This method is quite reliable for repetitive analyses involving stable matrices. However, if the matrix composition fluctuates, then this disturbs the equilibrium factors and reduces the precision of the result, by consequence of an irregular calibration. In this case cartridge-base extraction would be preferred.

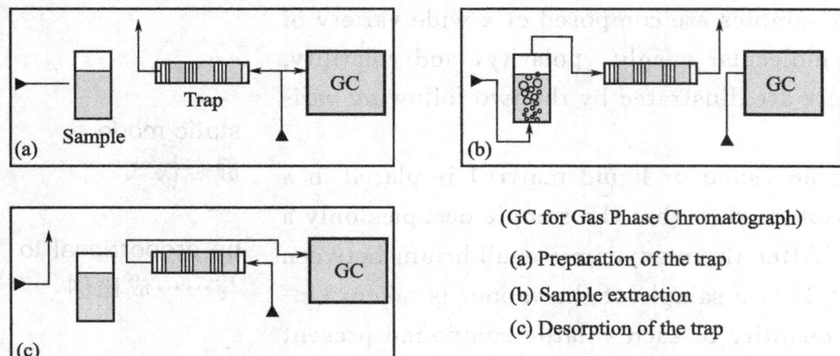

Figure 8.7 Headspace, dynamic mode. The sample is recovered by thermal desorption of a purge-and-trap cartridge, chosen as a function of the compound to be extracted

8.7 Supercritical fluid extraction (SFE)

Extraction using a fluid in the supercritical state is a well-known procedure employed in the food industry. Analytical extractors operate on the same principle. They incorporate a very resistant tubular container in which is placed the sample (in solid or semi-solid form) with the fluid (CO_2, N_2O or $CHClF_2$-Freon 22) in the supercritical state. There are two modes of operation:

Off-line mode consists of depressurising the supercritical fluid following extraction. By returning to the gaseous state, it leaves the analytes in a concentrated form with the undesirable impurities on the internal wall of the extraction vessel. The concentrate is dissolved using some classical solvent and submitted to a selective extraction on a solid-phase cartridge.

On-line mode consists of performing the direct analysis from the extract, while it is still under pressure, by passing the supercritical fluid through a chromatographic installation (SFC or HPLC). This procedure is applied only to those compounds for which interferences due to matrix are not likely to occur.

In summary, SFE acts on four parameters that can be modified to obtain a good selectivity: pressure, temperature, extraction period

depressurise
[di:'preʃəraɪz]
vt. 减压,降低

installation
['ɪnstə'leɪʃən]
n. 安装,装置

interference
[ˌɪntə'fɪərəns]
n. 干扰,冲突,干涉

and choice of modifier. It is also possible to change the solvation character of the fluid by the introduction of an organic modifier (e. g. methanol, acetone). Nonetheless, to separate the analyte from the matrix requires knowledge of the solubility and transfer rate of the solute in the solvent as well as of the physical and chemical interactions between solvent and matrix (Figure 8.8).

The use of microwaves combined with extraction by solvent in a pressurized container is another very efficient process for the rapid treatment of samples.

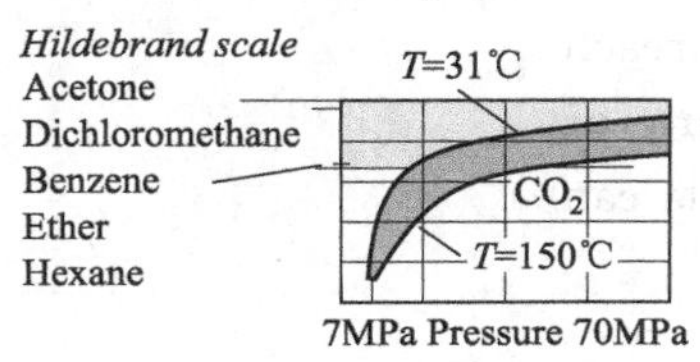

Supercritical Fluid Properties

Fluid	Critical Point(℃)	Dipole moment (Debye)
CO_2	32	0
N_2O	37	0.2
$CHClF_2$	96	1.4

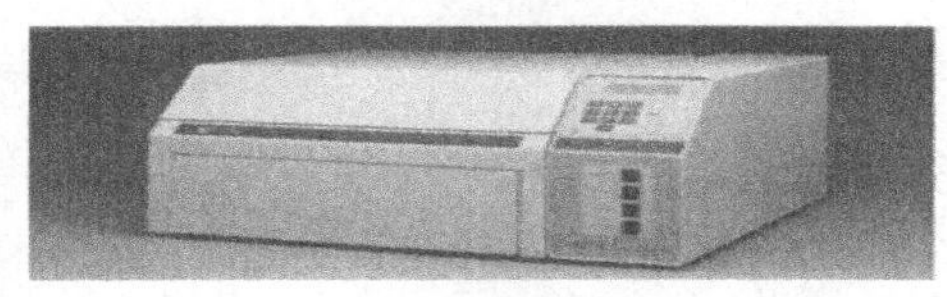

Figure 8.8 Supercritical fluid extraction. Comparison of the solvation strength of the CO_2 with respect to the usual solvents (Hildebrand scale) as a function of the temperature and pressure. The polarity of carbon dioxide in the supercritical state is comparable with that of hexane (for 100 atm and 35℃). SPE is a method for which automation becomes a justified investment when the sample throughput is large

8.8 Microwave reactors

The transformation to elements of samples containing organic material, known as mineralization, has motivated the development of many different approaches: dry procedures such as oven heating, combustion, while others by wet treatment such as with mineral acids, fluxes for fusion, etc. In the absence of a universal method applicable to all inorganic elements, the mineralization must be adapted to the sample. This indispensable step of the preparation of a large number of samples, in particular for those analyzed by AAS and OES, can be made easier by using of a "microwave digester."

Mineralization by strong acids conducted in open containers, in a fume-hood, with the ever-present risk of cross contamination and "bumping", is a long and fastidious operation which fundamentally has changed little over a century. Beyond the practical problems involved, some sample matrices are quite impossible to treat under these conditions (refractory or volatile materials, certain minerals, charcoals and heavy oils).

mineralization
[ˌmɪnərəlaɪˈzeɪʃən]
n. 矿化(作用)

flux
[flʌks]
n. 熔解

contamination
[kənˌtæmɪˈneɪʃən]
n. 污染

fastidious
[fæˈstɪdɪəs]
adj. 挑剔的，讲究的，苛求的

refractory
[rɪˈfræktərɪ]
adj. 难熔的；
n. 耐火物质

Microwave digesters consists of a Teflon "bomb" of a few mL volume in which the crude sample is placed along with the mineralization solution (necessary for carbonization or oxidation). The heating, induced by the microwaves, occurs via molecular agitation due to the dipoles of the water molecules. The closed calorimetric-type vessel avoids losses through droplets or aerosols, or evaporation of volatile components and splashing. The temperatures reached by the acidic solutions under pressure quickly exceed the boiling temperatures under normal conditions (the same effect is seen in a pressure-cooker). This gain in time has been known, in certain cases, to reach more than 90 per cent. In such conditions, digestions by sulfuric or perchloric acids, hydrolyses or mineralizations in oxidizing media can be performed. This system can be automated quite safely.

calorimetric
[kælərɪmetrɪk]
n. 量热

Problems

8.1 What is the purpose of sample pretreatment?

8.2 Describe the several kinds of sample pretreatment techniques.

第8章 样品制备技术

对于所有的化学分析，分析物（待测量的物质）必须具有足够的数量，并使用合适的仪器测量。大多数样品需要特定的预处理。根据所谓的取样程序，这个初步阶段一直是分析过程的“瓶颈”。样品前处理通常是化学分析中关键的一步，因为它对最终结果有影响。它与测量本身同样重要，会极大地影响分析过程的可靠性、准确性和成本。至少一半以上的分析任务都涉及痕量分析（天然样品中的分析物少于1mg/L），因此需要强有力的萃取方法。特殊的富集模式往往由仪器或所用的方法来决定。受益于现代机器人技术的发展，关于缩短预处理时间这个领域的研究开展得非常活跃。此外，大量小样品的分析促进了最先进的设备和最新程序的开发，其中一些专门用于色谱分析的样本将在本章中进行阐述。

8.1 样品前处理的必要性

开始分析时，需要研究能代表该批样品的样本。被研究的物质称为分析物，通常与大量可变的其他化合物一起称为基质。当在成千上万倍其他更高含量的化合物中测量痕量级的一种分析物时，常常担心分析物与其他化合物之间的干扰。甚至最好的分析技术也不能弥补由样品前处理效果不好产生的问题。这使得样品前处理必须得到高度重视。由于这项附加的工作往往乏味、耗时，因此已经开发了一些有关吸附、萃取或沉淀的方法（图 8.1），这将使样品前处理过程更简单，甚至自动化。

8.2 固相萃取

目前，在测量其中一种或几种成分之前，固相萃取（SPE）常用于纯化或浓缩粗提物。固相萃取是一种清洁分析方法，它大大改变了经典的在分液漏斗中进行的液/液萃取的溶剂萃取法，但液/液萃取法相对较慢，还有使用大量的有机溶剂等缺点。

固相萃取要在一个类似于注射器的开口小塑料卡盒内操作，卡盒内含适量的固体吸附剂（例如一盒含 100～300mg RP-18 反相填料），该过程包括将已知体积的液体样品通过（图 8.2 所示）。具体有以下两种方法。

目前最常用的方法　首先目标化合物被含吸附剂的 SPE 柱吸附，接着通过冲洗最大可能地除去不需要的杂质化合物，用于随后的测量。然后分析物富集到小体积合适的溶剂中。此萃取过程不仅分离了被分析物，而且预浓缩了被分析物，这对痕量分析特别有用。

第二种方法，尽管很简单，但不常用，样品过柱，不需要的杂质被吸附保留了，而分析物自由通过。这种方法很吸引人，因为它没有洗脱（解吸）步骤。另一方面，分析溶液被纯化，但没有被浓缩（富集系数为 1）。

固体吸附剂类似于液相色谱的固体固定相，键合的硅胶（反相极性）的颗粒粒径在 4～100μm 之间，允许更快的流速通过。其他吸附剂，包括石墨或其他共聚物，如苯乙烯-二乙烯基苯，这种有大官能团键合在其表面的吸附剂在酸性溶液中更稳定。

也有可作为化学过滤器使用的萃取盘片（厚度为 0.5mm，直径为 25～90mm），使用方法类似于将过滤纸垫在接有真空瓶的漏斗中。这种萃取盘片由键合的硅胶球组成，装载在多孔聚四氟乙烯或玻璃纤维基质中。它们保留了大量水溶液中存在的痕量有机化合物，但有些物质用上述的卡盒是不可能保留的。分析物在溶剂流过过滤器时被富集，这样痕量分析会更容易些，因为富集倍数可高达 100 倍。

从极性基质中用 C_{18} 衍生的硅胶柱或圆盘分离非极性化合物，极性化合物从被填满正相填料的非极性基质中分离。带电物质可以通过离子交换分离。这三种情况在 HPLC 分析中同样适用。

有两点需要注意：在萃取过程中，超出了体积限量的过量化合物将不再被卡盒（萃取小柱）萃取。另外在过滤过程中，所吸附化合物的回收率很少能完全达到 100%。如果测量技术采用质谱检测器，这个定量的问题可以通过在萃取之前，加入一个给定体积的样品来解决，加入已知数量的相同化合物采用稳定同位素来标记。将这一序列的两种相同形式的化合物信号相比较，建立它们的比率，从而测定未知浓度。这是等同于内标法，采用的内标物的相对响应系数为 1。

目前的研究主要涉及对特定类型的化合物具有显著亲和性的吸附剂，合成的聚合物以表面变形的方式在其表面形成识别位点，这更适合捕捉保留目标分子。原理类似于更具选择性的免疫萃取。

在传统的液/液萃取中，分析物在大量溶剂中被稀释。因此，当萃取溶液被浓缩时，非挥发性杂质也会包含在溶剂中。这往往使该方法不受欢迎，除非使用超纯溶剂。例如，如果 1mg 化合物与 1mL 99.9%纯溶剂混合，后者将带来与分析物本身相同质量的附加污染物。

8.3 免疫亲和萃取

用传统的固相萃取柱，被分析物的保留是基于疏水相互作用。如果基质含有浓度较大的干扰物，对目标分析物有可能产生干扰，因此这种方法缺乏选择性。在开发基于免疫吸附剂的新型小柱方面已经取得了很大的进展，能够从复杂的环境介质中有效富集单个分析物。这种分子识别是基于抗原/抗体类型间的相互作用。

通过共价键将抗体固定在一个合适载体上，将得到一种吸附剂，上面固定有相应的抗原作为分子印迹聚合物（图 8.3）。免疫亲和技术，现在广泛应用于制药和生物技术领域，可用于萃取环境样品中的有机小分子。有些萃取小柱含有除草剂或多环芳烃的抗体，适合通过快速渗滤的方法来萃取。

8.4 微萃取程序

8.4.1 固相微萃取（SPME）

固相微萃取将取样、萃取、浓缩和样品导入整合为无需溶剂的一个操作步骤中。该技术采用1～2cm长的多孔熔融石英纤维，涂上适当的吸附剂。这种纤维装在一个类似注射器的装置中，由钢针保护（图8.4）。该装置的柱塞压低使纤维暴露在样品中，然后在取样时间结束后收缩回来，该纤维可被引入水溶液中或位于其表面的空间（顶部空间）。这个过程不需要溶剂，而是对样品进行直接萃取，并将其浓缩在萃取纤维上。然后将纤维引入GC的特定注射器中，通过温度的作用，将感兴趣的化合物解析然后分离。SPME技术还可以与HPLC、HPCE和MS相结合使用。灵敏度通常为每万亿分之一，尽管同样的纤维可以使用大约50次，但这一过程的成本仍相当昂贵。

8.4.2 液相微萃取（LPME）

这种技术也称为单滴微萃取，以两相模式运行。将分析物从1～4mL含水样品中提取到25～50μL非常纯如己烷的有机溶剂中。它在传统的微注射器尖端使用微量的溶剂，而不是如上所述的脆性二氧化硅纤维。在注射器针头形成的微滴在几分钟内浸入到含有待萃取产物的水溶液中（图8.4b）。LPME一步可完成提取和浓缩。后续分析由GC进行。只要操作条件在所有校准程序中都保持不变，则可以进行定量处理。该方法已应用于食品和药品中的化学溶剂以及天然植物成分、水环境中挥发性和半挥发性污染物的分析。

传统上使用分液漏斗（液-液萃取）进行的少量水溶液的纯化可以很好地由含有强吸水剂的多孔化学惰性材料的小柱来完成。水溶液样品首先被吸收到柱中（尺寸必须能够使得所有溶液都被吸收）。在这个阶段，水会扩散到惰性材料中，并表现为一个水相的固定膜，它允许中和以及用非混相溶剂清洗水相。

8.5 在卡盒（饼）上进行气体提取

为了测量气态样品中存在的低浓度分子化合物，例如污染的大气，最好的方法是通过将已知体积的样品通过短管吸收材料或通过一次性SPE柱捕获它们（图8.5）。

捕捉保留柱的组成是可变的：可以是分子筛、石墨基炭黑、有机聚合物。捕捉器可以由一系列吸附剂组成。泵速率用于传送预先设定体积的气体，这与捕集器的容量（流速为0.1～1L/min）有关。

吸收的化合物通过溶剂（通常使用二硫化碳）提取或通过在载气中的热解吸而回收，这避免了分析物稀释和引入其他物质。

该技术可适用于气相色谱。将捕捉器插入特殊的烘箱中，并在几秒内可达350℃的温度梯度加热。置换洗脱的化合物直接引入GC注射器。作为这些特殊烘箱的替代品，可以使用直径足够小的提取管直接引入到一种改进过的GC注射器中。化合物的回收率达到60%时即认为是令人满意的，通常可使用该装置来定量。

填充有几层吸附剂的萃取管能够保留不同极性的很多分子。每一层吸附剂都保护活性更强的下一层吸附剂。第一层用于清除分子量大的化合物，较轻的化合物和气体被后续的吸附层所吸附。为了再生捕捉器，扫气管路反向，从而首先消除保留在管入口处的高分子量化合物（图 8.5）。

一系列新问题的出现影响了超过 50％的痕量分析样品的处理：容器壁吸收，大气氧气的作用，分析物的蒸发等。因此，提取时间更短的增强富集方法更受青睐。

可以应用相同的原理来开发检测器管，检测器管往往都含有对分析物特异性衍生化的试剂以及吸附剂。这种方法的灵活性是由于吸附剂的选择种类多，从而允许萃取方法的选择余地大。当分子在吸附剂上不稳定时，它们可以通过与吸附剂混合的试剂转化成特定的衍生物。

8.6 顶空萃取

这一相对简单的技术其灵敏度类似于上述的动态吹扫和捕集分析的灵敏度。顶空进样器是一种在气相色谱中（图 8.6）串联使用的进样装置。它用于分析基质中存在的挥发性化合物，其本身不能通过色谱法直接分析。一般情况下，分析样品大都由分布广泛、各种各样的化合物组成，其分子量、极性和挥发性有所差别。顶空萃取的基本原理有下面两种类型。

静态模式：样品（固体或液体基质）放置在一个封闭的玻璃小瓶中，且只占据瓶子的一部分体积。在相热力学平衡后（1/2～1h），得到样品蒸气。在这些条件下，每个挥发性化合物存在于顶空（液体以上的体积）的量将与基质中的浓度成比例。校准后（使用内标法或外标法），样品中的实际浓度与注入气相色谱仪中的蒸气浓度相匹配（图 8.6）。

动态模式：代替在封闭环境中的工作，载气如氦气或在试样表面吹过或在试样中鼓泡，使挥发性成分进入捕集管进行吸附和浓缩（图 8.7）。随后通过注入反向的气流对捕集管进行热脱附，使其进入色谱进样口。这种“吹扫和捕集”的原理是半定量的，而且没有样品残留。

这种方法对稳定基质的重复分析是相当可靠的。然而，如果基质组分发生波动，这将扰乱均衡状态的平衡因子，并通过一个不规则的校准结果降低结果的精确度。此时，柱萃取将是优选方法。

8.7 超临界萃取（SFE）

超临界状态下的液体萃取是食品工业中众所周知的一道工序，与分析萃取器的原理相同。它将样品（固体或半固体状）和超临界状态下的液体［CO_2、N_2O 或 $CHClF_2$（氟利昂 22）］放置在一个非常耐压的管状容器中。具体有两种操作模式。

离线模式是萃取后再降低超临界流体的压力。流体变为气体状态后，分析物被浓缩，不需要的杂质也留在了提取容器的内壁上。使用一些经典的溶剂溶解该浓缩物，可在一个固相小柱进行选择性萃取。

在线模式是指用超临界流体在加压状态下通过色谱装置（SFC 或 HPLC）直接分析萃取物。此过程仅适用于那些基质干扰不太可能发生的化合物。

总之，为获得一个良好的选择性，超临界流体萃取有四个参数可以进行优化：压力、温度、萃取周期和选择调节剂。另外，也可以引入一种有机改性剂（例如甲醇、丙酮），以改变流体的溶剂性质。尽管如此，分离基质中的分析物要求溶解度和溶剂传输速率以及溶剂和基质间的物理和化学相互作用的相关知识（图 8.8）。

在加压的容器内用有机溶剂与微波结合进行萃取是对样品进行快速处理的另一种非常有效的方法。

8.8 微波萃取

将含有机物的样品转换为单质的过程称为矿化，矿化有许多不同的方法：干法处理如烘箱加热、燃烧，湿法处理如与矿物酸作用、助溶剂熔化等。在没有适用于所有无机元素的通用方法的情况下，必须要用矿化来处理。这是大量样品预处理不可缺少的步骤，特别是采用原子吸收分光光度法（AAS）和发射光谱分析（AES）时，而通过“微波消解仪”会变得更容易些。

在通风橱内用强酸在敞开容器中进行矿化，永远存在交叉污染的风险。“硝化”是一个漫长而艰苦的操作，从根本上看一个世纪以来这个操作过程都没有多少进步。除了所涉及的实际问题，一些样本基质在这一条件下是不可能处理的（难熔性或挥发性物质，某些矿物质、木炭和重油）。

微波消解是将几毫升体积的粗品与矿化溶剂（必要进行碳化或氧化）混合液倒入聚四氟乙烯“管”中。通过微波加热，由于水分子偶极作用引起分子碰撞。封闭量热型容器可避免飞沫或气溶胶形式的损失，或挥发性成分的蒸发和飞溅损失。在加压下该酸性溶液达到的温度会很快超过在正常条件下的沸腾温度（相同的效果类似于一个高压锅内）。在一些情况下，这种方法可以节省 90%以上的处理时间。在这样的条件下，可以进行硫酸或高氯酸的消解，在氧化性介质中可以进行水解或矿化。该系统可以安全地自动运行。

Solutions

1.1 At equilibrium the 40mL of eluent contains $12 \times 40/10 = 48$mg of compound. There are therefore $100-48=52$mg in the stationary phase. If K represents the ratio of the masses present in 1mL of each phase in equilibrium, $K=(52/6)/(48/40)=7.2$.

1.2 When a molecule of the compound migrates along the column it passes alternately from the mobile phase, where it progresses at the speed of this phase, to the stationary phase where it becomes immobilized. The average velocity for the passage is then reduced in relation to the amount of time spent by the molecule in the stationary phase, increasing by consequence the value for t_S.

Now consider the total mass, m_T, of all of the identical molecules in the sample. Statistically at each moment, there will be a certain number of these molecules within the stationary phase while the rest remain in the mobile phase. The ratio of these molecules fixed in the stationary phase (m_S), and of those present in the mobile phase (m_M), will be the same as the ratio representing the sum of the intervals spent in each phase for a single isolated molecule; that is: $m_S/m_M=t_S/t_M$.

Thus, k will correspond to the ratio t_S/t_M and since $t_S=t_R-t_M$ then we can refind the classic expression for the retention factor (or capacity factor), as calculated from the chromatogram:

$$k=(t_R-t_M)/t_M$$

1.3

1. The separation factor (or selectivity factor), is generally calculated from the retention time t_R, yet in the question it is the retention volumes V_R while are given for the two compounds. However, V_R and t_R are linked by an expression while considers the flow of the mobile phase, D:

$$V_R=D\,t_R$$

Therefore, if for the expression in α below, each t_R (or t_M) is replaced by V_R (or V_M), we will have:

$$\alpha=[t_{R(2)}-t_M]/[t_{R(1)}-t_M]=[V_{R(2)}-V_M]/[V_{R(1)}-V_M]$$

which leads to $\alpha=1.2$.

Note: The expression in α is defined from the retention volumes and remains useful with a flow gradient but it is very rare that it is employed since the apparatus does not measure the instantaneous rate of the column. Therefore, the calculation of the retention volumes (other than by substituting in the relation, $V_R=Dt_R$) is not permissible.

The expression giving the retention factor can be written, $t_R=t_M(1+k)$, where t_M corresponds to the ratio between the length L of a column and the average velocity u of its mobile phase. Equally, $k=KV_S/V_M$, leading upon substitution to

$$t_R=(L/u)(1+KV_S/V_M)$$

In order to incorporate k into expression 1, the relation $t_M=t_R/(1+k)$ is used. Expression 1 then becomes:

$$N_{eff}=5.54\,\frac{\left(t_R-\dfrac{t_R}{1+k}\right)^2}{W_{1/2}^2}$$

Simplifying to

$$N_{\text{eff}}=5.54\times t_R^2\frac{\left(1-\frac{1}{1+k}\right)^2}{W_{1/2}^2}\text{ and then } N_{\text{eff}}=N\frac{k^2}{(1+k)^2}$$

The classic formula giving the resolution implies that the peaks are Gaussian. The known relationship between W, $W_{1/2}$ and σ permits the replacement of W by $4W_{1/2}/2.35$.

The basic formula then becomes,

$$R=2\frac{t_{R(2)}-t_{R(1)}}{4(W_{1/2})_1/2.35+4(W_{1/2})_2/2.35}=1.18\frac{t_{R(2)}-t_{R(1)}}{(W_{1/2})_1+(W_{1/2})_2}$$

The formula, proposed in the question, leads to an error of approximately 20%.

2. If these peaks are adjacent then the widths of their bases will generally be comparable. Allowing therefore $W=W_1=W_2$ and expressing W as a function of N_2 we will have:

$$R=\frac{t_{R(2)}-t_{R(1)}}{W_2}\text{, with }\frac{1}{W_2}=\frac{1}{4}\sqrt{N_2}\frac{1}{t_{R(2)}}$$

Leading, on combination, to:

$$R=\frac{1}{4}\sqrt{N_2}\frac{t_{R(2)}-t_{R(1)}}{t_{R(2)}}$$

k can then be introduced by incorporating the general expression, $t_R=t_M(1+k)$,

$$R=\frac{1}{4}\sqrt{N_2}\frac{k_2-k_1}{1+k_2}$$

then, finally by introducing $\alpha=k_2/k_1$:

$$R=\frac{1}{4}\sqrt{N_2}\frac{k_2-k_2/\alpha}{1+k_2}\text{ leading to }R=\frac{1}{4}\sqrt{N_2}\frac{k_2}{1+k_2}\times\frac{\alpha-1}{\alpha}$$

3. Then by introducing α and substituting, we arrive at expression 2.

1.4 In supposing that the following equation is true:

$$R=\frac{1}{4}\sqrt{N}\frac{k}{1+k}\times\frac{\alpha-1}{\alpha}$$

then,

$$N=16R^2\left(\frac{\alpha}{\alpha-1}\right)^2\left(\frac{k+1}{k}\right)^2\text{ but } N_{\text{eff}}=N\left(\frac{k}{1+k}\right)^2$$

Which on substitution will lead to the expression given in the question.

1.5 $R_1=(0.5\times 60)/(10+12)=1.36$; $R_2=1.47$ ($k_2=9.5$, $\alpha=1.056$ and $N_2=15270$) $R_3=1.5$ ($k_1=9$); $R_4=1.5$.

1.6 The solution of this problem us to the initial three lines of exercise 1.7, where we considered employing the retention volumes in place of the retention times.

2.1

1. $R_{F(A)}=27/60=0.45$; $R_{F(B)}=33/60=0.55$

$N_A=16\times 27/2=2916$; $N_B=16\times 33/2.5=2788$

$H_A=X_A/N_A=9.26\times 10\text{cm}$; $H_B=X_B/N_B=1.18\times 10\text{cm}$

2. $R=2(33-27)/(2+2.5)=2.67$

3. From the definition for the selectivity factor and from the relation linking R_F and k, we arrive at:

$$\alpha=[R_{F(B)}/R_{F(A)}][1-R_{F(A)}]/[1-R_{F(B)}]=1.49$$

2.2

1. Since it arises from a normal phase, the more polar compound will be retained the longest. In order of increasing migration distances we will have C, then B and finally A.

2. The order of elution will be A, B then C, reflecting the migration times.

3. Reverse order.

4. Approximate value

for $R_{F(A)}=22.5/50=0.45$. $HETP=L/N$ with $N=5.54x^2/d^2=5.54\times22.5^2/2^2=701$, leading to $HETP=50/701=0.07$mm.

3.1 In gas-liquid chromatography, the stationary phase is a liquid that is immobilized on a solid. Retention of sample constituents involves equilibria between a gaseous and a liquid phase. In gas-solid chromatography, the stationary phase is solid surface that retains by physical adsorption. Separation involves adsorption equilibria.

3.2 Gas-solid chromatography is used primarily for separating low molecular mass gaseous species, such as carbon dioxide, carbon monoxide, and oxides of nitrogen.

3.3 A chromatogram is a plot of detector response versus time. The peak position, retention time can reveal the identity of the compound eluting. The peak area is related to the concentration of the compound.

3.4 Sample injection volume, carrier gas flow rate, and column condition are among the parameter that must be controlled for highest precision quantitative GC. The use of an internal standard can minimize the impact of variations in these parameters.

3.5

(a) Advantages of thermal conductivity: general applicability, large linear range, simplicity, and nondestructive.

Disadvantage: low sensitivity.

(b) Advantages of flame ionization: high sensitivity, large linear range, low noise, ruggedness, ease of use, and response that is largely independent of flow rate. Disadvantage: destructive.

(c) Advantages of electron capture: high sensitivity selectivity toward halogen-containing compounds and several others and nondestructive.

Disadvantage: small linear range.

3.6 Liquid stationary phases are generally bonded and/or cross-linked in order to provide thermal stability and a more permanent stationary phase that will not leach off the column. Bonding involves attaching a monomolecular layer of the stationary phase to the packing surface by means of chemical bonds. Cross-linking involves treating the stationary phase while it is in the column with a chemical reagent that creates cross links between the molecules making up the stationary phase.

3.7

(a) Band broadening arises from very high or very low flow rates, large particles making up packing, thick layers of stationary phase, low temperature, and slow injection rates.

(b) Band separation is enhanced by maintaining conditions so that k lies in the range of 1 to 10, using small particles for packing, limiting the amount of stationary phase so that particle coatings are thin, and injecting the sample rapidly.

4.1

(a) Substances that are somewhat volatile and are thermally stable.

(b) Substances that are ionic.

(c) High molecular mass compounds that are soluble in nonpolar solvents.

(d) Chiral compounds (enantiomers).

4.2

(a) In an isocratic elution, the solvent composition is held constant throughout the elution.

(b) In a normal-phase packing, the stationary phase is quitepolar, and the mobile phase is relatively nonpolar.

(c) In a bonded-phase packing, the stationary phase liquid is held in place by chemically bonding it to the solid support.

(d) In ion-pair chromatography, a large organic counter-ion is added to the mobile phase as an ion-pairing reagent. Separation is achieved either through partitioning of the neutral ion pair or as a result of electrostatic interactions between the ions in solution and charges on the stationary phase resulting from adsorption of the organic counter ion.

(e) Gel filtration is a type of size-exclusion chromatography in which the packings are hydrophilic and the eluents are aqueous. It is used for separating high molecular mass polar compounds.

4.3 diethyl ether, benzene, *n*-hexane.

4.4 ethyl acetate, dimethylamine, acetic acid.

4.5 In adsorption chromatography, separations are based on adsorption equilibria between the components of the sample and a solid surface. In partition chromatography, separations are based on distribution equilibria between two immiscible liquids.

4.6 Gel filtration is a type of size-exclusion chromatography in which the packings are hydrophilic and the eluents are aqueous. It is used for separating high molecular mass polar compounds. Gel-permeation chromatography is a type of size-exclusion chromatography in which the packings are hydrophobic and the eluents are nonaqueous. It is used for separating high molecular mass nonpolar species.

4.7 In suppressor-column ion chromatography, the chromatographic column is followed by a column whose purpose is to convert the ions used for elution to molecular species that are largely nonionic and thus do not interfere with conduct ometric detection of the analyte species. In single-column ion chromatography, low capacity ion exchangers are used so that the concentrations of ions in the eluting solution can be kept low.

4.8

(a) For a reversed phase chromatographic separation of a steroid mixture, selectivity and, as a consequence, separation could be influenced by temperature-dependent changes in distribution coefficients.

(b) For an adsorption chromatographic separation of a mixture of isomers, selectivity and, as a consequence, separation could be influenced by temperature-dependent changes in distribution coefficients.

5.1 The stationary phase is of a cationic type. In the dry state it remains in a non-dissociated form but when water is added and in the presence of a high concentration of sodium ions, the following equilibrium is attained in which there is a liberation of H^+:

$$RSO_3H + Na^+ \longrightarrow R{-}SO_3Na + H^+$$

The calculation of the molar capacity of the stationary phase is as follows:

$$25.5 \times 0.105/1000 = 2.6775 \times 10^{-3}\ \text{mol/g}$$

The equilibrium above is long in establishing itself such that the addition of the sodium hydroxide solution removes all of the protons liberated. A certain time is required for them to reappear in solution.

Note: The helianthin $(Me)_2N(C_6H_4)N{=}N(C_6H_4){-}SO_3Na$ is another name for methyl orange, which colours red for pH<3.2 and yellow for pH>4.4 (λ_{max}=505nm).

5.2

1. Such a phase corresponds to a polymer ($250000 < M < 900000$) comprising acid groups, $-CH_2COOH$, and is therefore weakly cationic. The pH of the separation is inferior to the pI of proteins, thus their amino functionality will be fully protonated ($-NH_3^+$).

2. If the pH is increased then the ionic character of the column is reduced leading to a faster elution of the proteins. Here $pI_1 < pI_2 < pI_3$.

5.3

1. The concentration factor is 5 at the beginning since we have 1mL of plasma and, following treatment of this extract, there will be 200μL remaining of the solution containing the same amount of cyclosporin as the 1 ml at the beginning.

2. No, because we cannot weigh the collected cyclosporins A or D.

3. (a) Method of internal standard

Following convention we will give the letter A to cyclosporin A and D to cyclosporine D as used in the internal standard:

$$k_{A/D} = m_A A_D / m_D A_A = C_A A_A / C_D A_A$$

To calculate the relative response coefficient we will take for example the solution containing 400 ng/mL of cyclosporin A:

$$k_{A/D} = (400/250) \times (1/2.04) = 0.784$$

$$C'_A = C'_D k_{A/D} A'_A / A'_D$$

The chromatogram of the sample for which $A'_A / A'_D = 9/14$ (heights of the peaks measured from the original chromatogram), leads to

$$C'_A = 250 \times 0.784 \times 9/14 = 126\text{ng/mL}$$

(b) Method of calibration curve

Plot the graph of the ratios of the peak heights against concentration C of cyclosporine A.

By the method of least squares we find:

$$R_h = 0.005092C - 0.00411$$

Note: If we evaluate R_h incorrectly then we commit an error of some importance which is of course translated into the result. However, frequently, for very weak concentrations, we can estimate with respect to a limit below which the error holds less importance, e.g. in taking $R_h = 8/15$ rather than 9/14, the difference will be 20% and we will find that $C =$ 104ng/mL.

5.4 Through the application of the general expression we will have a total of four formulae giving the % mass of the various esters. Below is an example for the methyl butanoate ester:

$$x_i = 100\frac{A_{AE}\times 0.919}{A_{ME}\times 0.919 + A_{EE}\times 0.913 + A_{PE}\times 1.06 + A_{BE}}\times 100\%$$

We will solve this by taking and substituting the areas given in the question:

$$x_{ME} = 16.6\%$$

$$x_{EE} = 16.6\%$$

$$x_{PE} = 33.4\%$$

$$x_{BE} = 33.4\%$$

Note: The scale in mV corresponds to UV detection.

5.5

1. The addition of *N*-methylserotonin before the extraction is made removes the necessity of counting any eventual loss of product due to the different intermediate manipulations. We can suppose that the yield of the extraction is the same for the two compounds, which are structurally very similar.

2. To determine the relative response factor $k_{S/NMS}$ of the serotonin (S), with respect to its *N*-methyl derivative (NMS), the following relation is used:

$$k_{S/NMS} = \frac{m_S}{m_{NSM}}\times\frac{A_{NSM}}{A_S} = \frac{5}{5}\times\frac{30956727}{30885982} = 1.002$$

3. Measuring the sample:

$$m_S = m_{NSM}k_{S/NMS}\frac{A'_{NMS}}{A'_S} = 30\times 1.002\times\frac{2573822}{1719818} = 45\text{ng/mL}$$

which gives a concentration of approximately 45 ppb, for an aqueous solution.

6.1 In size exclusion chromatography $K \leqslant 1$, excepting cases where interactions develop between the solute and stationary phase since this creates a partition phenomenon which superimposes itself upon the diffusion in the pores.

6.2 By placing the following two columns end to end, first column C will separate the masses of 3×10^6 Da and 1.1×10^6 Da from the remaining two. These two masses will subsequently be separated by the column A. On the corresponding chromatogram, we will obtain four distinct peaks.

6.3

1. The total exclusion or interstitial volume is approximately 4.2mL. The intraparticle volume of the pore is close to: 7.9−4.4=3.5mL.

2. For the mass of 3250 Da, $K=(5.4-4.4)/(7.9-4.4)=0.29$.

7.1

1. Since it is known that $\mu_{app}=\mu_{EP}+\mu_{EOS}$, therefore $\mu_{EP}=\mu_{app}-\mu_{EOS}$, and since the experiment allows access to the velocities: v_{app} and v_{EOS}. Then by substitution $v_{app}=24.5/(60\times 2.5)=0.163\text{cm/s}$

and $v_{EOS}=24.5/(60\times 3)=0.136\text{cm/s}$

As a result: $v_{EP}=0.163-0.136=0.027\text{cm/s}$

$\mu_{EP}=v_{EP}/E=0.027\times 32/30000=2.88\times 10^{-5}\text{cm}^2/\text{s}$

2. $D=1^2/(2Nt_m)=24.5^2/(2\times 80000\times 150)=2.5\times 10^{-5}\text{cm}^2/\text{s}$

7.2

1. See the scheme in the book.

2. No, since it arises from a non-treated internal surface, which at the pH considered acts as a polyanion. This results in the creation of an electro-osmotic flow. The fact that the compound migrates towards the cathode does not necessarily mean that it is carrying a net positive charge. The possibility remains that it is being trained in that direction, even if carrying a negative charge. Following Convention:

3. $l = v_{app} \times t$ and $\mu_{app} = v_{app}/E = v_{app} \times L/V$ therefore $\mu_{app} = lL/Vt$

$$\mu_{app} = 7.5 \times 10^{-3}\,\text{cm}^2/(\text{s} \cdot \text{V})$$

4. Following the same reasoning, this leads to $\mu_{EOS} = lL/Vt_m$

$$\mu_{EOS} = 1.5 \times 10^{-3}\,\text{cm}^2/(\text{s} \cdot \text{V})$$

5. $\mu_{EP} = \mu_{app} - \mu_{EOS}$ thus $\mu_{EP} = -7.5 \times 10^{-4}\,\text{cm}^2/(\text{s} \cdot \text{V})$

6. The negative character of μ_{app} implies that it originates from a species carrying a net negative charge. The compound will migrate more slowly than a neutral marker.

7. If the internal lining is rendered neutral there would be no more electro-osmotic flux and as a result the compound will no longer migrate towards the cathode but will reappear in the anode compartment.

8. If the pI is 4 for all pH less than 4, then the compound will be in the form of a cation. In this case, the migration time will normally be shorter than for a neutral market.

9. In using the formula recalled in the question, we will find 337500 theoretical plates.

10. A small molecule diffuses faster than a larger one. Therefore the effectiveness is greater for molecules of greater mass.

7.3

Protein	p*I*	pH=3	pH=7.4	pH=10
Insulin	5.4	+	−	−
Pepsin	1	−	−	−
Cytochrome C	10	+	+	0
Haemoglobin	7.1	+	−	−
Albumin (serum)	4.8	+	−	−

8.1 The analyte in a sample usually contains the spices being mixed, in general, with a variable number of other compounds which constitute together what is called the matrix. When measuring traces of an analyte within other compounds thousands of times more abundant, there is a fear that interferences between analyte and other compounds may occur. So sample pretreatment is essential step in analysis for a sufficient quantity, suitable form for the instrument used, and the reliable, accurate results.

8.2 Liquid-liquid extraction, solid-phase extraction, solid-phase microextraction, gas extraction on a cartridge or a disc, headspace, Immunoaffinity extraction, supercritical phase extraction (SPE), microwave reactor.

参 考 文 献

1. 牟世芬等. 色谱技术丛书. 北京：化学工业出版社，2000.
2. Skoog Douglas A, West Donald M, Hollar James F. Fundamentals of Analytical Chemistry. 9th ed Belmout: Cengage learning, 2014.
3. David Harvey. Modern Analytical Chemistry. New York: McGraw-Hill, 1999.
4. Robards K., Haddad P. R., Jackson P. E., Principles and practice of modern chromatographic method. California: Elsevier academic Press, 2004.
5. Francis Rouessac, Annick Rouessac. Chemical analysis-modern instrumentation methods and techniques. London: John wiley & Sons, 2007.
6. James W Robinson, Eileen M Skelly Frame, George M. Frame Ⅱ. Undergraduate instrumental analysis. 6th ed. New York: Taylor & Francis, 2005.
7. Douglas A, Skoog, F. James Holler, Stanley R Crouch. Principles of Instrumental Analysis. 7th ed. Boston: Cengage learning, 2017.
8. Daniel C Harris. Quantitative Chemical Analysis. New York: W. H. Freeman, 2010.
9. Douglas A Skoog, Donald M West, F James Holler, Stanley R Crouch. Analytical Chemistry-An Introduction. 7th ed. New York: Saunders, 2000.
10. Robin Gill. Modern Analytical Geochemistry: An Introduction to Quantitative Chemical Analysis Techniques for Earth, Environmental and Materials Scientists (Longman Geochemistry Series). London: Routledge, 2014.
11. Ashutosh Sharma, Stephen G Schulman. Introduction to Fluorescence Spectroscopy (Techniques in Analytical Chemistry series), New York: Wiley-Interscience; 1999.
12. JM Andrade-Garda, A Carlosen-Zubieta, MP Gómez _ Garracedo, MA Macstro-Saavedra, MC Prieto-Bianco, RM Soto-Ferreiro. Problems of Instrumental Analytical Chemistry, New Jersay: World scientific, 2017.
13. Raymond P W Scott. Introduction to Analytical Gas Chromatography. 2nd ed. Floride: CRC pressing, 1997.
14. Stéphane Bouchonnet. Introduction to GC-MS Coupling. Floride: CRC pressing, 2013.
15. Robert E. Ardrey. Liquid Chromatography-Mass Spectrometry: An Introduction: Analytical Techniques in the Sciences (AnTs), London: John wiley & Sons, 2003.
16. Paul E Mix. Introduction to Nondestructive Testing: A Training Guide. London: John wiley & Sons, 2005.